T0253772

Technological Internationalism and World Order

Between 1920 and 1950, British and US internationalists called for aviation and atomic energy to be taken out of the hands of nation-states, and instead used by international organizations such as the League of Nations and the United Nations. An international air force was to enforce collective security and internationalized civil aviation was to bind the world together through trade and communication. The bomber and the atomic bomb, now associated with death and devastation, were to be instruments of world peace. Drawing on rich archival research and focusing on public and private discourse relating to the control of aviation and atomic energy, Waqar H. Zaidi highlights neglected technological and militaristic strands in twentieth-century liberal internationalism, and transforms our understanding of the place of science and technology in twentieth-century international relations.

Waqar H. Zaidi is Associate Professor of History at the Lahore University of Management Sciences. He is also a member of the Institute for Advanced Studies Princeton for 2020–21 and a research affiliate at the Future of Humanities Institute at the University of Oxford.

SCIENCE IN HISTORY

Series Editors

Simon J. Schaffer, University of Cambridge
James A. Secord, University of Cambridge

Science in History is a major series of ambitious books on the history of the sciences from the mid-eighteenth century through the mid-twentieth century, highlighting work that interprets the sciences from perspectives drawn from across the discipline of history. The focus on the major epoch of global economic, industrial and social transformations is intended to encourage the use of sophisticated historical models to make sense of the ways in which the sciences have developed and changed. The series encourages the exploration of a wide range of scientific traditions and the interrelations between them. It particularly welcomes work that takes seriously the material practices of the sciences and is broad in geographical scope.

Technological Internationalism and World Order

Aviation, Atomic Energy, and the Search for International Peace, 1920–1950

Waqar H. Zaidi

Lahore University of Management Sciences, Pakistan

CAMBRIDGE
UNIVERSITY PRESS

CAMBRIDGE
UNIVERSITY PRESS

Shaftesbury Road, Cambridge CB2 8EA, United Kingdom

One Liberty Plaza, 20th Floor, New York, NY 10006, USA

477 Williamstown Road, Port Melbourne, VIC 3207, Australia

314–321, 3rd Floor, Plot 3, Splendor Forum, Jasola District Centre, New Delhi – 110025, India

103 Penang Road, #05–06/07, Visioncrest Commercial, Singapore 238467

Cambridge University Press is part of Cambridge University Press & Assessment, a department of the University of Cambridge.

We share the University's mission to contribute to society through the pursuit of education, learning and research at the highest international levels of excellence.

www.cambridge.org
Information on this title: www.cambridge.org/9781108819190

DOI: 10.1017/9781108872416

First published 2021
First paperback edition 2023

A catalogue record for this publication is available from the British Library

ISBN   978-1-108-83678-4   Hardback
ISBN   978-1-108-81919-0   Paperback

For Zehra, Ali Akbar, and Yawar

Contents

Figures

Acknowledgements

I would like to begin by thanking Lucy Rhymer and Emily Sharp of Cambridge University Press, as well as Simon Schaffer and Jim Secord (General Editors of the Science in History series) for their guidance in shepherding the draft of this manuscript through to its final book form, and to the two anonymous readers for their insights and incisive suggestions.

This book began as a PhD project at the Centre for the History of Science, Technology, and Medicine at Imperial College London, and was completed during my tenure as Assistant and then Associate Professor at the Department of Humanities and Social Sciences at the Lahore University of Management Sciences. The faculty at Imperial College were particularly important to my intellectual and professional development. Amongst others, Serafina Cuomo, Mats Fridlund, Hannah Gay, Robert Iliffe, Andrew Mendelsohn, David Munns, Andrew Warwick, and Abigail Woods were instrumental in developing my understanding of the history and sociology of science and technology. I couldn't have asked for a better PhD supervisor than David Edgerton. He has been, without doubt, the most important intellectual guiding light during my academic career and remains a friend and mentor. He was kind enough to read more drafts of my book than perhaps anyone else and I continue to rely on him for his guidance and support.

Early in my career, I was fortunate to build an association with the Tensions of Europe research group (many of whom were at the Eindhoven University of Technology). Irene Anastasiadou, Alexander Badenoch, Vincent Lagendijk, Suzanne Lommers, Johan Schot, Frank Schipper, and Erik van der Vleuten: I fondly remember our many conversations and good times. Johan Schot's support, in particular, has been instrumental at crucial points in my career. At the Lahore University of Management Sciences I would like to thank Zubair Abbasi, Sadaf Ahmad, Shabbir Ahsen, Kamran Asdar Ali, Muhammad Azeem, Laila Bushra, Asma Faiz, Aurangzeb Haneef, Turab Hussain, Hasan Karrar, Ali Khan, Rashid Memon, Khalid Mir, Sameen Mohsin, Saba Pirzadeh,

Yunas Samad, and Mohammad Waseem, amongst many more, for their friendship, support, and guidance. Hasan and Spenta's warm company and encouragement helped this book along in many ways.

I have had the privilege to learn from a great number of students and academics from many other institutions and this book would not have been possible without their support. From the graduate programmes at Imperial College, I would like to acknowledge Sabine Clarke, Ralph Desmarais, Louise Devoy, Rebekah Higgitt, Max Stadler, and Michael Weatherburn. Others who have supported me along the way include Thomas Bottelier, John Krige, Daniel Laqua, Joe Maiolo, Kaveri Qureshi, Jessica Reinisch, Katharina Rietzler, Asif Siddiqi, Geert Somsen, Jessica Wang, and Peter Wilson. At the Meiji University Research Institute for the History of Global Arms Transfer: Tamara Enomoto, Kaori Takada, and Katsuhiko Yokoi. At the Centre for the Governance of AI of the Future of Humanity Institute at the University of Oxford: Allan Dafoe, Jeffrey Ding, Carrick Flynn, and Jade Leung.

Several generous fellowships and grants supported the research and writing of this book. Whilst at Imperial College London, I benefited from the AHRC Postgraduate Doctoral Award; the British Society for the History of Science Research Grant; the Hans Rausing Postgraduate Scholarship in the History of Science, Technology, and Medicine; the Royal Aeronautical Society Centennial Scholarship and Travel Bursary award; and the University of London Central Research Fund Grant. At the Lahore University of Management Sciences I was fortunate to have access to the Faculty Startup Grant, the Faculty Research Fund, and the Faculty Travel Grant.

This book would not have been possible without the generous time and assistance of archivists at the Churchill Archives Centre at Churchill College, the Library Manuscripts and Archives Division of the New York Public Library, the National Archives and Records Administration at College Park, the National Library of Wales, the Modern Records Centre at the University of Warwick, the Rare Books and Manuscript Library at Columbia University, the Special Collections at the University of Sussex Library, and the Special Collections Research Center of the University of Chicago. A big thank you also goes to the many librarians whose assistance and hard work made their libraries wonderful places to work and explore: the British Library, the British Library of Political and Economic Science at the London School of Economics and Political Science, the Central Library of Imperial College London, the Gad and Birgit Rausing Library at the Lahore University of Management Sciences, and the Senate House Library of the University of London.

An earlier version of Chapter 1 was published as 'Liberal Internationalist Approaches to Science and Technology in Interwar Britain and the U.S.', in *Internationalism Reconfigured: Transnational Ideas and Movements Between the World Wars*, ed. D. Laqua (London: I.B. Tauris, 2011), 17–43. Elements of Chapters 2, 3, 4, and 5 were published in '"Aviation Will Either Destroy or Save Our Civilization": Proposals for the International Control of Aviation, 1920–1945', *Journal of Contemporary History* 46, no. 1 (2011): 150–78. An earlier version of Chapter 7 was published in 'A Blessing in Disguise: Reconstructing International Relations Through Atomic Energy, 1945–1948', *Past and Present* 210, Supplement 6 (2011): 309–31. I thank the publishers for their permission to reuse material from these publications and the editors and anonymous readers of these chapters and papers for their critical feedback.

I owe everything to my parents: their love has supported me in all sorts of ways through my life in academia. My brothers and sister Ifty, Ammar, and Naveen have been with me from the beginning of this project and my in-laws with me towards the end. I am thankful to them all for the understanding and support. Tahir and Moona Kazmi, as well as Razia aunty, provided me with a home on my many research trips to London: my research relied in so many ways on their warmth and hospitality. Most of all, this book would not have been possible without the love and support of my immediate family: Zehra, Ali Akbar, and Yawar. I dedicate this book to them.

Abbreviations

ADA	Atomic Development Authority
AORES	Association of Oak Ridge Engineers and Scientists
AORES Records	Association of Oak Ridge Engineers and Scientists Records, Special Collections Research Center, University of Chicago Library
AScW	Association of Scientific Workers
AScW Records	Records of the Association of Scientific Workers, Modern Records Centre, University of Warwick
ATC	Air Transport Command
Atomic Scientists Records	Atomic Scientists of Chicago Records 1943–1955, Special Collections Research Center, University of Chicago Library
BOAC	British Overseas Airways Corporation
Carnegie Records	Carnegie Endowment for International Peace Records, 1910–1954, Rare Books and Manuscript Library, Columbia University
CFR	Council on Foreign Relations
CFR Records	Records of the Council on Foreign Relations, 1921–1951, British Library of Political and Economic Science, London School of Economics and Political Science

Chatham House Papers	Records of the Royal Institute of International Affairs, Chatham House, London
CSOP	Commission to Study the Organization of Peace
Davies Papers	Lord Davies of Llandinam Papers, National Library of Wales
Eichelberger Papers	Clark M. Eichelberger Papers, Manuscripts and Archives Division, The New York Public Library
FAS Records	Federation of American Scientists Records, Special Collections Research Center, University of Chicago Library
Hansard, HC	Parliamentary Debates, House of Commons, fifth series
Hansard, HL	Parliamentary Debates, House of Lords, fifth series
JCS	Joint Chiefs of Staff
JSSC	Joint Strategic Survey Committee
Mass Observation Archives	Mass Observation Archives, Special Collections, University of Sussex Library
MSC	United Nations Military Staff Committee
NARA	National Archives and Records Administration, National Archives at College Park, College Park, MD
Noel-Baker Papers	The Papers of Baron Noel-Baker, Churchill Archives Centre, Churchill College, University of Cambridge
PEP	Political and Economic Planning
TNA	The National Archives, UK
TVA	Tennessee Valley Authority
UN	United Nations
UNA	United Nations Association

UNAEC	United Nations Atomic Energy Commission
UNA Papers	Papers of the United Nations Association of Great Britain and Northern Ireland, Archives and Special Collections, British Library of Political and Economic Science, London School of Economics and Political Science
UNO	United Nations Organization
UNRRA	United Nations Relief and Rehabilitation Administration
WCA Central Committee Records	World Citizens Association Central Committee Records 1939–1953, Special Collections Research Center, University of Chicago Library

Introduction: Machines of Peace

In February 1963, not long after the Cuban Missile Crisis, David Lilienthal gave a series of lectures on nuclear weapons at Princeton University. As a leading US policymaker on atomic matters in the late 1940s (including as the first Chairman of the Atomic Energy Commission), Lilienthal had been instrumental in shaping early policy on the atomic bomb. Now, in the aftermath of the deepest nuclear crisis to envelop the United States, he conceded that his, and society's, earlier thinking on the bomb had turned out to be incorrect. 'We have been following a myth', he exclaimed, 'an illusion about the Atom':

What is the essence of this myth? To my mind, it is this: That because the development of the Atomic Bomb seemed to be the ultimate breakthrough in scientific achievement, in the control of physical matter, we could make a similarly radical departure in dealing with those problems in human affairs which the Bomb so greatly intensified. The Bomb was so colossal, a new force in the world, that we believed a new way must be found to meet its threat, an approach similarly sweeping, similarly radical and world-wide. In short, our obsession with the Atom drove us to seek a Grand Solution. We became committed to the concept of a total final settlement because nothing short of this would answer the tremendous threat. ... We became obsessed with the idea of a Single Solution for the Atom because we were obsessed with the revolutionary destructive power of the Atom itself.[1]

The 'Grand Solution' Lilienthal referred to was the international control of atomic energy. Taking atomic scientific research and development, as well as associated facilities, raw materials, and end products (such as the atomic bomb) out of the hands of nation-states, and placing them under the direct control of the United Nations would, it had been believed, provide the solution to the problem of atomic energy and bring with it international peace and security. International control had been widely discussed and supported in Britain and the United States in the aftermath of the atomic bombings of Hiroshima and Nagasaki, and

---

[1] David E. Lilienthal, *Change, Hope, and the Bomb* (Princeton, NJ: Princeton University Press, 1963), 20, 23.

1

was even adopted as official policy in the form of the Baruch Plan placed before the United Nations Atomic Energy Commission in June 1946.

These attempts at international control had failed. In retrospect, Lilienthal acknowledged they had a misguided approach to solving the problem of atomic energy. But Lilienthal had a deeper realization: that atomic energy had not turned out to be what he and other experts thought it would be in the late 1940s. They had proven to be wrong because their understandings of nuclear weapons had been tied to their understandings of social and world affairs at the time, and because they had understood atomic energy as presenting one problem with one solution. 'The Atom', he noted, 'seemed to present a whole new order of problems. It appeared to supersede and take precedence over every aspect of life that it touched, so that the grave problems of military and foreign policy and human relations were essentially transformed into a single monolithic problem: the Atom'. Atomic energy, he now conceded, was not, in fact, such a singular problem and had no single solution.[2]

This book explores such attempts at a technological solution to the problem of international peace and security, particularly through aviation and atomic energy. Both technologies appeared so powerful and destructive, yet so transformative that internationalists like Lilienthal did not try to ban them, but instead attempted to use them to bring about fundamental transformations in international relations. Proposals for the international control of aviation arose in Britain in the 1920s as part of wider attempts at disarmament and collective security, and reached a peak during the Geneva disarmament conference of the early 1930s. Driven largely by internationalists and supporters of the League of Nations, they were widely aired through the press and public gatherings, and within pressure groups, think tanks, international conferences, and state organizations. They re-emerged during the Second World War in Britain and even the United States as a part of conversations on post-war planning. Aviation, it was argued, was transforming international relations through both its integrative effects and as a powerful modern weapon of war. The spread of aviation and its social and political effects could not be halted or reversed, but could, it was thought, be controlled by international organization. Internationally minded bureaucrats, aviators, and other technical experts, working through international organizations, could be trusted to nurture this invention to fulfil its internationalist potential. Proposals centred on far-reaching international regulation or even outright ownership of aeroplanes and aerial facilities by a powerful international organization. Internationalized civil aviation would bind the world together

---

[2] Ibid., 23.

through trade and communication, and an international air force would enforce collective security.

After August 1945 internationalist interest shifted to atomic energy. It was now international control of atomic bombs and facilities which would prevent catastrophic warfare whilst strengthening the fledgling United Nations. Atomic energy appeared to offer the opportunity for peace, international order, and perhaps even international economic prosperity through cheap energy. Support for the international governance of atomic energy became widespread in both Britain and the United States by mid-1946. An official US proposal was eventually placed before the newly formed United Nations Atomic Energy Commission, though no agreement was reached.

These proposals were in fact not as monolithic as Lilienthal would later assume. They were multifaceted and diverse, reflecting the differing political and personal aims of their proponents. They were also profoundly shaped by the politics of their times, and reflected both the shifting power balances in international relations and, domestically, political rivalries, lobby group pressure, and election-year policies. They were also shaped by contemporary cultural, social, and economic currents; the differing states of British and US aviation and atomic industries had an impact too. The proposals pervaded public discourses on war, peace, and disarmament; and publications, informal networks, and formal gatherings and organizations transmitted these ideas across national boundaries. Although diplomatic attempts at international control ultimately failed, they nevertheless left their mark on popular culture, activism, and intellectual thought into the 1950s and beyond. Hi-tech international police forces, for example, continued as a staple of juvenile science fiction into the Cold War years. A rejuvenated scientists' internationalism in the 1950s and 1960s repeated many of the atomic internationalist refrains from 1946. The techno-globalist rhetoric which emerged so prominently in the 1980s and 1990s made strikingly similar arguments in relation to the newly emergent technologies of the time. Exploring these proposals, consequently, has much to tell us about the techno-politics of international relations and internationalism, and the wider social currents which supported them, of both that period and beyond.

Technological Internationalism

Proposals for international control were not simply about disarmament or security but about the creation of new liberal world orders built on and defended by these technologies. Internationalists argued that these machines necessitated new world organization, and that this organization was at last possible with the aid of these technologies themselves and their attendant experts. These proposals were, in that sense, manifestations of liberal internationalisms which swept Britain and the United States in the first half of the twentieth century. The belief that greater international cooperation and organization was required, and was in fact already emerging, dominated intellectual thought on international relations, as well as both popular and elite activism on foreign affairs. It was reflected also in state policy, with its most visible manifestation being the growth in international organizations in the first half of the twentieth century, especially the formation of the League of Nations in 1920 and the United Nations organization, and associated institutions, in 1944 and 1945.[3]

This impulse is central to our understanding of world politics in the middle decades of the twentieth century. Historians now contextualize the international organizations of the period within an arc of liberal internationalist thinking and advocacy dating back to the nineteenth century.[4] It is also now recognized that liberal internationalism's

---

[3] There is a large and growing literature on this. For example on US liberal internationalism: Robert A. Divine, *Second Chance: A Triumph of Internationalism in America During World War II* (New York: Atheneum, 1967); Andrew Johnstone, *Dilemmas of Internationalism: The American Association for the United Nations and US Foreign Policy, 1941–1948* (Farnham: Ashgate, 2009). On British liberal internationalism: Michael Pugh, *Liberal Internationalism: The Interwar Movement for Peace in Britain* (Houndmills: Palgrave Macmillan, 2012); Casper Sylvest, *British Liberal Internationalism, 1880–1930: Making Progress?* (Manchester: Manchester University Press, 2009). For overviews see: Mark Mazower, *Governing the World: The History of an Idea* (New York: Penguin, 2012); Daniel Gorman, *The Emergence of International Society in the 1920s* (Cambridge: Cambridge University Press, 2012); Glenda Sluga, *Internationalism in the Age of Nationalism* (Philadelphia: University of Pennsylvania Press, 2013); Jo-Anne Pemberton, *The Story of International Relations, parts one and two, Cold-Blooded Idealists* (Cham, Switzerland: Palgrave Macmillan, 2019, 2020). On the League: Patricia Clavin, *Securing the League of Nations: The Reinvention of the League of nations, 1920–1946* (Oxford: Oxford University Press, 2013); Susan Pedersen, *The Guardians: The League of Nations and the Crisis of Empire* (Oxford: Oxford University Press, 2015). On international organizations more broadly see: David MacKenzie, *A World Beyond Borders: An Introduction to the History of International Organizations* (Toronto: University of Toronto Press, 2010); Guy Fiti Sinclair, *To Reform the World: International Organizations and the Making of Modern States* (Oxford: Oxford University Press, 2017). Other significant case studies: Mark Mazower, *No Enchanted Palace: The End of Empire and the Ideological Origins of the United Nations* (Princeton, NJ: Princeton University Press, 2009); Daniel Laqua, *The Age of Internationalism and Belgium, 1880–1930: Peace, Progress and Prestige* (Oxford: Oxford University Press, 2015); Or Rosenboim, *The Emergence of Globalism: Visions of World Order in Britain and the United States, 1939–1950* (Princeton, NJ: Princeton University Press, 2017); Glenda Sluga and Patricia Clavin, eds., *Internationalisms: A Twentieth Century History* (Cambridge: Cambridge University Press, 2017); Simon Jackson and Alanna O'Malley, eds., *The Institution of International Order: From the League of Nations to the United Nations* (London: Routledge, 2018).

[4] Mazower, *Governing the World*; Sluga, *Internationalism in the Age of Nationalism*.

manifestations and impact spread far beyond the demand for international organization or cooperation. For Glenda Sluga there was
a widespread 'international turn' at the start of the twentieth century;
for Or Rosenboim a new 'globalism' emerged in the 1940s as a way of
imagining global space, politics, and society.[5] Erez Manela has shown
that there existed a powerful 'Wilsonian moment' in 1919 and 1920 in
which peoples from across the world couched their political visions in
terms of Wilsonian internationalism.[6] Internationalism is now acknowledged to be a significant motive force behind the creation of trans-
European infrastructure from the interwar period onwards.[7]

  Liberal internationalism became prominent (and in some ways prevalent) in national culture and politics too in the first half of the twentieth
century. That the emergent British and US international relations theorizing was largely liberal internationalist in nature has long been
recognized.[8] But liberal internationalism is now also known to be
a distinct sociopolitical movement with its own middle-class and liberal
aristocratic-led policies, programmes, and discourses. It is also recognized as being of some significance to mass mobilization and associational
life in 1930s Britain, and yet also intertwined with wider liberal militaristic culture and the political economy of the militaristic state.[9] Although
the interwar United States was outside the League of Nations, historians
have begun to explore the ways liberal internationalism continued to
flourish there – particularly through philanthropic foundations such as
the Ford Foundation or the Carnegie Endowment.[10] New work suggests
that US and British central bankers may also have operated within
a broadly liberal internationalist intellectual and institutional culture.[11]

---

[5] Sluga, *Internationalism in the Age of Nationalism*; Rosenboim, *The Emergence of Globalism*.
[6] Erez Manela, *The Wilsonian Moment: Self-Determination and the International Origins of Anticolonial Nationalism* (Oxford: Oxford University Press, 2007).
[7] Alexander Badenoch and Andreas Fickers, eds., *Materializing Europe: Transnational Infrastructures and the Project of Europe* (London: Palgrave Macmillan, 2010).
[8] Peter Wilson, 'Introduction: The Twenty Years Crisis and the Category of "Idealism" in International Relations', in *Thinkers of the Twenty Years' Crisis: Inter-war Idealism Reassessed*, eds. David Long and Peter Wilson (Oxford: Clarendon Press, 1995), 1–24.
[9] Pugh, *Liberal Internationalism*; Helen McCarthy, *The British People and the League of Nations: Democracy, Citizenship and Internationalism, c.1918–45* (Manchester: Manchester University Press, 2012); David Edgerton, *Warfare State: Britain 1920–1970* (Cambridge: Cambridge University Press, 2006); David Edgerton, *England and the Aeroplane: Militarism, Modernity and Machines*, 2nd ed. (London: Penguin, 2013).
[10] Katharina Elisabeth Rietzler, 'American Foundations and the "Scientific Study" of International Relations in Europe, 1910–1940' (PhD diss., University College London, 2009).
[11] Patricia Clavin, 'Men and Markets: Global Capital and the International Economy', in Sluga and Clavin, *Internationalisms*, 85–112.

We also now recognize liberal internationalism's diversities and multiplicities: that, for example, internationalists had differing (sometimes clashing) visions, aims, and understandings of the nature and trajectory of international relations, and operated in differing social, cultural, institutional, and political contexts. Liberal internationalism was not only a political ideology or a constellation of ideas and beliefs, but included activism and was embedded within organizations and institutions, and, in a more diffused way, national, international, and transnational societies, cultures, and norms.[12] Scholars recognize that it was not simply utopian or idealistic, and that nationalism, imperialism, and isolationism are not its antithesis. Imperialism and nationalism are indelibly intertwined with internationalism, and isolationism is more a problematic category tied to the internationalist rhetoric of the 1930s and 1940s than an opposing tendency.[13]

Our understanding of internationalism in the 1930s and 1940s remains chronologically restricted however, splintered by the Second World War. This is especially true of post-war atomic internationalism, which is generally seen as dominated by scientists and driven and shaped by the atomic bomb itself.[14] By bringing aviation and atomic energy together into the same historical and analytical frame, this book highlights the

---

[12] For a sense of diversity see: Long and Wilson, *Thinkers of the Twenty Years' Crisis*; Sluga, *Internationalism in the Age of Nationalism*; Glenda Sluga and Patricia Clavin, 'Rethinking the History of Internationalism', in *Internationalisms*, 3–14.

[13] On idealism and utopianism see: Peter Wilson, 'The Myth of the "First Great Debate"', in *The Eighty Years' Crisis: International Relations 1919–1999*, eds. T. Dunne, M. Cox, and K. Booth (Cambridge: Cambridge University Press, 1998), 1–15; Lucian M. Ashworth, 'Did the Realist-Idealist Great Debate Really Happen? A Revisionist History of International Relations', *International Relations* 16, no. 1 (2002): 33–51. On nationalism, imperialism, and isolationism: Andrew Johnstone, 'Isolationism and Internationalism in American Foreign Relations', *Journal of Transatlantic Studies* 9, no. 1 (March 2011): 7–20; Sluga, *Internationalism in the Age of Nationalism*; Holger Nehring, 'National Internationalists: British and West German Protests against Nuclear Weapons, the Politics of Transnational Communication and the Social History of the Cold War, 1957–1964', *Contemporary European History* 14, no. 4 (2005): 559–582; Brooke L. Blower, 'From Isolationism to Neutrality: A New Framework for Understanding American Political Culture, 1919–1941', *Diplomatic History* 38, no. 2 (2014): 345–376; David Long and Brian C. Schmidt, eds., *Imperialism and Internationalism in the Discipline of International Relations* (Albany: State University of New York Press, 2005); Miguel Bandeira Jerónimo, 'A League of Empires: Imperial Political Imagination and Interwar Internationalisms', in *Internationalism, Imperialism and the Formation of the Contemporary World*, eds. Miguel Bandeira Jerónimo and José Pedro Monteiro (Cham, Switzerland: Palgrave Macmillan, 2018), 87–126; Sluga and Clavin, 'Rethinking the History of Internationalism', 3–14; Stephen Alexander Wertheim, 'Tomorrow, the World: The Birth of U.S. Global Supremacy in World War II' (PhD diss., Columbia University, 2015), 1–26.

[14] For example: Alice Kimball Smith, *A Peril and a Hope: The Scientists' Movement in America: 1945–47* (Chicago: University of Chicago Press, 1965).

importance of pre-Second World War views about science, technology, and international relations to post-war internationalism and arms control. The aeroplane and bombing, I argue, laid the foundation for internationalist responses to the atomic bomb in the late 1940s.

Whilst the prominence of liberal internationalism was clearly a foundation for the emergence and popularity of proposals for the international control of aviation and atomic energy, it does not fully explain two prominent features of these ideas. The first is that their proposed governing organizations were conceived not merely as peace-loving bodies but as military organizations which would put down threats to global peace with force.[15] Technological internationalism incorporated a liberal militarism which put its faith in modern scientific weapons, seeing them as the ultimate arbiters of power and war. The use of these weapons abroad, rather than a reliance on manpower or supposedly traditional ways of warfare would, it was thought, avoid casualties and protect liberal democracy at home from militarism.[16] This liberal militarism unfolded by reproducing and expanding on the demand for collective security which emerged strongly in Britain and France after the First World War.[17] Bombers and atomic bombs, widely recognized and feared as weapons of mass destruction, were welcomed as potential instruments of peace, able to create and sustain peace and security. Their power, it was hoped, could be used to create new forms of collective security, defending liberal states against, in the 1930s, the revisionist powers, and in the late

---

[15] The militaristic side of liberal internationalism remains underexplored by historians, see for example: Sluga, *Internationalism in the Age of Nationalism* and Sluga and Clavin, *Internationalisms*. The overlap between pacifists and liberal internationalists has been problematic for historians: Martin Ceadel classified the League of Nations Union as part of the British 'peace movement' and identified militarists as the peace movement's 'ideological adversaries': Martin Ceadel, *Semi-detached Idealists: The British Peace Movement and International Relations, 1854–1945* (Oxford: Oxford University Press, 2000), 8–9. Liberal internationalism's militaristic edge has been explored in: Edgerton, *Warfare State*, 5, 54–58; Edgerton, *England and the Aeroplane*, chapter 3; Mazower, *Governing the World*, 166–171, 204–207; Pugh, *Liberal Internationalism*, chapter 5.

[16] Liberal militarism also emphasized civilian control of these weapons, which were to be directed against civilian, in addition to military, targets. David Edgerton, 'Liberal Militarism and the British State', *New Left Review* 185 (January–February 1991): 138–169; Edgerton, *Warfare State*, 7–14, 285; Bryan Mabee, 'From "Liberal War" to "Liberal Militarism": United States Security Policy as the Promotion of Military Modernity', *Critical Military Studies* 2, no. 3 (2016): 242–261; James Wood, 'Anglo-American Liberal Militarism and the Idea of the Citizen Soldier', *International Journal* 62, no. 2 (2007): 403–422.

[17] Peter Jackson, *Beyond the Balance of Power: France and the Politics of National Security in the Era of the First World War* (Cambridge: Cambridge University Press, 2013), 427–522; Thomas Bottelier, 'Associated Powers: Britain, France, the United States and the Defence of World Order, 1931–1943' (PhD diss., King's College London, 2018). More broadly: Sally Marks, *The Illusion of Peace: International Relations in Europe, 1918–1933*, 2nd ed. (Houndmills: Palgrave Macmillan, 2003), chapter 2.

1940s, the Soviet threat. Historians have pointed out that the aeroplane was used to scientifically order and govern peoples in far-flung colonial spaces; *Technological Internationalism and World Order* reveals how internationalists hoped to apply these imperial lessons to their metropolitan centres of civilization.[18] As well as providing military defence and governing outlying territories, international control was also to police these technologies themselves by governing the production, flow, and use of technical and scientific know-how and material between and within states. Prior to August 1945, aviation was to be regulated through an international police or an international air bureau, and after 1945 atomic energy was to be governed by an international atomic commission staffed by atomic experts and technocrats. Illegal national development of these technologies would be curtailed by new international conventions, and, if necessary, devastating force: an international air police, which after 1945 was to be armed with atomic bombs. The contradictions inherent in using fearsome weapons for policing regulatory and security regimes were managed through a recourse to prevailing ideas about the nature of modern technology and the new international liberal legalism of the time, which included domestic analogies, just law, and notions of anarchy, barbarism, and civilization.[19] These allowed internationalists to characterize the use of force not as warfare but as legal policing action sanctioned by international authority and carried out by an international police.

Second, proposals for international control voiced deeper and wider hopes and fears about modern scientific machines and their impact on world affairs. It was not just aviation and atomic energy that was thought to be transforming the world, but modern scientific inventions generally, and the science which was thought to underlie them. Diesel-powered ships, radio, and (earlier) the telegraph were thought to be bringing nations closer together through transport, communication, and trade, and so spreading peaceful relations. At the same time, mighty scientific weapons such as poison gas, tanks, and rockets threatened to not only

---

[18] David E. Omissi, *Air Power and Colonial Control: The Royal Air Force, 1919–1939* (Manchester: Manchester University Press, 1990); Priya Satia, *Spies in Arabia: The Great War and the Cultural Foundations of Britain's Covert Empire in the Middle East* (Oxford: Oxford University Press, 2008), chapter 8.

[19] Hidemi Suganami, *The Domestic Analogy and World Order Proposals* (Cambridge: Cambridge University Press, 1989); Martti Koskenniemi, *The Gentle Civilizer of Nations: The Rise and Fall of International Law 1870–1960* (Cambridge: Cambridge University Press, 2001); Hatsue Shinohara, *US International Lawyers in the Interwar Years: A Forgotten Crusade* (Cambridge: Cambridge University Press, 2012); Daniel Joyce, 'Liberal Internationalism', in *The Oxford Handbook of the Theory of International Law*, eds. Anne Orford, Florian Hoffmann, and Martin Clark (Oxford: Oxford University Press, 2016), 471–487.

devastate whole countries but possibly destroy civilization itself. For some this was a new 'Machine Age', and aviation and atomic bombs stood out as its most prominent exemplars and harbingers. Publics were in awe of these machines, transfixed by their spectacle, and certain that they portended epochal transformations, both constructive and destructive, in society and world affairs.[20] There was significant national enthusiasm too for their development: Britain and the United States poured substantial resources into their military and civilian deployment, hoping that their transformative properties could be used for military and geopolitical advantage.[21] These ideas, I argue, constituted a 'technological internationalism' which brought together prominent strands of liberal internationalism and beliefs about the efficacy of modern scientific machines and technical expertise. So ubiquitous and commonsensical (if a little naïve) did some of these ideas seem at the time, and to subsequent historians, that they have eluded study. Nor has their influence on our understanding of international relations or the impact of science and technology been properly grasped. This book is the first history of this remarkable phenomenon.

This technological internationalism was empowered by a growing consensus that the increasingly global and technical problems of international relations required expert technocratic solutions. International governance through technical expertise was demanded by intellectuals (such as H.G. Wells), bureaucrats in international organizations, imperial administrators, and internationalist policymakers. A plethora of international technical organizations, staffed with technical experts and bureaucracies, were created to deal with specific international issues, and even imperial governance took a technocratic turn both in British imperial governance and through the League's Mandates Commission.[22] In Europe plans for transnational rail, road, and electricity networks proliferated alongside

[20] Joseph J. Corn, *The Winged Gospel: America's Romance with Aviation, 1900–1950* (New York: Oxford University Press, 1983); Robert Wohl, *The Spectacle of Flight: Aviation and the Western Imagination, 1920–1950* (New Haven, CT: Yale University Press, 2005); Jenifer Van Vleck, *Empire of the Air: Aviation and the American Ascendancy* (Cambridge, MA: Harvard University Press, 2013); Paul Boyer, *By the Bomb's Early Light: American Thought and Culture at the Dawn of the Atomic Age* (New York: Pantheon Books, 1985); Allan M. Winkler, *Life Under a Cloud: American Anxiety about the Atom*, 2nd ed. (Urbana: University of Illinois Press, 1999).

[21] On Britain: Edgerton, *England and the Aeroplane*; Edgerton, *Warfare State*; G. C. Peden, *Arms, Economics and British Strategy: From Dreadnoughts to Hydrogen Bombs* (Cambridge: Cambridge University Press, 2007). On the United States: Michael S. Sherry, *The Rise of American Air Power: The Creation of Armageddon* (New Haven, CT: Yale University Press, 1987); Gregg Herken, *The Winning Weapon: The Atomic Bomb in the Cold War 1945–1950*, 2nd ed. (Princeton, NJ: Princeton University Press, 1988).

[22] For the interwar years: Pedersen, *The Guardians*; Clavin, *Securing the League of Nations*; Paul Weindling, ed., *International Health Organisations and Movements, 1918–1939*

schemes for European political union.[23] This impulse continued into the war and into the post-war period, where it found expression in the formation of the United Nations and the multitude of specialist international organizations associated with it.[24]

International control, though radical even for its time (and ultimately unsuccessful), was thus part of a wider scientific, technical, and technocratic intervention in international affairs. Through international control, internationalists (such as international relations experts, political scientists, or atomic scientists) promoted themselves as experts with scientific solutions, and sometimes scientific machines as *the* solution. Aeroplanes and atomic bombs came to symbolize technical and scientific control over world affairs, the ultimate triumph of modern science and expertise over anarchic nationalism and outdated diplomacy. These revolutionary new inventions, internationalists argued in 1925, in 1935, and then again in 1945, required revolutionary new expertise. Just as the International Labour Office of the League or United Nations Relief and Rehabilitation Administration (UNRRA) of the United Nations were solving specific transnational problems of workers and refugees, so, some hoped, would an Aerial Board of Control or an Atomic Development Authority solve the problems of international security. This book adds to the growing literature on the relationships between technocracy and liberal internationalism and shows that the two were closely intertwined, and that this connection could rest on widely-held assumptions about the nature of modern science and technology.[25]

In a broader sense, proposals for international control, and the technological internationalism surrounding them, reflected the wider

(Cambridge: Cambridge University Press, 1995). On the 1940s: Joseph Morgan Hodge, *Triumph of the Expert: Agrarian Doctrines of Development and the Legacies of British Colonialism* (Athens: Ohio University Press, 2007); Sabine Clarke, *Science at the End of Empire: Experts and the Development of the British Caribbean, 1940–62* (Manchester: Manchester University Press, 2018).

[23] Badenoch and Fickers, *Materializing Europe*; Jean-Luc Chabot, *Aux Origines Intellectuelles de L'Union Européenne: L'idée d'Europe unie de 1919 à 1939* (Grenoble: Presses Universitaires de Grenoble, 2005); Peter M. R. Stirk, ed., *European Unity in Context: The Interwar Period* (London: Pinter, 1989).

[24] Jessica Reinisch, 'Introduction: Relief in the Aftermath of War', *Journal of Contemporary History* 43, no. 3 (2008): 371–404; Frank Trentmann, A. B. Sum, and M. Riviera, eds., *Work in Progress. Economy and Environment in the Hands of Experts* (Munich: Oekom, 2018).

[25] See for example: Johan Schot and Vincent Lagendijk, 'Technocratic Internationalism in the Interwar Years: Building Europe on Motorways and Electricity Networks', *Journal of Modern European History* 6, no. 2 (2008): 196–217; Clavin, *Securing the League of Nations*; Jessica Reinisch, 'What Makes an Expert? The View from UNRRA, 1943–47', in Trentmann, Sum, and Riviera, *Work in Progress*, 103–130. Lynn Meskell, *A Future in Ruins: UNESCO, World Heritage, and the Dream of Peace* (New York: Oxford University Press, 2018), chapters 2 and 3.

technological modernity of the first half of the twentieth century. They were a family of 'sociotechnical imaginaries', that is 'collectively held, institutionally stabilized, and publicly performed visions of desirable futures, animated by shared understandings of forms of social life and social order attainable through, and supportive of, advances in science and technology'.[26] International control encapsulated the desire to order and control peoples and international affairs and defend civilization. Aeroplanes stood for space–time compression and allowed international-ists to imagine the world as a globe, interconnected and quickly reachable from any point. Atomic energy promised cheap industrial power and freedom from want, and war was thought to have been transformed into a new type of mechanized industrial warfare. The modernity of atomic weapons and aviation was constructed not only through the public spec-tacle of these technologies (as Bernhard Rieger, Robert Wohl, and others have shown) but also through discourses on international relations and arms control.[27] The proposed international organizations staffed by large cadres of technical experts mirrored the developing modernities of tech-nical state bureaucracies. Proposals for international control were sus-tained by continuous appeals to novelty and uniqueness; each proposal promised a jarring break with previous conditions of war and inter-national relations. These appeals reflected a wider modernist idiom which emphasized the power and possibilities of continuous transform-ation through science and technical expertise.

Yet if proposals for international control were about modernity and expertise, they were also about empire and imperialism. It is now well recognized that liberal internationalism and imperialism are not neces-sarily oppositional tendencies, but rather have overlapped, interacted, and sometimes reinforced each other in different ways through the twen-tieth century.[28] This book shows that, in similar vein, modernity and technology on the one hand, and imperialism on the other, were not

---

[26] Sheila Jasanoff, 'Future Imperfect: Science, Technology, and the Imaginations of Modernity', in *Dreamscapes of Modernity: Sociotechnical Imaginaries and the Fabrication of Power*, eds. Sheila Jasanoff and Sang-Hyun Kim (Chicago: University of Chicago Press, 2015), 4.

[27] Bernhard Rieger, *Technology and the Culture of Modernity in Britain and Germany, 1890–1945* (Cambridge: Cambridge University Press, 2005); Wohl, *The Spectacle of Flight*; Boyer, *By the Bomb's Early Light*.

[28] Long and Schmidt, *Imperialism and Internationalism in the Discipline of International Relations*; Jerónimo, 'A League of Empires', 87–126. On the confluence between liberalism and imperialism see: Uday Singh Mehta, *Liberalism and Empire: A Study in Nineteenth -Century British Liberal Thought* (Chicago: University of Chicago Press, 1999); Jennifer Pitts, *A Turn to Empire: The Rise of Imperial Liberalism in Britain and France* (Princeton, NJ: Princeton University Press, 2005); Duncan Bell, *Reordering the World: Essays on Liberalism and Empire* (Princeton, NJ: Princeton University Press, 2016).

mutually exclusive tendencies within technological internationalism, but in some ways mutually reinforcing ones. The use of aviation in the British colonies inspired liberal internationalists to imagine similar control over Europe or globally. Schemes for aviation usually incorporated imperial holdings without changing the nature of empire or the power relations between the metropole and colonial holdings. Moreover, whatever else they were, schemes for international control were also attempts to expand and solidify technological control and governance internationally – a sort of soft imperialism (for aviation what Jennifer Van Vleck has termed an 'Empire of the Air') which would set up military bases abroad, capture aerial and atomic markets, restrict other countries from developing aviation and atomic energy, and empower international organizations and the small number of great powers which controlled or influenced them.[29]

Ultimately, this book is an exploration and extension of David Lilienthal's realization that our understanding of science and technology, and its impact on society, is not stable or fixed, but rather is constructed and mediated through wider beliefs and discourses which can, and do, change over time. Thought of in this way the inventions explored in this book left behind their real-world complexity when they entered the liberal internationalist imagination and instead became products of their hopes and fears. They became liberal internationalist machines, causing great social and political change, but also capable of creating and sustaining liberal world orders. In a world of anarchy these machines of peace promised the creation and sustenance of rule of law. In a world that required rational management they were machines of rationality. Embedded in efficient effective international organizations managed by rational experts, planners, and technocrats, they promised the rational organization of global political, social, and economic affairs. They embodied the latest in technological and scientific research and development, and heralded much more to come. These machines, when insulated from militaristic impulses, could be used for, and even represent, the just use of force, which could end war, defend the righteous, and ultimately enforce international law and the liberal order. Efficient and cost-effective, international control assured economy, less competitive waste, and lower military spending. The machines held within them the promise of greater international trade, commerce, and communication, and through this promised greater international integration. Liberal

---

[29] Van Vleck, *Empire of the Air*. The critique of liberal internationalism as imperialism is of course long-standing. From the 1930s: E. H. Carr, *The Twenty Years' Crisis 1919–1939: An Introduction to the Study of International Relations* (London: Macmillan, 1939); Carl Schmitt, 'The Turn to the Discriminating Concept of War (1937)', in *Writings on War*, eds. Carl Schmitt and Timothy Nunan (Cambridge: Polity Press, 2011), 30–74.

democracy at home could be protected from enemies domestic and foreign. They promised not only a break with traditional and parochial approaches of the past to international affairs, but eventually the defeat of nationalism, militarism, fascism, and totalitarianism. They promised to extend the ascendency of the liberal powers who had won the World Wars, and the ongoing subjugation of their enemies. Although new and modern, they were the endpoint of historical development in international law, international institutions, and military and non-military technologies dating back centuries. They were imagined to be historical necessities which held the promise of a perfect, or perfectible, world order: a world that could be imagined in its global political and economic totality, that would be held together tightly bound and governed. Lilienthal pontificated on the myth of the single solution to war and peace but failed to realize that liberal machines had promised much more than that; they had promised the end of history itself.

Outline

The book begins with a chapter exploring the concepts that were used to link science and technology to international relations between 1920 and 1950. These ideas, explored through intellectual interventions on science, technology, and international relations, provide the conceptual backdrop to many of the later proposals and discourses explored in the book. The chapter highlights the pervasiveness of notions such as the technically driven global integration, the 'machine age', 'scientific warfare', and 'cultural lag' in the interwar years, and argues that internationalist engagement with science and technology was broader and deeper than historians have hitherto appreciated. It contextualizes these motifs within the international politics of their time, and shows how they traversed the Second World War, changed but still intact, into the post-war years, where they were incorporated into new theories of international relations.

Chapter 2 explores British discourses on disarmament and international organization in the 1920s. Its central argument is that conceptualizations of disarmament and collective security were transformed by internationalists searching for broader solutions to the problems of international affairs. Internationalists such as Philip Noel Baker and David Davies came to envisage armaments as products of modern science, and so to see disarmament itself as an increasingly technical and scientific project requiring the guidance of experts schooled in international relations or with a specialized understanding of armaments. Disarmament and collective security came to be seen also in broader political and institutional terms – intimately connected to and dependent on a powerful international organization to organize, operationalize, and police them. Aviation was central to this transformation. Apparently successful in policing and controlling vast imperial territories cheaply and effectively, to most observers aviation promised not only connectivity and internationality but military domination too. Thus for internationalists it came to represent the only weapon capable of giving international organization the teeth required to build legitimacy, enforce international law, and police arms control.

Chapter 3 argues that British thinking on aviation and internationalization became radicalized in the early 1930s, and remained so until the onset of the Second World War. By the start of the decade scientific disarmament was no longer enough for internationalists. They now pushed for a more comprehensive aerial transformation of international relations. The 1932 Geneva Disarmament Conference, and internationalist organizations such as the League of Nations Union and the New Commonwealth Society, emerged as sites for the discussion and propagation of these aerial visions. Arguments about the convertibility of civilian to military aviation were used to make the case for comprehensive international control of both military and civilian aviation. In the most radical proposals existing airlines and air forces were to be transferred to the League of Nations, which was to run them and use them to ensure international peace and security. By the mid-1930s, as the fascist threat loomed large, there emerged a more muscular internationalism willing to use bombing to bolster the fledgling League order. A central argument in this chapter is that these proposals were not simply a response to the rise of German aviation or fear of bombing (as Uri Bialer and others have argued) but instead reflected a national enthusiasm for aviation, as well as British aerial and scientific might.[30]

Chapter 4 explores proposals for the creation of a post-war United Nations air force in Britain and the United States during the Second World War. Although there is growing historical interest in wartime planning for a post-war United Nations, aviation (and collective security more broadly) remains missing from our histories.[31] Aviation, the chapter shows, emerged as a central component of internationalist campaigns for the creation of a post-war international security organization. Allied aerial

---

[30] Uri Bialer, *The Shadow of the Bomber: The Fear of Air Attack and British Politics 1932–1939* (London: Royal Historical Society, 1980).

[31] Mazower, *Governing the World*; Elizabeth Borgwardt, *A New Deal for the World: America's Vision for Human Rights* (Cambridge, MA: Belknap Press, 2005); Townsend Hoopes and Douglas Brinkley, *FDR and the Creation of the U.N.* (New Haven, CT: Yale University Press, 2000).

ascendancy appeared to offer the opportunity, and the aeroplane the perfect tool, for the creation of a new system of collective security and international governance. By examining discussions and proposals within internationalist organizations and state planning committees, the chapter examines the coalition of interests which formed in support of the force. In the United States it was supported by proponents of international law and of a strong United Nations Organization, and by a lobby keen to see the expanded wartime aviation industry used for post-war peacekeeping. In Britain support was built on visions of joint Anglo-American policing of the world and anticipation of British influence over an integrated post-war Europe. An international air force was incorporated into the blueprint for a United Nations Organization at the 1944 Dumbarton Oaks Conference, and although the force was never formed it remains a part of the United Nations Charter to this day.

In Chapter 5 I explore planning, debates, and rhetoric about post-war civil aviation during the Second World War in Britain and the United States. As well as public rhetoric, I focus on discussions at the 1944 Chicago conference on international aviation, and in state committees and internationalist organizations such as the Council on Foreign Relations (in the United States) and Chatham House (in Britain). The internationalism surrounding aviation was powerful enough, I argue, to manifest in a wide variety of visions for post-war aviation. This internationalism was not monolithic: I emphasize its fractured and contested nature by exploring the intermingling of political, commercial, and national interests within differing internationalist proposals. Although in both countries internationalists continued to see the aeroplane as a globalizing machine of prosperity, their proposals were also designed to safeguard national commercial interests. In Britain a post-war aerial regime managed by a powerful international organization (the Labour Party's policy pamphlet on the subject was titled 'Wings for Peace') was to safeguard British aviation and forestall the spread of US aviation around the world.[32] In the United States, on the other hand, most internationalists joined state officials and the aviation industry in a near-consensus that global nature of aviation necessitated minimal regulation, a so-called 'freedom of the air'.

Chapters 6 and 7 present new perspectives on internationalist reactions to the atomic bomb in Britain and the United States between 1945 and 1950. The chapters' central argument is that these reactions were too widespread and varied to be, as the existing literature contends, only or

---

[32] Labour Party, *Wings for Peace: Labour's Post-War Policy for Civil Flying* (London: Labour Party, April 1944).

largely driven by scientists.[33] Instead, the chapters suggest that scientists were part of a wider reinvigorated effort by British and US internationalists to deal with the problems of atomic energy, to incorporate the bomb in their internationalist worldviews, and to strengthen international organization. Although historians have depicted this reaction as a novel reaction to a novel problem, this book suggests that this impulse echoed interwar thinking about science and aviation.

Chapter 6 makes this point by highlighting the complex and contested nature of atomic internationalism in the United States between 1945 and 1950. It conceptualizes the introduction of atomic energy into the public sphere as the opening of a new discursive space which was fought over by contending activists and their associated organizations. Each struggled to have its vision of international atomic energy governance and world order accepted by the public and the state as official policy. In this chapter the well-known Baruch and Acheson-Lilienthal Plans emerge as just two of a number of contending schemes for international control. These proposals, I demonstrate, were as much shaped by their authors' visions of international relations, notions of expertise, and in response to each other, as they were by the desire to tackle an existential threat.[34]

Chapter 7 focuses on post-war Britain and highlights the liberal nature of post-war atomic internationalism by examining responses to the bomb from across the political spectrum. The chapter explores media coverage (the centre/centre-left *News Chronicle* newspaper welcomed the atomic bomb as a 'blessing in disguise') as well as positions taken by prominent intellectuals; particularly liberals Arthur Salter and Bertrand Russell, realists such as Lord Hankey, and the leftist perspective developed by physicist Patrick Blackett.[35] The chapter also argues that long-established liberal internationalist organizations such as the New Commonwealth Society and the United Nations Association were more crucial to the propagation of proposals for international control than the scientists' associations emphasized in existing histories of the period.[36]

---

[33] Most influentially in: Smith, *A Peril and a Hope*.

[34] In particular I build on the understanding that scientists' expertise is (at least partially) socially constructed and is intertwined with their political and social interventions. See: Thomas F. Gieryn, *Cultural Boundaries of Science: Credibility on the Line* (Chicago: University of Chicago, 1999), 1–36, 65–114; Harry Collins and Robert Evans, *Rethinking Expertise* (Chicago: University of Chicago, 2007), chapter 1.

[35] 'An End or a Beginning?', *News Chronicle*, 8 August 1945.

[36] For example in: Greta Jones, 'The Mushroom-Shaped Cloud: British Scientists' Opposition to Nuclear Weapons Policy, 1945–57', *Annals of Science* 43, no. 1 (1986): 1–26.

Terminologies and Scope

Science, technology, and technical expertise intermingled in liberal internationalist discourses in ways which defy our current understandings of these categories. The word 'technology' was hardly used – instead writers tended to directly name the artefacts, systems, and industries they referred to or use terms such as 'mechanical inventions', 'modern inventions', and 'means' or 'forms' of 'modern communication'.[37] This book also, consequently, uses a mix of such terms. 'Science' was often taken to mean something much broader than what we understand it to be today. One interwar British writer on international relations, in a major text on the League of Nations, defined it as 'organized and cumulative knowledge', but it often also included scientists, engineers, and their associated facilities.[38] Modern complex machines and technologies were often assumed to be direct outcomes of science, so for example it was often assumed that 'aviation science' lay behind developments in aviation.[39] When using the term 'aviation', internationalists generally meant not only aeroplanes but also the infrastructure and expertise surrounding them: the pilots, airports, and construction, repair, and research and development facilities. Discussions on the atomic, meanwhile, often used the term 'atomic energy' which referred not only to the energies unleased through fission but also to both the practice and materiality of the release of atomic energy. It thus included the atomic bomb and the engineering behind its development, other sciences and engineering associated with fission, and atomic facilities and their associated scientists.[40]

I use the term 'liberal internationalism' to mean the promotion of international cooperation and intercourse in order to bring the peoples of the world closer together culturally, economically, and politically, and secure a more just and peaceful world. It was 'liberal' in that it was grounded in liberal approaches to politics and the economy (particularly a commitment to lowered trade and transport barriers), and eschewed discussions of more

---

[37] On the emergence of the word technology see: Eric Schatzberg, *Technology: Critical History of a Concept* (Chicago: Chicago University Press, 2018), chapters 10, 11, and 12.

[38] C. Howard-Ellis, *The Origin, Structure, and Working of the League of Nations* (London: George Allen & Unwin, 1928), 23. On understandings of science see: Frank M. Turner, 'Public Science in Britain, 1880–1919', *Isis* 71, no. 4 (December 1980): 589–608.

[39] For example: F. H. Sykes, *Aviation in Peace and War* (London: Edward Arnold, 1922), 7, 16.

[40] Thus for example the Baruch Plan for international control was published as: Department of State, *United States Atomic Energy Proposals: Statement of the United States Policy on Control of Atomic Energy as Presented by Bernard M. Baruch, Esq., to the United Nations Atomic Energy Commission June 14, 1946* (Washington, DC: Government Printing Office, June 1946).

leftist categories such as class. I use the term to include, amongst others, those the historical literature terms the British 'new liberal internationalists' and those, for the US case, who have often been called 'Wilsonian internationalists'.[41]

Although the book is largely limited to exploring British and US technological internationalism, the ideas and world orders imagined in these countries had global impact. They were part of a wider Anglo-American shaping of international organization and international life, including the League of Nations and the United Nations. This techno-logical internationalism was not restricted to Britain and the United States, and, as the book notes, could be found in other countries (e.g. France) where liberal internationalism was prominent. Although there is some discussion of them in this book, there is unfortunately not enough space to explore these geographies in depth. Nevertheless, the book suggests that there is a wider world of technological internationalism waiting to be explored. Lastly, it is worth keeping in mind that there were, in fact, multiple internationalisms and conceptualizations of global order from across the political, class, race, and gender spectrum in Britain, the United States, and across the world. This book should in no way be taken to mean a diminishment of their significance and impact on world affairs; indeed, recognizing technological internationalism should alert us to the multiplicity of internationalisms, the ideas within them, and the impulses driving them.

---

[41] For broad definitions of liberal internationalism: Fred Halliday, 'Three Concepts of Internationalism', *International Affairs* 64, no. 2 (Spring 1988): 187–198; Beate Jahn, *Liberal Internationalism: Theory, History, Practice* (Houndmills: Palgrave Macmillan, 2013), chapter 1; Christer Jönsson, 'Classical Liberal Internationalism', in *International Organization and Global Governance*, eds. Thomas G. Weiss and Rorden Wilkinson (Abingdon: Routledge, 2014), 105–117. For an understanding specific to interwar Britain: David Long, *Towards a New Internationalism: The International Theory of J.A. Hobson* (Cambridge: Cambridge University Press, 1996), 66; Sylvest, *British Liberal Internationalism*, 1–22. Although a wider variety of internationalisms has been used for the American case (Warren F. Kuehl, 'Preface', in *Biographical Dictionary of Internationalists*, ed. Warren F. Kuehl [Westport, CT: Greenwood Press, 1983], ix–xii), the term liberal internationalism is nevertheless widely used and recognized: Warren F. Kuehl, 'Concepts of Internationalism in History', *Peace and Change* 11, no. 2 (1986): 1–10. See also: Tony Smith, *Why Wilson Matters: The Origin of American Liberal Internationalism and its Crisis Today* (Princeton, NJ: Princeton University Press, 2017), 1–30. The term 'internationalism' is not, however, without its issues: Johnstone, 'Isolationism and Internationalism in American Foreign Relations'.

1 Invention, Interdependence, and the Lag
Conceptualizing International Relations in the Age of the Machine

> The great fact that the world is a unit rests upon the underlying conditions of modern invention and science
>
> – Paul S. Reinsch, *Public International Unions* (1911)[1]

In 1934, the American National Committee on Intellectual Cooperation of the League of Nations, an internationalist pressure group, published a report on the state of the discipline of international relations. In the introduction, the eminent historian and political scientist James T. Shotwell made an impassioned plea for the scientific study of international relations. This was required, he reasoned, because in the modern era, 'radically different from any that has gone before', science was leading not only to 'the conquest of time and space but a complete readjustment of the activities of mankind'.[2] In making this argument Shotwell expressed a sentiment widespread amongst thinkers on international relations. At a time widely characterized as a modern 'machine age', science, machines, and technical expertise were seen as decisive motive forces affecting not only society and culture but also international relations and war. Their primary effects on world affairs were thought to be a growing global integration and interdependence, as well as the transformation of war through scientific weapons. Mankind's organizations and social forms, however, lagged behind technical development, leading to imbalances in world affairs, international tension, and ultimately warfare. International organization and global integration, the solution to the problems of modern science, were held back by the intransigence of nationally minded leaders and backward traditional institutions.

This chapter charts the development of these ideas from 1910 onwards to 1950, and argues that liberal internationalist thinking on science and

---

[1] Paul S. Reinsch, *Public International Unions* (Boston: Ginn, 1911), 8.
[2] James T. Shotwell, 'Introduction', in *The Study of International Relations in the United States: Survey for 1934*, ed. Edith W. Ware (New York: Columbia University Press, 1934), 3–20.

machines in Britain and the United States was much broader in scope, and more important to theory, than has generally been recognized. This theorizing on science and machines was driven by disciplinary developments within the maturing field of international relations, international events, and changing perceptions of international organizations and international relations. It also built on widespread perceptions about the impact of various machines and devices on society. Aviation and the atomic bomb, in particular, informed, energized, and shaped this thinking in a multitude of ways. This thinking was shared between Britain and the United States, although there were notable differences in the prominence and context of these ideas and when they were expressed. Although some of the concepts developed and used by internationalist thinkers, particularly 'cultural lag', have subsequently disappeared, others live on and are still with us today.

Interdependence

The origins of liberal internationalist thinking on international relations date back to the seventeenth and eighteenth centuries. By the late 1800s earlier beliefs about the peaceful effects of free trade and liberal republican systems of government had been incorporated into a fully formed and influential British internationalist ideology.[3] The ideas of philosophers Herbert Spencer and Henry Sidgwick typify this development. In *Political Institutions* (1882) Spencer argued that the key emergent distinction in international affairs was between industrial and militant societies. The urge to produce, trade, and make profit meant that industrial societies preferred individual freedom and peace, whereas warfare and militarism were inimical to the growth of international prosperity. Sidgwick argued that the search for international security was pushing nations into larger agglomerations, and characterized the British Empire as part of this natural evolution.[4] By the beginning of the twentieth century liberal internationalists could point to growing international organizations and global interconnections as evidence of increasing international integration.[5] The growth of

---

[3] *Sylvest, British Liberal Internationalism*, 35–45; Duncan Bell, 'Victorian Visions of Global Order: An Introduction', in *Victorian Visions of Global Order: Empire and International Relations in Nineteenth-Century Political Thought* (Cambridge: Cambridge University Press, 2007), 1–25.

[4] Sylvest, *British Liberal Internationalism*, 101–147; Herbert Spencer, *Political Institutions, being Part V of the Principles of Sociology (The Concluding Portion of Vol. II)* (London: Williams and Norgate, 1882).

[5] Sluga, *Internationalism in the Age of Nationalism*, 11–32; Emily S. Rosenberg, 'Transnational Currents in a Shrinking World', in *A World Connecting 1870–1945*, ed. Emily S. Rosenberg (Cambridge, MA: Belknap Press, 2012), 813–996.

international law held up the possibility of international peace, and a series of international treaties and conventions appeared to demonstrate that this was not a pipe dream.[6] By the end of the nineteenth century science and modern means of transport and communication were increasingly coming to be seen as drivers of global interdependence. New mechanical and electrical devices, alongside transnational scientific and technical cooperation, were, it was believed, having an integrative effect on the world.[7] The telegraph, perhaps the most prominent and celebrated of the new means of communication of the late eighteenth century, attracted significant comment, with many claiming that long distances were no longer a barrier to the formation of common political and cultural identities.[8]

The 'new liberal internationalists' developed these notions further in the first two decades of the twentieth century whilst retaining the Cobdenite antithesis between industrial and commercial activity on the one hand and militarism on the other. This is particularly clear in the intellectual output of Norman Angell, an influential and widely read writer on international affairs at that time. By 1914 Angell's insistence that growing 'interdependence' and trade made war an increasingly 'diminishing factor' and irrational choice in international relations was predicated not only on an international division of labour but also on the 'the improvement of communication and the cheapening of transport'. For him, as for other internationalists, it was not the changing patterns of transport and communications over many hundreds of years that counted, but rather the introduction of a small number of 'mechanical' inventions in the nineteenth century: 'the compound steam-engine, the railway, the telegraph' as well as 'printing, gunpowder, steam, electricity'.[9] The Cobdenite faith in modern mechanical industry remained. Conflict could be eradicated by replacing it with mechanical work: 'Machinery and the steam-engine have done something more than make fortunes for manufacturers ... the more man succeeds in his

---

[6] Sylvest, *British Liberal Internationalism*, 61–100; Koskenniemi, *The Gentle Civilizer of Nations*, 11–97.

[7] Bell, 'Victorian Visions of Global Order'; Duncan Bell, 'Dissolving Distance: Technology, Space, and Empire in British Political Thought, 1770–1900', *The Journal of Modern History* 77, no. 3 (September 2005): 523–562. A significant US example were the statements of Representative William L. Wilson during the 1888 tariff debates. For example, National Democratic Committee, *The Campaign Text Book of the Democratic Party of the United States, For the Presidential Election of 1888* (New York: Brentanos, 1888), 570.

[8] Simone M. Müller, *Wiring the World: The Social and Cultural Creation of Global Telegraph Networks* (New York: Columbia University Press, 2016), chapter 3.

[9] Norman Angell, *Arms and Industry: A Study of the Foundations of International Polity* (London: G. P. Putnam & Sons, 1914), xx–xxi, xxii; Norman Angell, *The Great Illusion: A Study of the Relation of Military Power to National Advantage*, 4th ed. (New York: G. P. Putnam & Sons, 1913), 335, 142.

struggle with nature, the less must be the role of physical force between men, for the reason that human society has become, with each success in the struggle against nature, a completer organism'.[10]

Communications and transport-driven interdependence emerged as an important motif in international relations writing in the first decade of the twentieth century and was often used to make the argument for international governance. In Britain, the most influential thinkers to use interdependence in this way were Leonard Woolf, particularly through his 1916 Fabian Society pamphlet *International Government*, and J. A. Hobson through his *Towards International Government* (1915) and *Democracy after the War* (1917).[11] In the United States the most detailed arguments were penned by diplomat and political scientist Paul S. Reinsch in a series of articles published in the first decade of the twentieth century, and compiled in his influential 1911 volume *Public International Unions*.[12] They were also widely disseminated through Pitman B. Potter's books, especially his popular *An Introduction to the Study of International Organization*, which was first published in 1925 and went through four editions to 1935.[13]

These works took Victorian-era ideas about modern means of transport and communications bringing the world closer together and inserted them into more systematically liberal internationalist analyses of international relations and interdependence. Reinsch labelled them 'instrumentalities', and for Hobson they were the 'vast and complex machinery of communications and transport'.[14] The most commonly cited examples were the telegraph, railways, and the post, though the steamship and the telephone also often made an appearance. These, along with science and sometimes other technical problems of an international nature (usually relating to health or crime), required international governance through

---

[10] Angell, *The Great Illusion*, 277–278.
[11] Leonard Woolf, *International Government: Two Reports by L.S. Woolf Prepared for the Fabian Research Department, Together with a Project by a Committee for Supernational Authority that will Prevent War* (London: Fabian Society, 1916); J. A. Hobson, *Towards International Government* (London: George Allen & Unwin, 1915); J. A. Hobson, *Democracy after the War* (London: George Allen & Unwin, 1917).
[12] Reinsch, *Public International Unions*.
[13] Pitman B. Potter, *An Introduction to the Study of International Organization* (New York: Century, 1922), 308–309. Also Pitman B. Potter and Roscoe L. West, *International Civics: The Community of Nations* (New York: Macmillan, 1927), v, 38–46. Other prominent works which referenced interdependence include Raymond Leslie Buell, *International Relations* (London: Sir Isaac Pitway & Sons, 1926), 139–140; Stephen Haley Allen, *International Relations* (Princeton, NJ: Princeton University Press, 1920), 2; Edmund A. Walsh, ed., *History and Nature of International Relations* (New York: Macmillan, 1922), 96.
[14] Reinsch, *Public International Unions*, 176; Hobson, *Towards International Government*, 116–117.

either smaller specialist technical organizations (e.g. for Woolf, or Reinsch, who labelled them 'Unions') or one large international organization (as in Hobson). In these works, as in earlier ones, existing technical organizations (the Telegraphic Union and the Universal Postal Union came up the most) were held up as examples of how this 'internationalism' was already coming into being through necessity.[15]

By the beginning of the 1930s interdependence was firmly established as an indisputable fact of international relations. Although introduced as a 'platitude' by one book on the topic, it was still nevertheless thought novel enough to warrant several popular works which took it as their central motif.[16] In Britain, economic historian George N. Clark's 1920 *Unifying the World* pointed to 'modern methods of communication' as driving integration, though alongside the usual devices such as the telegraph and 'wireless machines' he included inventions which transmitted ideas ('ideal communications'): printing, photography, cinematography, and the typewriter.[17] Similarly, the Liberal Party intellectual Ramsay Muir organized his *The Interdependent World and its Problems* (1932) into seven chapters which closely followed the then commonly held arguments relating to interdependence. The first chapter described the 'Interdependent World' and how it had come about, whereas the second emphasized the continuing existence and power of nationalism. The third explained the 'Perils of Interdependence' arising from the interaction between nationalism and growing interdependence, followed by three chapters on solutions to these perils: the limitation of state sovereignty, the abolition of war, and economic cooperation. It concluded with a chapter on the inadequacy of national approaches for dealing with these problems.[18] These arguments were commonplace enough for Alfred Zimmern to welcome the book as 'a useful restatement ... of conditions familiar to most students of international affairs'.[19] William L. Langer's review for *Foreign Affairs* was blunter: 'A

---

[15] Woolf, *International Government*, 99, 116, 117, 197–200; Hobson, *Towards International Government*, 116–117; Hobson, *Democracy after the War*, 196–197; Reinsch, *Public International Unions*, 4, 6, 7, 12–76, 176. On Reinsch, see Jan Klabbers, 'The Emergence of Functionalism in International Institutional Law: Colonial Inspirations', *The European Journal of International Law* 25, no. 3 (2014): 645–675. On Hobson's and Woolf's wartime writing, see Long, *Towards a New Internationalism*, 121–172; Peter Wilson, *The International Theory of Leonard Woolf: A Study in Twentieth-Century Idealism* (London: Palgrave Macmillan, 2003), 23–81.

[16] Ramsay Muir, *The Interdependent World and its Problems* (London: Constable, 1932), 1.

[17] G. N. Clark, *Unifying the World* (London: Swarthmore Press, 1920), 9–36.

[18] Muir, *The Interdependent World and its Problems*, chapters 1, 3, 7.

[19] Alfred Zimmern, 'Review of *The Interdependent World and its Problems*, by Ramsay Muir', *International Affairs* 12, no. 2 (March 1933): 247–248.

survey of present day political and economic problems, offering little that is new'.[20]

In the United States, political scientist Pitman B. Potter's *The World of Nations*, published in 1929 for a non-academic audience, struck a peculiarly optimistic note, symptomatic perhaps of the heights of exuberance prior to the Great Crash. He predicted a 'complete unification of world markets and world supplies of material goods, and a complete unification of world information and experience in another seventy years ... something approaching a real world state'.[21] James T. Shotwell's thought in the first three decades of the century, meanwhile, emphasized that communications-driven interdependence was a manifestation of the transformative effects of modern science and industrialization.[22] Shotwell had stressed the importance of science for the creation of modern society from his earliest writings and, by the twenties (in essays such as 'Mechanism and Culture'), was characterizing the modern epoch as one characteristically shaped by science-based machines.[23] The connection to international relations arrived in his 1929 *War as an Instrument of National Policy*, which explained spreading industrialization and growing communications-driven interdependence as characteristic of a new 'scientific era' for international affairs.[24] The impact of science, working through industrial machines and transport and communication inventions, remained important for his subsequent writings, particularly his widely read 1936 internationalist tract *On the Rim of the Abyss*. 'Mechanism and Culture' was meanwhile incorporated into a 1942 collection on *Science and Man*, which included essays by well-known scientists.[25]

By the mid-twenties internationalists increasingly linked the transformative effects of modern communications and transport to the League of Nations. This was especially so in Britain, where the League

---

[20] William L. Langer, 'Some Recent Books on International Relations', *Foreign Affairs* 11, no. 4 (July 1933): 720–732.

[21] Pitman B. Potter, *The World of Nations: Foundations, Institutions, Practices* (New York: Macmillan, 1929), 339–340.

[22] On Shotwell see Harold Josephson, *James T. Shotwell and the Rise of Internationalism in America* (Cranbury, NJ: Associated University Presses, 1975).

[23] James T. Shotwell, 'History', in *The Encyclopaedia Britannica: A Dictionary of Arts, Sciences, Literature and General Information*, 11th ed. vol. XIII, ed. Hugh Chisholm (New York: Encyclopaedia Britannica, 1910), 527–533; James T. Shotwell, *The Religious Revolution of Today* (Boston: Houghton Mifflin, 1913); James T. Shotwell, 'Mechanism and Culture', *The Historical Outlook* 16 (January 1925): 7–11.

[24] James T. Shotwell, *War as an Instrument of National Policy and its Renunciation in the Pact of Paris* (New York: Harcourt, Brace, 1929), 27–31.

[25] James T. Shotwell, *On the Rim of the Abyss* (New York: Macmillan, 1936), 42–43; James T. Shotwell, 'Mechanism and Culture', in *Science and Man*, ed. Ruth Nanda Anshen (New York: Harcourt, Brace, 1942), 151–162.

had more salience for internationalists. For supporters of the League its Covenant and political functions were the institutional equivalent to the processes of global interdependence already underway. In a 1930 survey of the success of the League, Ramsay Muir noted that 'the world had become a single political and a single economic system, and that all peoples must henceforward be interdependent. This idea dictated the institution of the League of Nations'.[26] For the first Sir Ernest Cassel Professor of International Relations at the University of London, Philip Noel Baker (later Noel-Baker), the League Covenant was exactly the political instrument required to deal with the 'virtual destruction of the international barriers of time and space' unleashed by 'scientific discovery and invention' in the nineteenth century. The integration initiated by 'the steamship, the railway and the telegraph' decades ago was in the process of being renewed by 'aircraft and wireless telegraphy'.[27] Geneva-based journalist C. Howard-Ellis in his 1928 *Origin, Structure, and Working of the League of Nations* attributed three world-changing effects to 'science' (which he defined as 'organized and cumulative knowledge') over the past two hundred years. He presented 'the telegraph, telephone, railway, steamship and cheap printing' as products of the 'application' of science which had led the 'globe' to 'shrink'. Because of this, 'mankind is culturally and economically becoming one interdependent society'. The League of Nations was formed in order to 'build on the peace-organizing tendency' and to give 'formal and binding expression to the interdependence of modern nations' – the 'next step' in the organization of international relations. The Concert of Europe, the Hague Conferences, and public international unions were part of a previous 'unconscious' development of international organization which the League Charter had transformed into a 'conscious' one. The Covenant was 'the turning-point in the evolution of the world toward international organization'.[28] In the United States, textbooks on international organization positioned the League as the rational culmination of decades of development in international institutions. For political scientists E.C. Mower and Clyde Eagleton, it was the next phase in the development of 'international government', whereas for Potter the League was the

---

[26] Ramsay Muir, *Political Consequences of the Great War* (London: Thornton Butterworth Limited, 1930), 178. Similarly, William E. Rappard, *International Relations as Viewed from Geneva* (New Haven, CT: Yale University Press, 1925), 4–5.

[27] P. J. Noel Baker, 'The Growth of International Society', *Economica* 4, no. 12 (November 1924): 262–277.

[28] Howard-Ellis, *The Origin, Structure, and Working of the League of Nations*, 24, 25–26, 60, 67, 485. On the enigmatic Howard-Ellis, see James Cotton, '"The Standard Work in English on the League" and its Authorship: Charles Howard Ellis, an Unlikely Australian Internationalist', *History of European Ideas* 42, no. 8 (2016): 1089–1104.

'culminating event in the development of international organization' and the germ of the 'international federation' required to manage growing interdependence ('cosmopolitanism').[29]

By the early 1930s internationalist supporters of the League (once again, more so in Britain than the United States) had come to see the League's 'technical' organs and not the Covenant or its political activities as the new significant factor in international relations. Surveys of the League noted the difficulties faced in its political work – as early as 1926 academic (and League diplomat) William E. Rappard's stock-take of the League's progress expressed disappointment with its attempts at the 'prevention of war', but judged its 'promotion of international co-operation', particularly in the economic sphere, a success.[30] International relations scholar Charles K. Webster's 1933 *The League of Nations in Theory and Practice* blamed the 'instability of the League' on great power politics and economic depression (the 'instability of the age').[31] Political scientist Harold Laski's 1931 lecture at the Geneva Institute of International Affairs argued that the League needed to develop its 'technical bodies' in order to meet the needs of an increasingly interdependent society.[32]

Political scientist H.R.G. Greaves' 1931 *The League Committees and World Order* was the most significant work to emphasize the League's technical activities.[33] Greaves highlighted the success of 'international technical co-operation' through League organs such as its committees, the International Labour Organization, and the disarmament commissions. This technical work, he argued, mirrored the increasingly technical functions carried out by national governments, and as it progressed would eventually obviate the need for national governments. Not only was the League's best contribution to peace 'technical and administrative', but

---

[29] Edmund C. Mower, *International Government* (Boston: D. C. Heath, 1931); Clyde Eagleton, *International Government* (New York: Ronald Press, 1932); Potter and West, *International Civics*, 198; Potter, *An Introduction to the Study of International Organization*, 455.

[30] William E. Rappard, 'The Evolution of the League of Nations', *The American Political Science Review* 21, no. 4 (November 1927): 792–826. Similarly, see William E. Rappard, 'The League of Nations as a Historical Fact', *International Conciliation* 11 (June 1927): 270–322.

[31] Charles Webster and Sydney Herbert, *The League of Nations in Theory and Practice* (Boston: Houghton Mifflin, 1933), 304.

[32] H. J. Laski, 'The Theory of an International Society', in *Problems of Peace*, 6th Series, eds. H. J. Laski, et al. (London: George Allen & Unwin, 1932), 188–209.

[33] H. R. G. Greaves, *The League Committees and World Order: A Study of the Permanent Expert Committees of the League of Nations as an Instrument of International Government* (Oxford: Oxford University Press, 1931). A corresponding work published in the United States: Norman Hill, *International Administration* (New York: McGraw-Hill, 1931).

nothing is more significant than the League of Nations and the technical experiments it is carrying on. Not only are these co-ordinating the technical organization of the world along the various functions of society, and improving the conditions of life at the same time, but they are also developing a habit of co-operation and mutual confidence.[34]

This 'international technical government' worked because, just as in the domestic sphere, technical committees were increasingly staffed by 'experts' and 'technicians' who focused on technical rather than political issues.[35]

The emphasis on the League's role in engendering international technical cooperation died away by the mid-1930s. The growing belief that the League was failing in its political endeavours also coloured assessments of its administrative and (so-called) technical activities. More broadly the global depression led many to question the efficacy of economic and social experts and their organizations. Leonard Woolf's 1931 review of Greaves' *The League Committees and World Order* faulted him for taking 'a too optimistic view of the achievements of some of the committees'.[36] Georg Schwarzenberger's 1936 *The League of Nations and World Order* ignored technical expertise, and instead explained international cooperation on technical issues in terms of nations' common interests outweighing their opposing interests.[37] Stephen King-Hall's 1937 history of interwar international relations recognized the League's 'important experiments in international co-operation', but ultimately dismissed them as having made 'little progress' towards the 'substitution of world order for world chaos'.[38] George Keeton's 1939 *National Sovereignty and International Order*, largely about the League, devoted barely a sentence to its 'organs of international administration', and yet still dwelled on the advent of such organizations prior to 1914.[39] The fourth (1935) edition of Potter's *An Introduction to the Study of International Organization* ended with a new section questioning the 'Effectiveness and Value of the League', including in the 'fields of conference and administration'.[40] Alfred Zimmern, in his 1936 *The League of*

---

[34] Greaves, *The League Committees and World Order*, vii, 6, 244.   [35] Ibid., 5, 6.
[36] Leonard Woolf, 'Review of *The League Committees and World Order*, by H. R. G. Greaves', *Economica* 11, no. 34 (November 1931): 485–487.
[37] Georg Schwarzenberger, *The League of Nations and World Order* (London: Constable, 1936), 124–128.
[38] Stephen King-Hall, *The World Since the War* (London: Thomas Nelson and Sons, 1937), 95.
[39] George Keeton, *National Sovereignty and International Order* (London: Peace Book, 1939), 61–62.
[40] Pitman B. Potter, *An Introduction to the Study of International Organization*, 4th ed. (New York: D. Appleton-Century, 1935), 481–494.

Nations and the Rule of Law, distanced himself from his own earlier belief in the increasing power of the League of Nations expert.[41] Now, in 1936, he noted that many were asking if expert cooperation:

extended indefinitely throughout the field of public affairs? Why should not one problem after another be detached from the complex of 'high politics' and subjected to scientific treatment in the new atmosphere of international co-operation? And why should not this lead, in the long run, to the elimination of the causes of war?[42]

This sort of thinking, he noted, emerged 'in League circles' and was a 'curious combination of Fabianism and Cobdenism': 'Little by little, so it began to be believed, the morass of "high politics" would dry up along its edges, as one issue after another was drained off to Geneva'. These hopes, however, had turned out to be 'unfounded'.[43]

Yet, notwithstanding growing disillusionment with the League, inter-dependence remained a prominent motif in liberal international relations writing through to the end of the 1930s. Schwarzenberger, for example, still used it to argue that a universal League was a necessity in the modern world, and King-Hall asserted that modern international relations was moulded by the effects of the two industrial revolutions through the mechanization and the speeding up of processes, most prominently communications.[44] Zimmern, who had been writing of the effects of the industrial revolution on international relations since 1914, did so again in his 1936 *The League of Nations and the Rule of Law*, where he talked of the 'industrial revolution' creating a 'material internationalism' and pre-sented as examples "five different fields of communication: Posts, Telegraphs, Wireless Telegraphy, Railways and Motor-cars".[45] Interdependence was so omnipresent that it could be found in avowedly anti-liberal internationalist works on international relations, such as the prominent English Marxist analysis of international relations, Palme Dutt's 1936 *World Politics*, which took as axiomatic that communications and trade-driven unification was the fate of the current nation-state system.[46]

---

[41] For example: Alfred Zimmern, 'The Prospects of Democracy', *Journal of the Royal Institute of International Affairs* 7, no. 3 (May 1928): 153–191.

[42] Alfred Zimmern, *The League of Nations and the Rule of Law 1918–1935* (London: Macmillan, 1936), 321–322.

[43] Ibid., 322–323.

[44] Schwarzenberger, *The League of Nations and World Order*, 175; King-Hall, *The World Since the War*, 85–87.

[45] Alfred Zimmern, 'German Culture and the British Commonwealth', in *The War and Democracy*, eds. R. W. Seton-Watson, et al. (London: Macmillan, 1914), 348–382; Zimmern, *The League of Nations and the Rule of Law*, 40–41.

[46] Palme R. Dutt, *World Politics 1918–1936* (London: Victor Gollancz, 1936), 21, 351–352.

Once the Second World War began, liberal internationalist intellectuals added their voices to the growing call for the United Nations to formalize their alliance into a United Nations Organization (UNO). Significantly, however, interdependence was not prominent in justifications for the formation of such an organization. It was instead eclipsed by the new language of the Atlantic Charter (1941) and the United Nations Declaration (1942). At the April 1943 Annual Meeting of the American Academy of Political and Social Science, for example, international legal scholar Quincy Wright was the only one to make the interdependence argument in support of the UNO. However, even Wright failed to address in any detail how the UNO was better suited than the League to deal with 'the new conditions of technology and economy'. Although he offered some explanation in terms of allowing fairer international trade, he rooted this in the values of the proposed organization, which were to flow from the four freedoms of the Atlantic Charter and the United Nations Declaration. The ideals embodied in these declarations were certainly liberal internationalist in nature but were not explicitly connected to interdependence. Instead, for internationalists such as Wright, the freedoms in the Atlantic Charter were rooted further back in time. They had 'spread throughout the world since the Renaissance' and were 'accepted as the standards of world civilization'.[47] The other major reasons for the UNO, offered by fellow internationalist Clark Eichelberger, were the need for cooperative post-war reconstruction and collective security.[48] These reasons were disseminated widely by internationalist lobbies. Grayson Kirk and Walter R. Sharp's 1942 Headline booklet, *United Today for Tomorrow*, for example, argued that a UNO would be an expression of the Four Freedoms, and was grounded in the need for post-war reconstruction. There was only a brief mention of interdependence in the final paragraph.[49]

The rhetoric of technological interdependence and integration was heard during the war but was linked to the plethora of Allied international technical and specialist organizations created from 1942 onwards. These wartime organizations, internationalists argued, portended a new type of technocratic organization which was necessary not only for the prosecution of the new technical and scientific warfare now being waged but also for the

---

[47] Quincy Wright, 'United Nations-Phrase or Reality?', *Annals of the American Academy of Political and Social Science* 228, no. 1 (July 1943): 1–10. On the new language of the Charter see: Borgwardt, *A New Deal for the World*, chapters 1, 2.

[48] Clark M. Eichelberger, 'Next Steps in the Organization of the United Nations', *Annals of the American Academy of Political and Social Science* 228, no. 1 (July 1943): 34–39.

[49] Grayson Kirk and Walter R. Sharp, *United Today for Tomorrow: The United Nations in War and Peace* (New York: Foreign Policy Association, 1942).

transformed era of international relations after the war ended. Such claims were particularly prominent in Britain. Historian and international relations expert E.H. Carr, once a critic of liberal internationalist thinking, noted in *Nationalism and After* (1945) the existence of 'modern technological developments which have made the nation obsolescent as the unit of military and economic organization and are rapidly concentrating effective decision and control in the hands of great multi-national units'. As examples of such 'functional' organizations, he pointed to interwar technical organizations linked to the League, as well as to wartime organizations such as UNRRA, the FAO and the Middle East Supply Centre.[50] Harold Laski too (in 1943) called for the formation of separate specialist technical international organizations to deal with global technical 'functions' (aviation, railways, road transport, and currency).[51]

The pacifying and integrative effects of international technical cooperation were captured and elucidated in their most sophisticated form by political scientist David Mitrany. His influential 1943 paper on *A Working Peace System* took as its central problematique the inability of the nation-state system to solve the 'technical' problems created by modern 'technical inventions'. Nation-states, he argued, were increasingly unable to deal with the problems created by modern transport and communications, or to ensure that they achieve their fullest potential. Drawing inspiration from many areas, including the US Tennessee Valley Authority (TVA) project and wartime Allied cooperation and post-war planning, he argued that the most successful trans-border organizations were those pragmatically set up to tackle specific trans-border problems – whose powers were limited to specific areas sufficient to fulfil their limited functions. Such technical organizations (be they global or regional in scope), he argued, should have remits limited to particular economic or technical areas, and have sufficient authority and jurisdiction to fulfil their functions. Fruitful areas for such international cooperation could include water and natural resource management, power generation, transport and communication, healthcare, and even scientific and technological research. He singled out 'railway systems', 'shipping', 'aviation', and 'broadcasting' to be particularly amenable to this 'functional' approach.[52] Apolitical technocrats would ensure the success of

---

[50] E. H. Carr, *Nationalism and After* (London: Macmillan, 1945), 38, 48. For a proposed 'European Reconstruction Corporation' and a 'European Planning Authority' see E. H. Carr, *Conditions of Peace* (London: Macmillan, 1942), 236–275.

[51] Harold Laski, *Reflections on the Revolution of Our Time* (London: Allen and Unwin, 1943), 7, 234–239.

[52] David Mitrany, *A Working Peace System: An Argument for the Functional Development of International Organization* (London: Royal Institute of International Affairs, 1943), 20, 33. On his functionalism see: Cornelia Navari, 'David Mitrany and International Functionalism', in Long and Wilson, *Thinkers of the Twenty Years' Crisis*, 214–246;

these organizations and convince nation-states to further empower them to deal with other technical areas. As the number and functions of these organizations increases, he argued, the increasing web of technical and economic integration will lead to political integration. National sovereignty, and eventually the nation-state itself, would wither away. This process, he argued, was necessarily to be a gradual one: history had shown that attempts to create, from scratch, international authorities with wide-ranging political powers would fail. Technical international organizations would also help in the prevention of war; a 'joint European transport organization such as the new European Central Inland Transport Organization', he argued in 1946, 'should be able to plan the railways and canals of Europe with a view to improving civilian communications and facilitating trade, but prevent the construction of railways and roads primarily for strategic purposes'.[53]

Fuelled by expectations raised by formation of the United Nations and its associated agencies, the functional approach came to be celebrated amongst international relations and government policy circles through to the early 1950s. By that time, however, it became clear that the United Nations, even in its supposedly functional organs (most prominently UNESCO), had become bogged down in political wrangling. International relations theorists began to rework Mitrany's functionalist approach within a regional context – foremost amongst them the political scientist Ernst Haas, whose *The Uniting of Europe* (1958) modified the functionalist approach to de-emphasize the role apolitical internationally minded technocrats.[54] Haas's neofunctionalist approach was closely linked to European integration – as this integration appeared to slow down in the 1960s many, including Haas himself, came to doubt the applicability of neo-functionalism, and both neo-functionalist and functionalist approaches declined in influence within the discipline of international relations.[55]

Scientific War

The widespread perception that the First World War had inaugurated a new type of scientific warfare found its way quickly and easily into international relations writing by the end of the twenties. The new

Lucian M. Ashworth, *Creating International Studies: Angell, Mitrany and the Liberal Tradition* (London: Taylor and Francis, 2017), 76–105.

[53] David Mitrany, 'The Growth of World Organisation', *Common Wealth Review* 3, no. 8 (June 1946): 12–13.

[54] Ernst B. Haas, *The Uniting of Europe: Political, Social and Economic Forces 1950–1957* (Stanford: Stanford University Press, 1958); Ernst B. Haas, *Beyond the Nation-State: Functionalism and International Organization* (Stanford: Stanford University Press, 1964).

[55] Stefan Borg, *European Integration and the Problem of the State: A Critique of the Bordering of Europe* (Houndmills: Palgrave Macmillan, 2015), 43–61.

scientific warfare was attributed to the invention of new armaments, which were assumed to have their origins in civilian science. Internationalists emphasized the power of this new warfare by arguing that it gave an over-riding advantage to the aggressor, adding further instability to the already anarchic state of international relations. Ideas about scientific warfare and its impact first became important for British international relations writers in the 1920s, but they quickly transferred over to the USA where they resonated much more strongly in the late 1930s and during the Second World War.

In Britain, the most developed versions of these arguments were put forward by the industrial magnate and Liberal MP David Davies (later a Baron), particularly in his *The Problem of the Twentieth Century: A Study in International Relationships*, published in 1930.[56] In it Davies presented a lengthy articulation of the argument that the First World War had ushered in a new era of destructive warfare based on modern, science-based weapons, the most powerful of which were 'aeroplanes and poison gas'.[57] Davies did not see new warfare as substantially altering the system of international relations – warfare was still driven by imperialist or nationalist motives, and remained a function of the anarchic nature of international relations. It did, however, make wars more deadly, and so made the prevention of warfare more important than ever before. As the next chapter shows, Davies developed these ideas into what he called the principle of the 'differentiation of weapons' which became part of the intellectual foundation of his attempts to reconstruct international relations through collective security and international organization. By ceding these new scientific weapons to international organization, leaving the older and less effective ones in the hands of nations, it was now possible, he reasoned, to abolish war.[58]

In the United States these ideas were explored in most depth by the academics and intellectuals Quincy Wright and James T. Shotwell, and the popular writer Ely Culbertson, though they found a wide audience through textbooks as well. For these internationalists the destructiveness of modern warfare was the end point of the progressive development of weapons through the ages. The industrial revolution accelerated this process, culminating in the modern scientific armaments which now threatened, in the words of Shotwell's 1936 *On the Rim of the Abyss*, to return civilization 'to the Dark Ages'.[59] Widely used international relations textbooks such as Frederick Schuman's 1933 *International Politics*

[56] David Davies, *The Problem of the Twentieth Century: A Study in International Relationships* (London: Ernest Benn, 1930).
[57] Ibid., 275, 326–327.   [58] See Chapter 2.   [59] Shotwell, *On the Rim of the Abyss*, viii.

and Walter R. Sharp's and Grayson Kirk's 1940 *Contemporary International Politics* argued that the most important transformation in warfare arose from post-1800 development of industrialized weapons which lead to the birth of modern 'mechanized armies' many times more powerful than those which had gone before.[60] Shotwell similarly argued for a transformation of warfare in the 1800s in his 1929 *War as an Instrument of National Policy*, a warfare which in the new 'scientific era' mobilized new inventions, industry, and the 'entire economic structure of the belligerent nations'.[61] His 1944 *The Great Decision* used the then increasingly popular terms 'total war' and 'totalitarian war' (interchangeably) to describe this new type of warfare.[62] These histories allowed the authors to argue that modern war was immensely more destructive than previous wars, and so its abolishment needed to be a priority for diplomats and politicians.

The most prominent liberal internationalist theorist of warfare was the international legal expert Quincy Wright. In 1926 Wright was placed at the head of a large interdisciplinary research project on war at the University of Chicago. Wright published some of his early thinking on war in the 1930s, but the fullest exposition of his ideas arrived in 1942 with the publication of the 1,500-page *Study of War*.[63] Both an assimilationist culmination of the Chicago war project as well as the most developed elucidation of Wright's own thinking, *Study* was the most thorough liberal internationalist study of war to date. Wright presented a schematic for the development of warfare through the ages, culminating in the age of 'industrialization and nationalistic wars' (1789–1914) and finally from 1914 onwards the age of the 'airplane and totalitarian war'.[64] Modern current-day warfare, he argued, had six distinguishing characteristics: mechanization, increased army size, militarization of population, nationalization of war effort, total war, and 'intensification of operations'. These taken together he termed 'totalitarian war' – the logical end-point of a centuries-old arms race which manifested through the cyclical dominance of defensive and then offensive armaments.[65] In the totalitarian

---

[60] Frederick L. Schuman, *International Politics: An Introduction to the Western State System* (New York: McGraw Hill, 1933), 644–649; Grayson Kirk and Walter R. Sharp, *Contemporary International Politics* (New York: Farrar & Rinehart, 1940), 397–427.

[61] Shotwell, *War as an Instrument of National Policy*, 32–38.

[62] James T. Shotwell, *The Great Decision* (New York: Macmillan, 1944), 3–15. On the terms 'total war' and 'totalitarian war' see: Hew Strachan, 'Essay and Reflection: On Total War and Modern War', *The International History Review* 22, no. 2 (2000): 341–370.

[63] Quincy Wright, *The Causes of War and the Conditions of Peace* (London: Longmans, Green, 1935); Quincy Wright, *A Study of War*, 2 vols. (Chicago: University of Chicago Press, 1942).

[64] Wright, *A Study of War*, vol. 1, 291–312.   [65] Ibid., 303–312.

phase, it was the offensive armaments which were in the ascendant. This warfare was totalitarian because it suited totalitarian states (who would be better than democracies at waging it), and because non-totalitarian states would have to adopt some totalitarian characteristics in order to wage it effectively. The impact on international relations was thus to give an advantage to aggressive totalitarian states, and to put liberal democracies at a disadvantage. To make matters worse these modern wars had a tendency to spread because of growing interdependence.[66]

Study did not reach a wide audience, instead the US reading public were exposed to internationalist thinking on war and international relations through the writings of the celebrity bridge player Ely Culbertson. His widely read *Total Peace: What makes Wars and How to Organize Peace* (1943) used a Davies-like typology of modern armaments to explain the impact of war on international affairs. Like others, Culbertson saw the industrial revolution as introducing a 'machine age' which 'revolutionized the structure and the technique of modern weapons, creating a new military age of heavy fighting machines'. Like Davies, he envisaged that this had led to an increasing divergence between two classes of weapons: 'heavy', scientific, complex mechanical weapons, and cheaper, lighter, simpler ones. Although initially only the USSR, Britain, and the USA would have such weapons, spreading industrialization would mean that other countries, especially China and Japan, could eventually develop them too. Larger countries would retain an advantage in industrialization, though, leading to a dangerously unstable system of international relations in which a small number of large powerful industrialized states confronted each other, with smaller states 'satellized around protector states'. As later chapters show Culbertson, like Davies, then went on to argue that this differentiation of weapons allowed for the formation of an effective international political organization armed with an international police force.[67]

The internationalist view of interdependence and war was contested by opponents of liberal internationalist causes. Although few denied that transport and communications were interconnecting the world, opponents argued that its significance was overstated. In the United States these arguments emerged most forcefully during debates surrounding arms-embargo legislation tabled before the Senate in 1933.[68] John Bassett

---

[66] Ibid., 300–304, 313–314. For more on the internationalist nature of *A Study of War* see: Waqar Zaidi, 'Stages of War, Stages of Man: Quincy Wright and the Liberal Internationalist Study of War', *The International History Review* 40, no. 2 (2018): 416–435.

[67] Ely Culbertson, *Total Peace: What makes Wars and How to Organize Peace* (Garden City, NY: Doubleday, Doran, 1943), 21, 23–31, 42.

[68] Shinohara, *US International Lawyers in the Interwar Years*, 123–131.

Moore, an opponent of this legislation and a senior authority on international law (and once a judge on the Court of International Justice),
made his case for non-intervention in European affairs by attacking the
'remarkably unfounded' internationalist argument that 'improved means
of communication' were now causing previously localized conflicts to
spread more widely. His 'Appeal to Reason' (published in *Foreign
Affairs* in 1933) pointed out that the First World War 'did not begin as
a local war' and 'did not exceed the spread of all previous wars, or equal
that of some of them'. Moreover, 'The numerous local wars that have
since occurred, but have remained local, clearly demonstrate that the
supposed greater likelihood of spread is fanciful'.[69] In fact, Moore
argued, speedier communications and transport empowered the state,
and would allow the United States government to more effectively
enforce its neutrality. He also attacked the 'hasty supposition' that 'by
various modern devices ... discordant races and peoples have been har-
moniously united in thought and in action and in brotherly love'. Indeed,
'The French and the Germans have for centuries lived side by side. No
artificial device is needed to enable them quickly to come into contact'.[70]
In Britain, by contrast, there was much less overt criticism of interdependence, perhaps because of perceptions of empire and because
debates about intervention in Europe did not rest on connectivity to
Europe, as economic and other connections could not be denied.[71]
Internationalists claimed, as a riposte, that their opponents displayed
ossified and out-dated worldviews. In *On the Rim of the Abyss* Shotwell
maintained that Moore's opinions were part of a wider 'conservative'
worldview of international relations which failed to recognize the transformations science had wrought on society and international affairs. Only
the 'post-war peace movement' appreciated the impact of the 'interdependence of nations' and the fact that it had been brought about
'through the inventions and discoveries of science'. Shrugging off the
criticism that peace movements were idealistic, he claimed that their

---

[69] John Bassett Moore, 'An Appeal to Reason', *Foreign Affairs* 11, no. 4 (1933): 547–588.
For more on Moore's view: Justus D. Doenecke, 'Edwin M. Borchard, John Bassett
Moore, and Opposition to American Intervention in World War II', *The Journal of
Libertarian Studies* 6, no. 1 (1982): 1–34.
[70] Moore, 'An Appeal to Reason'.
[71] Perhaps the most significant came from E. H. Carr in his 1939 attack on 'utopianism',
The Twenty Years' Crisis. He, however, only focused on 'instruments' of mass media,
which he argued carried no agency themselves. Instead of any inherent internationalism,
their effects reflected the vested interests of those that controlled them: in 'totalitarian
countries' the government and in 'democratic countries ... immense corporations' which
collaborated with government. Carr, *The Twenty Years' Crisis*, 171–172.

activism was in fact grounded in an awareness of this new material reality.[72]

Cultural Lag

Interdependence and scientific war were drawn together through the concept of a 'cultural' or 'social lag' to construct a wider view of international relations. This master trope framed the various other claims about modern science, invention, and society, and helped to position social scientists and other commentators as experts on the impact of science on society. The lag argument was used both in relation to national society and in explanations of international relations – in the latter allowing internationalists to rebut more effectively the claim that there was precious little internationalism to be found in relations between states. The argument could be found in the writings of almost every British and American liberal internationalist one might care to name from the 1920s onwards.

Although there were several versions of the lag argument, in its essence it asserted that science and scientific inventions had advanced beyond mankind's understanding or control, leading to detrimental effects on society or international affairs. Typical arguments were that modern science-based armaments had advanced beyond society's ability to control their effects; or that man's social and/or governmental institutions were unable to deal with the consequences of the new transport-driven interdependence. In these arguments nationalism, nation-states and traditional militaristic thinking were depicted as increasingly redundant relics of an earlier era now holding back scientific and technical advance. Old-fashioned diplomats and politicians were often characterized as embodying the worst of this backwardness. Nationalism was more dangerous now than ever before because modern communications were putting peoples into increasing contact with each other, causing their nationalist impulses to clash, and leading to war. Sometimes the argument was made that man's understanding of social affairs lagged behind his understanding of scientific and technical matters. In some cases this argument was distilled into a specific concern with the state of the political and social sciences – that it was these which lagged professionally, methodologically and financially behind the natural sciences. If the social sciences could be advanced through the scientific study of society and politics, it was argued, the social effects of modern science and its applications could be understood and controlled.

---

[72] Shotwell, *On the Rim of the Abyss*, 41–44.

Although their earliest use in international relations writing was in Britain (for example by Norman Angell in 1913, Alfred Zimmern in 1914, and Arthur Greenwood in 1916), lag arguments were elaborated in the greatest detail in the United States.[73] In Britain lag arguments were tied more closely to calls for the empowerment of the League of Nations and the radical (Wells-esque) technocratic ordering of international relations, and were one of several motifs used to support these aims. Internationalists such as C. Delisle Burns and Noel Baker argued that the League was required for society and nations to deal with the effects of modern science and science-based machines.[74] They were also prominent calls for the technical and technocratic study and ordering of society – as in for example economist Josiah Stamp's 1936 Presidential address to the British Association for the Advancement of Science.[75] But it was through the writings of H.G. Wells that these arguments were introduced to wider audiences, particularly in relation to technocratic ordering. Although prominent in his earlier writings on the need for international organization, Wells' articulation of the lag probably reached its widest audience through his 1933 best-selling novel *The Shape of Things to Come*.[76] Through the novel Wells observed that man's 'social invention' lagged behind 'mechanical invention', but used the lag most forcefully to explain the onset of the war: 'new means of communication and transport, and the new economic life ... were necessitating the reorganization of human affairs as a World-State', however the world was 'already parcelled up', leading to 'steadily intensified mutual pressure to develop into more or less thinly disguised attempts at world conquest'. The destructiveness of war reflected the fact that armaments had become too advanced for the 'small and antiquated disputes' for which they were used. Wells' solution was technocratic rule by a small group of aviators committed to technical advancement and free trade.[77]

---

[73] Zimmern, 'German culture and the British Commonwealth'; Angell, *The Great Illusion*, xiii; Arthur Greenwood, 'International economic relations', in Grant et al., *Introduction to the Study of International Relations* (London: Macmillan, 1916), 66–112.

[74] For example: C. Delisle Burns, *Modern Civilization on Trial* (New York: Macmillan, 1931), 178; Noel Baker, 'The Growth of International Society'; Howard-Ellis, *The Origin, Structure, and Working of the League of Nations*, 24, 63; C. Delisle Burns, *A Short History of International Intercourse* (London: George Allen & Unwin, 1924); C. Delisle Burns, *International Politics* (London: Methuen, 1920), 8, 11.

[75] Josiah Stamp, 'The Impact of Science Upon Society', *Science* 84, no. 2176 (September 1936): 235–239. Later expanded as: Josiah Stamp, *The Science of Social Adjustment* (London: Macmillan, 1937).

[76] H. G. Wells, *In the Fourth Year: Anticipations of a World Peace* (New York: Macmillan, 1918); H. G. Wells, *The Idea of a League of Nations* (Boston: Atlantic Monthly Press, 1919); H. G. Wells, *The Shape of Things to Come* (London: Hutchinson, 1933; London: Penguin, 1993). Citations henceforth refer to the Penguin edition.

[77] Wells, *The Shape of Things to Come*, 36, 55–56. These aviators are discussed in Chapter 2.

Lag arguments were more prominent in the United States by the early 1930s, especially in popular books where they often provided over-arching framing arguments. The internationalist activists and academics used them to make a case for the scientific study of international relations (especially interdependence), and for the strengthening of international organization. One prominent exposition was by noted internationalist Raymond B. Fosdick in a collection of his lectures, titled, tellingly, *The Old Savage in the New Civilization* (1928). Fosdick argued that man would not be able to control the great powers that the 'scientific revolution' had put at his disposal unless the social sciences were developed too. He lamented the 'divergence between the natural sciences and the social sciences' caused by the gap between 'the brilliant development of scientific knowledge on the one hand and the almost stationary position of our knowledge of man on the other'. He called for the social sciences to inculcate 'the same technique that characterize our treatment of physics and chemistry'. Such scientific study of international affairs would confirm, he claimed, the need for a League of Nations to 'handle the common interests of mankind that overflow national boundaries'.[78] In a letter to Paul Mantoux (the co-founder of the Geneva-based Graduate Institute of International Studies), he noted that:

It is a platitude that the recent war, together with our stupendous scientific advances, such as transcontinental trains, fast steamers, airplanes, wireless, etc., have brought the nations practically to each other's door-steps without having provided an adequate corresponding advance in their methods of dealing with each other. The relationships between nations since 1914 have been so fundamentally revolutionized that practically all the pre-war studies and theories have been swept by the board and a wholly new set of difficulties created.[79]

John Herman Randall, Sr., founder of the World Unity Foundation, made similar arguments in his 1930 *A World Community*. Randall held that 'the new means of communication which science has devised' were the crucial driver behind the creation of a single global consciousness. Citing the works of prominent British and American internationalists Randall asserted that these new means join the world together into 'one physical neighbourhood' and 'one geographic community'.[80] Repeating

[78] Raymond Blaine Fosdick, *The Old Savage in the New Civilization* (Garden City, NY: Doubleday, Doran, 1928), 36–37, 40, 44. The first lecture in this collection is titled 'Our Machine Civilization'.
[79] Raymond Blaine Fosdick, Memorandum, 'A Proposal to Establish an Institute of International Research', n.d., c. 1926, folder 11, box 154, Paul Mantoux Papers, Archives of the League of Nations, Geneva.
[80] John Herman Randall, *A World Community: The Supreme Task of the Twentieth Century* (New York: Frederick A. Stokes, 1930), 9, 21.

Fosdick's arguments almost word for word, he noted that warfare occurred because of the gap between the physical sciences and the lagging 'knowledge of man'. To stop society from 'cracking under the strain' he called for the 'increasing development of the same scientific method and spirit in the social sciences that has already found expression in the physical sciences, and a frank recognition that it is only though the scientific, rather than the older political methods, that the desirable changes can be brought about'.[81]

In the United States lag arguments were in fact well known beyond writings on international relations, with a rich literature in particular in academic sociology. Notably sociologist William F. Ogburn carved out a niche as the leading theorist of cultural lag arguments in relation to national society, and his work came to be widely debated in US sociology circles, and more broadly, by the 1930s.[82] The greater prominence of lag arguments in the United States as compared to Britain was due to a greater fascination with mechanization and industrialization, and their impact on society. The sense of rapid mechanical change was accentuated by faster national communications and connectivity, rapid electrification and mechanization in homes, and the growth of huge industrial and civil engineering enterprises.[83] Lag arguments also drew off a deeper well of engineers', social scientists', and social reformers' calls for more technical expertise in national policymaking.[84] And so Randall was able to announce that in relation to international relations 'social engineers and technicians must be recognized and accepted for what they are – the trained and competent experts in their particular field of social control and social reorganization, to whom rulers and statesmen must look for light on the new problems that old methods and formulas have proved themselves unable to solve'.[85]

Lag arguments continued well into the Second World War and were once again most noticeably reproduced in popular works. Wells in his 1940 *The New World Order* noted that the League of Nations 'broke down'

---

[81] Ibid., 53, 50, 52, 87–88.
[82] For an example of Ogburn's lag arguments see: William F. Ogburn, *Social Change with Respect to Culture and Original Nature* (New York: B. W. Huebsch, 1922). In 1957 Ogburn wrote that he had first used the term 'cultural lag' in 1914 and had developed a fully articulated theory by 1915: William F. Ogburn, *On Culture and Social Change* (Chicago: University of Chicago Press, 1964), 87. For a review of critiques of Ogburn's lag arguments see: Joseph Schneider, 'Cultural Lag: What Is It?', *American Sociological Review* 10, no. 6 (December 1945): 786–791.
[83] John M. Jordan, *Machine-Age Ideology: Social Engineering & American Liberalism, 1911–1939* (Chapel Hill: University of North Carolina Press, 1994), 3.
[84] Ibid., 255–279.   [85] Randall, *A World Community*, 87–88.

because it ignored the 'vast disorganisation of human life by technical revolutions, big business and modern finance that was going on'. He criticized the British government for failing to look beyond Hitler and to plan for the required 'new world order'; the war was 'delaying and preventing an overdue world adjustment'.[86] By the end of the war US world federalists were mobilizing lag argument in their increasingly popular calls for world federation. Books by publishers William B. Ziff, Sr. (whose magazines included technology-orientated *Popular Aviation, Radio News, Amazing Stories, Air Adventures,* and *Popular Electronics*) and Emery Reves claimed that the lag between modern science and the structure of international relations could only be resolved through international (in Ziff's case regional) federation. Ziff updated the usual list of world-changing modern sciences and machines by including 'chemistry, light metals and electronics'.[87] Reves emphasized industrialization more broadly. Echoing a motif first used by Harold Laski in 1931, he framed his best-selling (pre-Hiroshima) 1945 *The Anatomy of Peace* through the rubric of outdated 'Ptolemaic' political institutions versus a scientific 'Copernican' world. Reves explained why 'Our Ptolemaic political conceptions in a Copernican industrial world are bankrupt': 'scientific and technological developments achieved by the industrial revolution' had given rise to new problems of an international nature, including the need to promote international trade and communication, and the problem of modern industrialized war. The solution to the 'clash between industrialism and political nationalism' was world federation. The 'false notion of Inter-nationalism' and its international machinery such as the 'San Francisco League' would not work as it left national sovereignties intact.[88]

Lag arguments persisted in international relations writing too. Leonard Woolf's 1940 response to Carr's *Twenty Years' Crisis* framed an explanation for the war in terms of the lagging 'mediaeval' thinking of the 'European ruling classes' combined with the 'nationalism' of the 'ordinary people'.[89] Although the lag argument played an important part in Mitrany's functionalist approach as expressed in 1943, it was only briefly stated. The 1946 edition of this essay made it more explicit. The

---

[86] H. G. Wells, *The New World Order: Whether It Is Attainable, How It Can Be Attained, and What Sort of World a World at Peace Will Have to Be* (London: Secker and Warburg, 1940), 36. See also page 30.

[87] William B. Ziff, Sr., *The Gentlemen Talk of Peace* (London: John Lane the Bodley Head, 1945), 107–108.

[88] Emery Reves, *The Anatomy of Peace* (New York: Harper Brothers, 1945), 27, 29, 42, 175–176, 184, 268, 274; H. J. Laski, 'The Theory of an International Society', in Laski, *Problems of Peace*, 6th Series, 188–209.

[89] Leonard Woolf, *The War for Peace* (London: George Routledge & Sons, 1940), 74–77.

nineteenth century 'saw the rise of national states' but at the same time new factors of 'communications, of new sources of power, of new materials, of the opening up of new lands and the rise of mass production' broke down barriers and 'bound peoples increasingly together'. This, then, is what had led to the central 'paradox' of the twentieth century, that 'social life has a highly integrated organic unity, but politically our outlook is bound to a mosaic of separate national units'.[90]

Textbooks

Much of this thinking on science and machines was reproduced in and popularized by leading US international relations textbooks in the 30s and 40s. They were also instrumental in popularizing the word 'technology' in international relations writing, helping it to become the standard term for modern world-changing scientific machines by the mid-forties. Professor of Government Clyde Eagleton's *International Government* was the more conventional of the two major internationalist textbooks of the 1930s. The 1932 edition (there were revised editions in 1948 and 1957) was essentially a description and history of the development of particular forms of international cooperation, including the League of Nations. These were explained as manifestations of modern international relations which, Eagleton insisted, needed to be envisaged as a 'community' of interdependent nations, formed on the basis of 'new inventions' ('steam and electronic railways and ships, telegraphs and telephone, newspapers, and now aviation, radio, and moving pictures') flowing out of the industrial revolution. Like other internationalists, he saw these inventions as the product of international 'science' ('the joint produce of men of various nationalities who combined their knowledge for the benefit of mankind') which had 'decreased the size of the earth, made mankind one interdependent community socially and economically, and has changed the material basis of civilization faster than habits and traditions can keep pace'.[91]

Internationalist political scientist Frederick L. Schuman's *International Politics* was by far the most widely read, and probably the best-selling, textbook of the 1930s. First published in 1933, it was revised and republished in 1937 and 1941, and then four times thereafter. *International Politics* suggested that just as all 'state systems' had in the past evolved to

---

[90] David Mitrany, 'A Working Peace System (1943)', in *The Functional Theory of Politics* (London: LSE & Political Science, 1975), 123–132.

[91] Eagleton, *International Government*, 10, 15. On 1930s textbooks see: Warren F. Kuehl, 'Webs of Common Interests Revisited: Nationalism, Internationalism, and Historians of American Foreign Relations', *Diplomatic History* 10, no. 2 (April 1986): 107–120.

some form of world state, so would the current 'Western state system'. The first three editions incorporated broadly the same arguments regarding science, 'technology' and this progressing international organization. These began with Reinsch's claim that International Public Unions marked the beginnings of a modern process of international organization. They were formed as a direct result of the 'industrial revolution' through which 'the new technology of the machine age introduced greater changes in the techniques of production, distribution, transportation, and communication than had occurred during the previous millennium'.[92] International organization consequently arose not because of the 'agitation of pacifists and reformers' or the need to deal with 'great issues of international politics', but rather out of the 'urgent necessity of international action in dealing with technical and routines matters'.[93] This was only the beginning of mankind's march to global integration, 'The world government of the future is already technologically possible':

Machine technology facilitates the surmounting of such barriers and makes possible an extension of State power over distances once regarded as fantastic. The world empires of today are existing realities made administratively possible by the new technology.[94]

Yet international organization was not progressing as fast as it should due to the lag between the 'the impact of science, technology, and the Industrial Revolution' and 'people unable or unwilling to adapt old habits to the stubborn facts of a changed world'. The word 'technology' was used singularly to refer to the complex of industrial and mechanical inventions emerging from the industrial revolution. This new 'system of technology' or 'machine industry' was created through the application of science to war, production, transportation, and communication, and had resulted in the 'unification of the world ... with no commensurate unification of the separate sovereignties of the World State System into a world polity. The Industrial Revolution changed a world of isolated, independent societies into a world of integrated, interdependent societies'. This lag required, Schuman suggested, a reformation of the system of international relations towards 'world unity'.[95]

Of the internationalist textbooks published during the war, two stand out for their emphasis on science and technology. The more popular of these was *Contemporary International Politics* (1940) by the political scientists Grayson Kirk and Walter Sharp. The book is notable for its widespread use of the term 'technology', particularly in a chapter titled

[92] Schuman, *International Politics*, 242.   [93] Ibid., 242.   [94] Ibid., 506–507.
[95] Ibid., 93. Frederick L. Schuman, *International Politics: The Western State System in Transition*, 3rd ed. (New York: McGraw-Hill, 1941), 689.

'Technology and Communications', and was the first textbook to fully reflect the developing thinking on science and technology within the wider discipline of the social sciences.[96] By the late 1930s social scientists had begun to articulate the relationship between the term 'technology' and science. John C. Merriam's article on 'The Relation of Science to Technological Trends' in the 1937 report of the Subcommittee on Technology of the National Resources Committee on *Technological Trends and National Policy*, for example, noted that, although science was distinct from technology, its contribution to the latter was self-evident, though 'only in part direct'. Instead 'inventive genius' mediated by applying the results of 'research' to transform technology. A 'relatively large percentage' of 'recent advances' were a result of the application of 'the contribution of science carried to application by engineering', including new developments in transport and communications such as the automobile and radio.[97] This understanding was expounded in *Contemporary International Politics*, which, like earlier textbooks, presented a handful of transport and communication 'inventions' (presumed to be civilian) as evidence of the increasing effect of technology on international relations.[98] 'The airplane, the motion picture, the talking picture, and the radio', the authors claimed, would be serving the cause of internationalism were they not being perverted by the forces of nationalism. On radio, for example, they concluded that 'so long as the world is rent by such deep political cleavages as have marked the 1930s, the instrumentality of radio is not likely to contribute anything of importance to the organization of peaceful international relations. Indeed, by perverting the radio to aggressive propaganda, fascism has pursued a diametrically opposite course'. Similarly, 'Because of the close connection of civil and military aviation, the progress of commercial air transport has been influenced, if not handicapped, by national military considerations.' They counterposed a future of globalized 'constructive internationalism' against a future of misused technology, of 'political isolationism, national self-sufficiency, and competitive militarism'.[99]

University of Washington-based political scientist Linden A. Mander's *Foundations of Modern World Society* (1941), the other prominent wartime internationalist textbook, chose instead to emphasize international technical organization. The book was arranged around international problems which 'nations cannot by themselves adequately organize' and so

---

[96] Kirk and Sharp, *Contemporary International Politics*.

[97] John C. Merriam, 'The Relation of Science to Technological Trends', in *Technological Trends and National Policy*, ed. National Resources Committee (Washington, DC: Government Printing Office, 1937), 91–92.

[98] Kirk and Sharp, *Contemporary International Politics*, 146–191. [99] Ibid., 166, 190, 191.

required 'international government'. Individual chapters tackled, amongst others, international health, crime, monetary issues, trade, population and resources. Rather than suggesting a Mitranian solution based on the integrative activities of international technical organizations, he suggested instead the formation of an international political organization: 'it would be the height of folly to restore the small nations as sovereign entities ... of what use to re-establish even a sovereign Britain or France?'[100]

Into the Post-war Period: Theorizing War and the Bomb

These approaches to science and machines persisted after the Second World War and were prominent enough to elicit a gathering at the University of Chicago in May 1948 on 'Technology and International Relations'. The assembled social scientists (mostly) repeated assumptions about science and technology noted in the sections above, but imported the atomic bomb into their Theorizing. William F. Ogburn's summing up of the papers concluded that both 'modern war inventions' (the atomic bomb, the air bomber, and the tank) and 'transportation inventions' (ship, railroad, the automobile, and most importantly the aeroplane) tended to amalgamate nations. Even if 'durable world state seems remote, the forces which help to produce the very large durable state are in existence'.[101] He espoused his lag approach in papers on 'Aviation and International Relations' and 'The Process of Adjustment to New Inventions'. Quincy Wright talked about the effects of 'war inventions' and 'peace inventions' and the consequent need for world federation in his paper on 'Modern Technology and the World Order'; the political scientist Robert Leigh reminded readers of the global integrative effects of 'mass communication inventions'; and historian A.P. Usher suggested that the 'present crisis in international relations' was 'primarily due' to the Industrial Revolution in his paper on 'The Steam and Steel Complex and International Relations'.[102]

[100] Linden A. Mander, *Foundations of Modern World Society* (Stanford: Stanford University Press, 1941), vi, vii, 882.

[101] William F. Ogburn, 'Introductory Ideas on Inventions and the State', in *Technology and International Relations*, ed. William F. Ogburn (Chicago: University of Chicago Press, 1949), 1–15.

[102] The only contrary approaches were by realist political scientists Bernard Brodie and William T. R. Fox who did not see international political integration as an outcome or response to the atomic bomb. William T. R. Fox, 'Atomic Energy and International Relations', in Ogburn, *Technology and International Relations*, 102–125; Bernard Brodie, 'New Techniques of War and National Policies', in Ogburn, *Technology and International Relations*, 144–173.

Certain themes emerged more strongly than others in post-war writings on international relations. Interdependence, though still referred to in internationalist texts, lost its previous prominent place in their narratives. The fourth (1948) edition of Schuman's *International Politics* barely mentioned interdependence.[103] Quincy Wright's last major internationalist text, published in 1955, de-emphasized the role of interdependence, though still made reference to it. The historical development of global systems of international relations ('the Hague System, through the League of Nations, to the United Nations'), he informed his readers, had progressed in step with the development of communications ('steady progress from the slow and infrequent sailboats, horseback riders ... through the more rapid and abundant steamboats and railroad trains ... telegraph, cable, and radio').[104] Both led to ever-increasing international government, though counter currents remained, even as 'material and technological tendencies make for world unity ... moral and sentimental tendencies make for national societies'.[105] New textbooks such as Norman Palmer and Howard Perkins' 1953 *International Relations* noted the existence and importance of interdependence, but characterized it as an economic phenomena, and did not highlight any mechanical or scientific foundation.[106] One significant exception was the 1948 edition of Eagleton's *International Government* which continued to dedicate a section to interdependence, only reducing it in the 1957 edition with a footnote explaining that it was no longer necessary to 'fully illustrate' the 'effects of interdependence'.[107]

Although the terms 'machine' and 'machine age' disappeared by the late forties, 'science' and 'technology' continued to be used. Schuman's fourth edition of *International Politics* (1948) continued to use 'technology' as a collective term for advanced machines and armaments, and framed the era of modern international relations as a technological age in which technology brought both benefits and threats. In this edition this point was illustrated through detailed studies of the use of modern armaments in the Second World War and failed attempts at the international control of the atomic bomb.[108] Quincy Wright, who had barely used the

[103] F. L. Schuman, *International Politics: The Destiny of the Western State System*, 4th ed. (New York: McGraw-Hill, 1948).

[104] Quincy Wright, *The Study of International Relations* (New York: Appleton-Century-Crofts, 1955), 275–276.

[105] Ibid., 276.

[106] Norman D. Palmer and Howard Perkins, *International Relations: The World Community in Transition* (Boston: Houghton Mifflin, 1953).

[107] Clyde Eagleton, *International Government*, 2nd ed. (New York: Ronald Press, 1948), 8–13; Clyde Eagleton, *International Government*, 3rd ed. (New York: Ronald Press, 1957), 8.

[108] Schuman, *International Politics*, 4th ed., 16, 61, 87, 98, 104, 381–391, 414, 925–937.

word 'technology' in the 1930s and 40s, now concluded in *The Study of International Relations* that:

Technological advance tends to lead to the dilemma of one world or none. Until there is a sufficiently general and simultaneous will to make international organization work one world can give little security. But so long as each state looks to itself alone for defense, no state will enjoy security in a technologically united world. Technological advance has left man with the necessity of understanding himself in society, as well as he understands nature and its control, if he is to solve his political problems.[109]

The political scientist John Herz, more than anyone else, put the bomb at the heart of his post-war theorizing on international relations. By doing so he echoed many of the themes already developed by liberal internationalists. By the late forties Herz had come to believe in the need for a synthesis of 'political realism' and 'political idealism', a synthesis which advocated achievable liberal aims.[110] Although his 'realist liberalism' called for a radically transformed international relations in the face of recent science-based invention, it actually restated long-standing thinking. He contended that from the interwar period onwards the 'territoriality' of modern states was being denationalized by various factors, the most important of which were 'air warfare' and 'atomic warfare'. These weapons, however, were not the end of it; the 'processes of scientific invention and technological discovery' would continue to create more deadly 'innovations'. Before the bomb, he concluded, internationalists who called for world government were 'utopians'. Now, however, the world had to move beyond national sovereignty in order to deal with these newly invented weapons. He expressed his preference for some form of international government, which, quoting prominent physicist J. Robert Oppenheimer, he described as a 'radical solution' rather than a 'conventional one'.[111]

Cultural lag theories were also given a new (but short-lived) lease of life by the bomb. Ogburn turned to apply his well-established sociological analysis to international relations after the war. A United Airlines-funded study *The Social Effects of Aviation*, published in 1946, developed a lag-based model of the relationship between aviation and nationalism. Ogburn assumed aviation to be inherently civilian in nature, and

---

[109] Wright, *The Study of International Relations*, 385.

[110] See in particular: John Herz, 'Idealist Internationalism and the Security Dilemma', *World Politics* 2, no. 2 (January 1950): 157–180; and John Herz, *Political Realism and Political Idealism: A Study in Theories and Realities* (Chicago: University of Chicago Press, 1951).

[111] John Herz, 'The Rise and Demise of the Territorial State', *World Politics* 9, no. 4 (July 1957): 473–93. See also: John Herz, *International Politics in the Atomic Age* (New York: Columbia University Press, 1959).

significant for international relations in that respect. Military aviation's effect was a deviation from aviation's natural influence, which was to allow for the formation of larger states within the world. But, and here is where Ogburn added his own twist, aviation could further either nationalism or internationalism within each state, depending on where the nation stood in terms of its nationalistic maturity – where it sat on a 'curve of nationalism'. Countries with differing languages, poor communications, and large sizes tended to have become 'over-expanded', and so there nationalism and national cohesion has some way to develop. In such states, which had low 'cohesiveness', aviation would bind the nation closer together, strengthening nationalism. In small states, which would already have high cohesiveness (where the curve of increasing nationalism had reached a plateau), civil aviation was more likely to inculcate internationalism. Military aviation, meanwhile, furthered nationalism. His conclusion was that aviation would strengthen regional groupings of states and eventually develop 'co-operation in world government'.[112] But, like others, Ogburn called for further study of the lag created by the invention of the atomic bomb, and suggested international control and the dispersion of the populations from America's largest cities as solutions to this lag.[113]

Duke University-based sociologist Hornell Hart produced the most articulated arguments relating the atomic bomb to lag. Hart presented his lag thesis through a paper on 'Technology and the Growth of Political Areas' at the Chicago institute mentioned earlier, and in papers published in the late forties and early fifties.[114] Like Ogburn he called for social science to be 'applied to the problems of directing international co-operation toward the protection of mankind from destruction by physical science applied to military technology'.[115] Modern international relations, he noted, was in crisis due to 'technological acceleration' which was manifesting through the invention of increasingly deadly weapons. This was leading to an 'atomic crisis' which was 'the result of the lag of social sciences behind the accelerating evolution of physical sciences'. Social sciences had, however, solved such lags in the past – he cited reduced instances of lynching, typhoid deaths, air fatalities, tuberculosis deaths, diarrhoea and

[112] William Fielding Ogburn, *The Social Effects of Aviation* (Boston: Houghton Mifflin, 1946), 686–706, 720, 723. Also: William Fielding Ogburn, 'Aviation and Society', *Air Affairs* 1, no. 1 (September 1946): 10–20.
[113] William Fielding Ogburn, 'Sociology and the Atom', *The American Journal of Sociology* 51, no. 4 (January 1946): 267–275.
[114] Hornell Hart, 'Technology and the Growth of Political Areas', in Ogburn, *Technology and International Relations*, 29–57.
[115] Hornell Hart, 'Technological Acceleration and the Atomic Bomb', *American Sociological Review* 11, no. 3 (June 1946): 277–293.

enteritis, and railway fatalities.[116] He also plotted graphs of the earlier developments of various bodies of knowledge as they became increasingly scientized to demonstrate that the social sciences should also, in theory, be able to close the current gap with the physical sciences. Once this was done, mankind would have the instruments to be able to control 'technological acceleration' and atomic energy.[117] Hart's work was to be the last gasp for lag arguments in international relations. They were soon to disappear from sociology as well: the final significant outing was a 1964 reprint of some of Ogburn's research.[118]

As the liberal internationalist colour within the discipline of international relations faded into the early Cold War, so too did many of these long-standing approaches to science and technology. International relations theorists and textbooks no longer referred to lags or technologically driven interdependence as revolutionary transformations in international relations. New works focused on power dynamics made science and machines appear secondary to the understanding of world order. Political scientist Hans Morgenthau, in his assault on 'scientific' attempts to solve the problem of war (*Scientific Man Versus Power Politics*, 1946), went as far as to explicitly challenge interdependence arguments directly. He pointed out that the experience from the domestic realm was that 'modern communications' had not created new types of political unity but had instead strengthened those that 'existed before and independently of the development of modern technology'. So, although 'this is "one world" technologically', it would not develop into 'one world politically', because the world was deeply divided politically.[119] Rather than scientific or technologically driven change, the great 'moving force' within Morgenthau's conceptualization of international relations was, as he put it in his *Politics Among Nations* (1948), 'the aspiration for power of sovereign nations'.[120] The notion of a scientifically transformed warfare survived longer thanks to atomic weapons, but it disappeared as the discipline accommodated itself to the strategy of mutually assured destruction.[121]

[116] Hornell Hart, 'Some Cultural-Lag Problems Which Social Science Has Solved', *American Sociological Review* 16, no. 2 (April 1951): 223–227.
[117] Ibid. See also: Hornell Hart, 'Atomic Cultural Lag: I. The Value Frame', *Sociology and Social Research* 32 (March 1948): 768–775; and Hornell Hart, 'Atomic Cultural Lag: II. Its Measurement', *Sociology and Social Research* 32 (May 1948): 845–55.
[118] Ogburn, *On Culture and Social Change*.
[119] Hans Morgenthau, *Scientific Man Versus Power Politics* (Chicago: University of Chicago Press, 1946), 79–80.
[120] Hans Morgenthau, *Politics Among Nations* (New York: Alfred A. Knopf, 1948), 8.
[121] Lawrence Freedman, 'The First Two Generations of Nuclear Strategists', in *Makers of Modern Strategy from Machiavelli to the Nuclear Age*, ed. Peter Paret (Princeton, NJ: Princeton University Press, 1986), 735–778.

2 Controlling Scientific War
International Air Police and the Reinvention of Disarmament

In 1930 the industrialist and one-time Liberal MP David Davies published *Problem of the Twentieth Century*, a 795-page rumination on the problems of war and anarchy in international relations and how they might be solved. The only solution, he argued, was the formation of a powerful international police force controlled by an international organization. Science-based 'modern weapons', prominently submarines, heavy artillery, bombers, and poison gas, were to be removed from national military forces and instead monopolized by this international force. The air arm was to be the dominant component of his international police: the focus of investment and research and development, it was to have its own factories for the manufacture of fighters and bombers. The force was to usher in a new era of international law and security by enforcing the decisions of new and powerful international judicial and arbitral tribunals.[1]

David Davies was not a writer of science fiction, and nor was he a marginal figure in British intellectual and political circles. A wealthy coal industrialist, a Liberal MP and then Peer, Davies was a regular speaker on issues of war and peace after the First World War who endowed the first ever chair in international relations in 1919, in Aberystwyth in his native Wales. Named in honour of Woodrow Wilson, the first holder was the leading internationalist scholar Alfred Zimmern. Davies remained through to the 1930s a powerful force in Liberal political circles. Although his proposal for an international police force was radical, it was by the late 1920s not uncommon in proposals for international organization and collective security.[2]

This chapter examines the growth of schemes such as Davies's, and how they became central to internationalist attempts at disarmament. In the 1920s aeroplanes emerged at the centre of a rejuvenated internationalism,

---

[1] Davies, *The Problem of the Twentieth Century*, 318, 360–382, 390, 425–502.
[2] On Davies see: Neil D. Bauernfeind, 'Lord Davies and the New Commonwealth Society 1932–1944' (MPhil diss., University of Wales at Aberystwyth, 1990); Brian Porter, 'David Davies and the Enforcement of Peace', in Long and Wilson, *Thinkers of the Twenty Years' Crisis*, 58–78.

which looked to science and modern machines as the linchpin of collective security and international law. These schemes were based on new conceptions of disarmament, war, armaments, and aviation, which emerged and spread during and after the First World War. As the conflict came to be conceptualized as a scientific war, and modern armaments as products of modern science, liberal internationalists came to see disarmament itself as a scientific project requiring guidance from technical experts schooled in armaments or international relations. Military aviation, in particular, came to be conceptualized as the most powerful application of science and technical expertise in the military sphere. These ideas converged in proposals for the international air force: for internationalists such as Davies aviation came to represent the ultimate application of science to international policing, and the only weapon capable of policing disarmament. At a time of a wider technocratic turn in international governance, aviation's apparent success in policing vast imperial territories cheaply and effectively made it a doubly promising tool for international governance.

Scientific Warfare and the First World War

Although intellectuals had been warning of a new age of industrial warfare since the Crimean War, the First World War altered as well as intensified these concerns.[3] The war appeared to inaugurate a radically new type of modern warfare in which civilian science and industry were for the first time mobilized for the creation of modern scientific armaments, and in which the nation itself was mobilized, and war organized, in a scientific manner.[4] On the one hand, this new understanding focused on the widespread application of science and the novelty and destructiveness of the weapons it created. It was commonly argued, as John Fletcher Moulton (head of British explosives production during the War) did in his 1919 Rede Lecture, that the Great War represented 'the results of the totality of scientific progress'. It was 'the realization of all that which the accumulated powers with which Science has endowed mankind can effect when used for destruction'.[5] Although explosives and aviation were also seen as products of the application of civilian science, it was poison gas which stood out for contemporaries both in its scientific foundations and

---

[3] Christopher Coker, *The Future of War: The Re-Enchantment of War in the Twenty-First Century* (Oxford: Blackwell, 2014), 2–6.

[4] For the disjuncture between this belief and reality, see: Anne Rasmussen, 'Science and Technology', in *A Companion to World War I*, ed. John Horne (Chichester: Wiley-Blackwell, 2010), 307–322.

[5] John Fletcher Moulton, *Science and War: The Rede Lecture* (Cambridge: Cambridge University Press, 1919), 7.

its destructiveness.[6] Chemical weapons expert Victor Lefebure explained in a lecture to the Grotius Society in 1921 that the 'applications of science to warfare' occurred in two ways. Gradual 'technical advances', for example, in range-finding and machine-gun improvements, did not 'vastly outdistance similar measures or counter-measures on the part of a rival'. 'Revolutionizing' scientific applications, such as chemical warfare, however, could not be defended against.[7]

Civil servants, politicians, scientists, and industrialists, on the other hand, stressed that civilian science's contribution to the war had been beneficial both during the war and would continue to be during peacetime as well.[8] In journalist and editor George A.B. Dewar's 1921 history of wartime munitions production, the 'great munition feat' was a lesson in the application of science and rational organization that would carry through to the peacetime years and allow for more efficient industrial production across civilian and military industry.[9] Physiologist J.B.S. Haldane, in his defence of chemical warfare during the war, argued that greater scientific education and scientific study of war was needed so that the public and the military could appreciate that poison gas was more humane than 'cruel and obsolete killing machines'.[10] Eugenicist Julian Huxley's 1934 *Scientific Research and Social Needs* stressed that, for as long as the threat of war existed, scientists should continue to make 'warfare as efficient as possible' in order to lower costs and shorten wars. This would require a 'scientific' prosecution of warfare, for which he suggested the formation of a 'War Services Research Council' akin to the Department of Scientific and Industrial Research. Huxley highlighted the peacetime applications of scientific developments made during the war, including being able to fly from Europe to Africa or Asia, new industrial alloys, caterpillar tractors, optical instruments, and modern psychology.[11]

The scientific development of warfare appeared inevitable, powered by the necessity to press all national resources for the war effort. 'Nations

---

[6] Marion Girard, *A Strange and Formidable Weapon: British Responses to World War I Poison Gas* (Lincoln: University of Nebraska Press, 2008), 157–190.

[7] Victor Lefebure, 'Chemical Warfare: The Possibility of Control', *Transactions of the Grotius Society* 7 (1921): 153–166.

[8] David Edgerton, 'British Scientific Intellectuals and the Relations of Science, Technology and War', in *National Military Establishments and the Advancement of Science and Technology Studies in the 20th Century*, eds. P. Forman and J. M. Sanchez-Ron (Dordrecht: Kluwer Academic, 1996), 1–35.

[9] George A. B. Dewar, *The Great Munition Feat 1914–1918* (London: Constable, 1921), vii.

[10] J. B. S. Haldane, *Callinicus: A Defence of Chemical Warfare* (London: Kegan Paul, Trench, Trubner, 1925), 28.

[11] Julian Huxley, *Scientific Research and Social Needs* (London: Watts, 1934), 154–160, 167.

cannot be blamed for arming themselves with all the resources of science and invention', noted the Liberal/Labour MP J.M. Kenworthy in his 1927 work on international relations, *Will Civilization Crash?* 'Once war has broken out the best brains of the chemists and scientists will be prostituted to its service'. His chapter on 'The Application of Science in Battle' described how the two most important military applications of science, aviation and poison gas, would be the 'determining factors in the next war'.[12] Communist crystallographer J.D. Bernal's *The Social Function of Science* (1939) traced science's contribution to war back to ancient times but argued that it was during the First World War that scientists 'found themselves for the first time not a luxury but a necessity to their respective governments'. Mechanization transformed warfare, whilst science itself was transformed through the institution of state direction. He singled out 'aerial and chemical warfare' as the two most significant products of this revolution. Bernal was critical of this 'prostitution' of scientists and hoped for an eventual separation of the 'constructive and destructive aspects of science'.[13]

The scientization of armaments argument fitted neatly into long-standing liberal apprehensions over accelerating arms races and their increasing financial burden.[14] The First World War and broader concerns over military spending led economists such as Arthur Cecil Pigou to produce analyses which emphasized the prohibitive cost of increasingly complex armaments, as well as the cost of the withdrawal of scientists and 'ingenious machines' from the national economy.[15] F.W. Hirst (one-time editor of the *Economist* and member of the Cobden Club) highlighted 'new tools and machinery for the building of more and more bombing aeroplanes, new and enlarged laboratories for the preparation of poisonous gases, and other devilish horrors of modern warfare' as a central aspect of the 1930s arms race.[16] The scientific nature of armaments destabilized international affairs for other reasons too: scientific weapons increased the 'potential swiftness and deadliness of an attack', argued

---

[12] J. M. Kenworthy, *Will Civilization Crash?* (London: Ernest Benn, 1927), 238. Similarly: Philip Kerr and Lionel Curtis, *The Prevention of War* (New Haven, CT: Yale University Press, 1923), 11–12.

[13] J. D. Bernal, *The Social Function of Science* (London: George Routledge and Sons, 1939), 29–32, 165–190.

[14] Joseph Maiolo, 'Introduction', in *Arms Races in International Politics: From the Nineteenth to the Twenty-First Century*, eds. Thomas Mahnken, Joseph Maiolo, and David Stevenson (Oxford: Oxford University Press, 2016), 1–2; Pugh, *Liberal Internationalism*, 51–52.

[15] A. C. Pigou, *The Political Economy of War* (London: Macmillan, 1921), 4. See also Arthur Guy Enock, *The Problem of Armaments: A Book for Every Citizen of Every Country* (London: Macmillan, 1923).

[16] Francis W. Hirst, *Armaments: The Race and the Crisis* (London: Cobden-Sanderson, 1937), 72.

disarmament expert (and Professor of International Relations at the London School of Economics) Philip Noel Baker, leading to an increased 'sense of insecurity' and ultimately the primacy of militaristic thinking in foreign affairs.[17]

The focus on new armaments and their identification with civilian science was echoed by those calling for the British military to support new types of technological warfare. Military strategist Major-General J.F.C. Fuller highlighted the importance of tanks for the British war effort in support of his calls for greater mechanization of the British army. His 1920 *Tanks in the Great War* characterized tanks as part of a wider turn from 'hand-weapons to machine-weapons', which mirrored the change from 'hand-tools' to 'machine-tools' in industry. These new machine weapons (artillery, explosives, machine guns, poison gas, mechanical detecting devices, aeroplanes, lorries, railways, and tanks) were the work of science: his book was dedicated to the 'modern military scientist, that small company of gentlemen who, imbued with a great idea, were willing to set all personal interest aside in order to design a machine destined to revolutionize the science of war'.[18]

Later writings developed these ideas further. Fuller's 1923 *The Reformation of War* argued that the First World War demonstrated that conservative militaries had not realized that warfare had evolved due to these new weapons. *Reformation* advocated their further use so as to bring war to a quicker and so less bloodier end. This required new tactical and strategic doctrines, as well as scientific research and development. Indeed, 'invention' was now an 'important branch of strategy' itself, and 'that side which gains supremacy in invention and design' would win the next war. He advocated the creation of a specific 'weapon-producing department' for the military with separate research, design, and experiment divisions manned by both military and civilian scientists.[19] Fuller's programmatic *The Foundations of the Science of War* (1926) was, partially at least, a rational framework for the deployment of these new weapons.[20] Fuller was not alone in connecting science, weapons, and the Great War in this way. Winston Churchill's history of the First World War, *The World Crisis, 1916–1918* (1927), presented modern 'war machinery' as crucial to the win on the Western Front: artillery, aviation, 'railway or mechanical transport', mortars, tanks, and poison gas. In subsequent writings Churchill would often muse over the scientific transformation of the

---

[17] P. J. Noel Baker, *Disarmament* (London: Hogarth Press, 1926), 21.
[18] J. F. C. Fuller, *Tanks in the Great War, 1914–1918* (London: John Murray, 1920), ix, xviii, xix, 299.
[19] J. F. C. Fuller, *The Reformation of War* (London: Hutchinson, 1923), 134, 168–169, 234.
[20] J. F. C. Fuller, *The Foundations of the Science of War* (London: Hutchinson, 1926).

world through faster communications and production, and deadlier warfare.[21]

Scientific Disarmament

The belief that new scientific armaments needed scientific disarmament had its origins in the growing movement for arms control during the war. By 1919 support for general disarmament was strong enough for the Preamble to the Paris Treaty's Disarmament Clauses to state that Germany was being disarmed as a prelude to the 'general limitation of the armaments of all nations'. Under Article 8 of the Covenant of the League of Nations, members of the League committed themselves to 'the reduction of national armaments to the lowest point consistent with national safety'.[22] Various schemes for disarmament were presented at the League in the early 1920s, and in 1926 a Preparatory Commission was tasked with the production of a draft agreement for final debate at a future disarmament conference. A successful agreement on the reduction of naval armaments in 1921–22, and the signing of the Kellogg-Briand Pact of 1928, which condemned aggressive war, appeared to demonstrate that agreement was not a pipe dream.[23] These initiatives spurred considerable interest in disarmament amongst British internationalists, for whom it emerged as a major policy goal. The lesson they learnt from these attempts was that, although agreement on arms limitations was possible, it could only be achieved as part of a wider international political settlement – one realized through the strengthening of international organization and collective security. In the words of internationalist Salvador de Madariaga (Section Head of the Disarmament Section of the Secretariat of the League of Nations, 1922 to 1928): 'Questions of disarmament resolve themselves into questions of security and the questions of security into questions of international organization'.[24]

---

[21] Winston Churchill, *The World Crisis*, vol. 3, *1916–1918* (London: Thornton Butterworth, 1927), 307; Justin D. Lyons, 'Strength without Mercy: Winston Churchill on Technology and the Fate of Civilization', *Perspectives on Political Science* 43, no. 2 (2014): 102–108.

[22] League of Nations, 'The Covenant of the League of Nations (Including Amendments adopted to December, 1924)', Avalon Project, accessed 2 August 2020, https://avalon.law.yale.edu/20th_century/leagcov.asp#art8.

[23] Andrew Webster, '"Absolutely Irresponsible Amateurs": The Temporary Mixed Commission on Armaments, 1921–1924', *Australian Journal of Politics and History* 54, no. 3 (September 2008): 373–388; Thomas Richard Davies, *The Possibilities of Transnational Activism: The Campaign for Disarmament between the Two World Wars* (Leiden: Martinus Nijhoff, 2007), 55–84; David Macfadyen et al., *Eric Drummond and His Legacies: The League of Nations and the Beginnings of Global Governance* (Cham, Switzerland: Palgrave Macmillan, 2019), 133–134.

[24] Salvador de Madariaga, *Disarmament* (New York: Coward-McCann, 1929), 194. See also: Andrew Webster, 'From Versailles to Geneva: The Many Forms of Interwar Disarmament', *The Journal of Strategic Studies* 29, no. 2 (April 2006): 225–246. On the

From the late 1920s onwards internationalists linked general disarmament to international organization and collective security. Robert Cecil (Viscount Cecil of Chelwood, a leading supporter of disarmament and the League in Britain, and then chairman of the Executive Committee of the League of Nations Union), informed the Foreign Policy Association in New York in 1923 that 'if you are to carry out a general scheme of disarmament, you must have an international organization to supervise it'. More than that, 'material disarmament' also required 'moral disarmament' – a willingness to co-operate internationally. That once again could only be achieved through an over-arching international organization.[25] Similarly, for League official Arthur Salter, disarmament and collective security were 'an inseparable part of a double problem ... they must come together ... Neither can advance far without the other'.[26] Noel Baker welcomed the 1924 Geneva Protocol for the Settlement of International Disputes as 'not only a project for disarmament, but the foundation of a system of legal justice among nations, backed by the rudimentary beginnings of an international police'.[27] Looking back in 1927, John Wheeler-Bennett (editor of the Royal Institute of International Affairs' *Bulletin of International News*) believed that it was during debates on the Esher Plan for disarmament at the Temporary Mixed Commission on Armaments in 1922 when it became clear that 'security must precede disarmament', and both became 'integral parts of one formula'.[28]

For most internationalists this holistic approach required a range of armaments to be tackled, rather than just one. Simply focusing on the navy, argued Noel Baker after the failure of the 1927 Geneva Naval Conference, was pointless as the greatest danger to Britain and her navy came from Air Forces, 'a danger against which there is no adequate defence'.[29] Some internationalists elevated disarmament to an existential issue for the League. When addressing the Geneva Institute of International

centrality of disarmament for interwar liberal internationalism see: Andrew Webster, 'The League of Nations, Disarmament and Internationalism', in Sluga and Clavin, *Internationalisms*, 139–169.

[25] Robert Cecil, 'Disarmament and the League', in *The Way of Peace: Essays and Addresses* (London: Philip Allan, 1928), 211–232. On Cecil and disarmament see: Gaynor Johnson, *Lord Robert Cecil: Politician and Internationalist* (Abingdon: Routledge, 2016), 121–142, 159; Peter J. Yearwood, *Guarantee of Peace: The League of Nations in British Policy 1914–1925* (Oxford: Oxford University Press, 2009), chapter 5.

[26] Arthur Salter, 'Disarmament – the Prospects', *Listener*, 26 October 1932, 577–579.

[27] P. J. Noel Baker, *The Geneva Protocol for the Pacific Settlement of International Disputes* (London: PS King & Son, 1925), 2.

[28] J. W. Wheeler-Bennett, *Information on the Problem of Security* (London: George Allen & Unwin, 1927), 91–94.

[29] P. J. Noel Baker, *Disarmament and the Coolidge Conference* (London: Hogarth Press, 1927), 40–43. See also Noel Baker, *Disarmament*, 48–49.

Relations in 1927 on preparation for the planned Geneva disarmament conference, Madariaga highlighted the Conference's importance by noting that disarmament was the 'first in chronology and the first in importance of the tasks which the League must carry out if it is to survive'.[30]

Internationalists buttressed their proposals by arguing that only radical solutions would bring real advances in international relations. David Davies, writing to Noel Baker in 1926, argued that disarmament would not be achieved if it was envisaged simply as a 'negative' policy of 'reduction in numbers, a reduction in equipment, the lopping-off here and there of certain units of military and naval establishments'. He instead stressed the need for a 'positive policy', a 'constructive proposal' that would achieve a 'real advance' in international relations and 'at the same time almost unconsciously to bring about an enormous reduction in existing armaments'.[31] For Noel Baker, writing soon after the failure of the 1932 Geneva disarmament proposal became apparent, the snail-paced and ultimately deadlocked discussions over the technical details of disarmament, as well as 'political' disagreements, were best overcome 'by drastic, rather than by moderate disarmament plans'.[32]

Proponents of world federation, such as H.G. Wells, took the disarmament conferences of the 1920s as an opportunity to argue that arms limitations would not stop war. Disarmament and international peace first required closer political integration between the leading world powers.[33] Wells had, in fact, incorporated scientific weapons into his earliest calls for international organization. His 1919 article in the *Atlantic Monthly*, 'The Idea of a League of Nations', presented this new warfare as the primary reason why such a League was now required. Modern 'unlimited warfare' had now expanded beyond all historical social boundaries, forcing nations to press every resource into service, including science: 'The outbreak of that struggle forced upon the belligerents, in spite of the natural conservatism of all professional soldiers, a rapid and logical utilization of the still largely neglected resources of mechanical and chemical science; they are compelled to take up every device that offered, however costly it might be; they could not resist the

---

[30] Salvador de Madariaga, 'The Preparation of the First General Disarmament Conference', in *Problems of Peace*, 2nd Series: Lectures Delivered at the Geneva Institute of International Relations, eds. William E. Rappard, et al. (London: Oxford University Press, 1928), 124–142.

[31] David Davies, 'Memorandum on Disarmament', to Philip Noel Baker, 5/67–1, The Papers of Baron Noel-Baker, Churchill Archives Centre, Churchill College, University of Cambridge (hereafter cited as Noel-Baker Papers).

[32] Philip Noel Baker, 'Peace and the Official Mind', in *Challenge to Death*, eds. Noel Baker, et al. (London: Constable, 1934), 87.

[33] H. G. Wells, *Washington and the Riddle of Peace* (New York: Macmillan, 1922), 25, 295.

drive toward scientific war which they had themselves released'. The most prominent examples Wells dwelled upon were the air force and the tank (an idea which was 'an old and very obvious one', but which had been 'disliked and resisted by military people for many years'). He emphasized, however, that this was a broader process, involving '[e]normous new organizations of anti-submarine flotillas, of mine-field material and its production, of poison-gas manufacture' and transportation such as the 'motor-lorry'. This usurpation of science and scientists for military purposes had meanwhile denuded the civilian science left behind. University education had 'practically ceased' and 'the laboratories of the English public schools are no longer making the scientific men of the future, they are making munitions'.[34] Important here was the notion (widespread amongst British scientific intellectuals) that the military was inherently conservative and not modern, and that the usurpation of science by the military occurred not because of the natural inclinations of the military, but against them. The military had only reluctantly been forced to turn to science by the needs of war.[35]

One new direction for disarmament was signposted by the new understandings of scientific warfare. Internationalists argued that this new scientific warfare required new scientific approaches to disarmament by technical specialists. Noel Baker located the armaments revolution as having occurred before the Great War, during the last quarter of the nineteenth century, when 'modern' weapons were born due to the 'application of modern science to war', particularly 'steam and electricity and chemistry'. The continuous 'invention' of scientific weapons was now so rapid that simply outlawing certain types of warfare, such as the bombing of cities, was bound to fail as such laws would become outdated rapidly. Even 'fixed limitations' on armaments was inadequate, and the 'old international law of war' was itself unable to control or limit modern warfare.[36] These modern armaments had made disarmament such a technical matter that it required specialized experts (such as himself) for its analysis, negotiation, and implementation.[37] It was not possible, claimed political scientist C. Delisle Burns, to reverse the movement of 'research and invention of new weapons' from private arms manufacturers into government departments, where it was securely lodged.[38] Classicist G. Lowes Dickinson believed that disarmament could at least

---

[34] Wells, *The Idea of a League of Nations*, 8, 9–12, 21.
[35] See: Edgerton, 'British Scientific Intellectuals'.
[36] P. J. Noel Baker, 'Disarmament', *Economica* 6, no. 16 (March 1926): 1–15. Frederick Maurice, 'Disarmament', *Journal of the Royal Institute of International Affairs* 5, no. 3 (May 1926): 117–133.
[37] Noel Baker, *Disarmament*, 42.   [38] Burns, *Modern Civilization on Trial*, 167.

partially be achieved if scientists withheld their services from the state.[39] H.G. Wells, commenting on the Washington conference, thought it useless to focus on particular armaments, as technical expertise was having a much broader effect on the business of war-making. Abolishing this expertise was consequently undesirable: 'We do not want the extinction of this great body of business, metallurgical, chemical, engineering and disciplined activities, but we do want its rapid diversion from all too easily obtained destructive ends to creative purposes now'. Modern 'financial and engineering energy' could be unleashed through world government projects for the improvement of transport and communications, urban reconstruction, and irrigation.[40]

Perhaps the best-known calls for a scientific approach to disarmament in the 1920s and early 1930s came from the industrial chemist and military advisor Victor Lefebure, whose lectures and writings became an important source of support for internationalists and scientists calling for more experts to be involved in disarmament negotiations.[41] Starting from the premise that negotiations had failed because diplomats and politicians had been unable to work and think dispassionately, Lefebure called for a more rational and technical approach to arms control. His 1931 *Scientific Disarmament* explained that, because 'the study of armament is a branch of applied science' and 'disarmament is a technical subject based on facts surrounding armament', arms control required the application of a 'technical approach' which took into consideration the 'scientific and technical characteristics of armament'. He emphasized that the new modern nature of warfare rested on a country's industrial might, and stressed the convertibility of civilian to military technologies and industries, especially in relation to chemicals and aviation. But he then argued for the existence of a 'conversion lag', being the time taken to convert any particular civilian industry to military use. This lag, which varied from industry to industry and country to country, he reasoned, meant that the abolition or internationalization of whole industries and technologies was unnecessary. Disarmament regulations could instead focus on making the conversation of civilian facilities to military armaments as difficult as possible. A scientifically determined mix of measures, including supervision of research and development, limitations on aerial personnel, and the

---

[39] G. Lowes Dickinson, *War: Its Nature, Cause and Cure* (London: George Allen and Unwin, 1923), 38–39.

[40] Wells, *Washington and the Riddle of Peace*, 31.

[41] Victor Lefebure, *The Riddle of the Rhine: Chemical Strategy in Peace and War* (London: Collins, 1921); Lefebure, 'Chemical Warfare'; Victor Lefebure, *Common Sense about Disarmament* (London: Victor Gollancz, 1932); Editorial, 'A Scientific Approach to Peace', *Nature* 134, no. 3394 (17 November 1934): 749–751. On Lefebure see: Girard, *A Strange and Formidable Weapon*, 165.

banning of bombardment, would protect national civilian industries from immediate misuse by the state for military purposes.[42]

Chariots of the Air

Although scientific warfare would become central to the concept of the international air force, the idea of international policing itself had distinctly liberal origins dating back to the nineteenth century, when the Royal Navy was widely considered to be the protector of the world's sea lanes and the guarantor of international law and free trade. By the end of the century, with the rise of other power navies, calls emerged for the formation of a multinational naval police force. These calls were particularly prominent in the growing movements for international law in the years leading up to the First World War, and were rooted more broadly in the growth of liberal internationalism in Britain from the latter half of the nineteenth century onwards.[43] They were also a manifestation of growing US naval power: one influential call was made in 1910 by ex-President Theodore Roosevelt (including in his Nobel Prize Lecture).[44] Roosevelt and those he inspired envisaged a naval force composed largely of British and American ships; for junior naval officer Thomas Kinkaid (later Admiral), the force was to be composed of 144 battleships, 144 destroyers, and 36 scout ships, and be stationed in 'neutralized harbours and islands' around the world.[45]

The First World War accelerated calls for an international force, both in the United States and in Europe, and proposals were increasingly linked to the growing calls for a comprehensive international organization to keep peace after the war. Prominent British liberal internationalist thinkers on international relations such as Norman Angell and J.A. Hobson called for such a force, as did internationalist legal scholars such as Cornelis van Vollenhoven.[46] In Britain it was envisaged to be an

---

[42] Victor Lefebure, *Scientific Disarmament* (London: Mundanus, 1931), 17, 21.

[43] Roger Beaumont, *Right Backed by Might: The International Air Force Concept* (Westport, CT: Praeger, 2001), 4–7; Sylvest, *British Liberal Internationalism*, 37–38.

[44] Theodore Roosevelt, 'Acceptance Speech', The Nobel Prize, accessed 2 August 2020, www.nobelprize.org/prizes/peace/1906/roosevelt/acceptance-speech.

[45] T. W. Kinkaid, [untitled], in *War Obviated by an International Police: A Series of Essays, Written in Various Countries*, eds. C. van Vollenhoven et al. (The Hague: Martinus Nijhoff, 1915), 158–162. Also: C. F. Goodrich, 'Wanted – an International Police', in Vollenhoven, *War Obviated by an International Police*, 172. See also: James R. Holmes, *Theodore Roosevelt and World Order: Police Power in International Relations* (Washington, DC: Potomac Books, 2006), 122–127.

[46] Hobson, *Towards International Government*, 72–89; Norman Angell, 'Introduction', in *Nationalism, War and Society*, ed. Edward Krehbiel (New York: Macmillan, 1916), xiii–xxxv; Vollenhoven, *War Obviated by an International Police*. For pre-World War One

Anglo-American force, representative of their global power and responsi-bility – Norman Angell's internationalist tract *The World's Highway* (1915), for example, called for the 'internationalisation of sea control' through combined Anglo-American naval power. The suitability of naval forces for policing was emphasized through the global reach of the war-ship. For journalist and writer H.N. Brailsford naval forces offered a new and unique ability to 'strike in all the four quarters of the earth' and were to be the power behind his suggested League of Nations.[47] In the United States the League to Enforce Peace, formed in 1915, was perhaps the most influential of the several organizations calling for an international police to enforce the decisions of a new international judicial court.[48] These ideas were widespread enough to enter fiction. New York salesman Allen Robert Dodd's novel *Captain Gardiner of the International Police*, published in 1916 but set a few decades after the First World War, envisaged a world ruled by an 'international federation', armed with an international police consisting of an 800,000-strong army and a navy of forty-eight battleships alongside cruisers, destroyers, and submarines.[49]

In the United States growing fear of foreign entanglements, crowned by the Senate's refusal to ratify the League of Nations Charter in 1919, led to the post-war disappearance of proposals for an international force. The League to Enforce Peace dissolved in 1923, and internationalists instead focused on less radical aims, including deepening ties with the League, pushing for more American participation in international conferences and the International Court of Justice, and increasing international non-governmental interaction. The Carnegie Endowment, which had always been against an international force (the Director of its Division for International Law, speaking against the League to Enforce Peace in 1917, called instead for a 'campaign of systematic instruction' on inter-national law and justice) emerged, alongside other philanthropic founda-tions, as a powerful focal point for US internationalist engagement in the 1920s and 1930s.[50]

proposals see: Davies, *The Problem of the Twentieth Century*, 110–115; J. M. Spaight, *An International Air Police* (London: Gale and Polden, 1932), 29–45; and Hans Wehberg, 'La Police Internationale', *Recueil des Cours* 48 (1934 part II): 1–132.

[47] Norman Angell, *The World's Highway* (New York: George H. Doran, 1915), 331; H. N. Brailsford, *A League of Nations*, 2nd ed. (New York: Macmillan, 1917), 201–202.

[48] Ruhl Jacob Bartlett, *The League to Enforce Peace* (Chapel Hill: University of North Carolina Press, 1944); Martin David Dubin, 'Elihu Root and the Advocacy of a League of Nations, 1914–1917', *The Western Political Quarterly* 19, no. 3 (September 1966): 439–455.

[49] Allen Robert Dodd [Robert Allen], *Captain Gardiner of the International Police* (New York: Dodd, Mead, 1916), 16.

[50] James Brown Scott, 'International Organization – Executive and Administrative', *Advocate of Peace through Justice* 83, no. 4 (April 1921): 133–136. For more on this see:

In Britain, on the other hand, calls for an international police force continued to grow, becoming a staple of internationalist demands on international relations by the late 1920s. There were publications, discussions within internationalist organizations, letters to the national press, petitions to Government, and proposals put before the League Assembly. The Grotius Society emerged as an early venue for debates on such proposals, and the League to Abolish War, formed in 1915 to push for an international police force, sent a delegation to meet the prime minister in June 1920.[51] Most envisaged that the force would be controlled by the League, empowering the organization by enforcing its decisions. The force also came to be seen as an indispensable requirement for general disarmament and, for some, the solution to the anarchy in international relations. Only through such a force, it came to be believed, would there be sufficient confidence in international organization and in a peaceful future for nations to consider arms reductions. It was to impose justice, rationalism, and order into international affairs, and promote and protect open international trade, commerce, communication, and travel.

The concepts of scientific warfare and weapons became central to calls for an international force, and particularly so for David Davies, who emerged as its most forceful proponent in the interwar years. Davies served in the First World War and returned with a deep sense of the scientific and mechanical transformation of warfare. He laid out his early thoughts at a Grotius Society meeting in November 1919, and then in a letter to *The Times*, published in April 1920. Like other internationalists, his rhetoric emphasized the centrality of scientists and invention to the creation of new weapons which had caused such bloodshed and suffering during the war, and were now fuelling an international arms race. In 1919 he pointed broadly to 'bigger guns, tanks, warplanes, poison gas and submarines' as the new inventions, but by the time of his 1930 *The*

Warren F. Kuehl and Lynne K. Dunn, *Keeping the Covenant: American Internationalists and the League of Nations, 1920–1939* (Kent, OH: The Kent State University Press, 1997); Inderjeet Parmar, *Foundations of the American Century: The Ford, Carnegie and Rockefeller Foundations in the Rise of American Power* (New York: Columbia University Press, 2012); Katharina Rietzler, 'Fortunes of a Profession: American Foundations and International Law, 1910–1939', *Global Society* 28, no. 1 (2014): 8–23.

[51] David Davies, 'Disarmament', *Transaction of the Grotius Society* 5 (1919): 109–118; David Davies, letter to the editor, 'More Armaments: A Menace to Future Peace, An International Police', *The Times*, 1 April 1920; League of Nations, *Arbitration, Security and Reduction of Armaments: Extracts from Debates of the Fifth Assembly* C.708.1924.IX (C.C.O.1.) (Geneva: League of Nations, 1924), 155, 271–274. 'International Police: Deputation to the Prime Minister', *The Times*, 17 June 1920; 'No League Police: Prime Minister on US Withdrawal, "Pressure" on Poland and Russia', *The Times*, 19 June 1920. For more on this see: Waqar H. Zaidi, '"Aviation Will Either Destroy or Save Our Civilization": Proposals for the International Control of Aviation, 1920–45', *Journal of Contemporary History* 46, no. 150 (2011): 150–178.

Problem of the Twentieth Century, he was singling out 'aeroplanes and poison gas' as the most dangerous of the new scientific weapons.[52]

Davies connected his proposals for disarmament and an international police force to modern scientific weapons through a classification of modern armaments which he developed from 1919 onwards. Civilian discoveries and inventions originally intended for peaceful purposes had been, for the first time, converted into weapons of war. These new scientific weapons were subsequently a divergent development from traditional weaponry, which was less scientific and not based on peaceful inventions.[53] The crucial dividing line was 1914:

> The era of intense application of scientific discoveries to the art of war may be said to have begun in 1914. During the four years which followed it reached its highest pitch of intensity, culminating in the gigantic preparations for the anticipated campaign of 1919. ... The year 1914 may therefore be taken as the dividing line which, broadly speaking, marks the vital change in the character of weapons. The years which followed may be described as the golden age of military scientists.[54]

This new science-based warfare was one of the central problems of modern international relations, he reasoned, but could be solved through the very post-1914 weapons which threatened mankind. This principle of the 'differentiation of weapons' became the cornerstone of Davies's attempts, over the next ten years, at the reconstruction of international relations through disarmament, collective security, and international organization. By ceding these new scientific weapons to international organization, leaving the older and less effective ones in the hands of nations, it was now possible, he declared, to abolish war.

Davies outlined a particularly detailed scheme for the creation of an international police force. It was to consist of a land, sea, chemical, and air arm, to be created under a new international authority. Each arm was to constitute a separate department managed by a Board which initially was to be composed of retired military officers and civilians nominated by member states, but eventually was to be composed of retirees from the international police itself, nominated by the international organization. The heads of these departments were themselves to sit on a higher Board, chaired by a 'High Constable' answerable to the controlling international organization. The international police was to be based around the world in internationalized territories run by the international organization and headquartered in Palestine. The objectives of the force would be to 'repel aggression on the part of a member or non-member State, and to enforce the decision of the judicial and arbitral tribunals of the international

[52] Davies, 'More Armaments'; Davies, *The Problem of the Twentieth Century*, 275.
[53] Ibid., 297–339.   [54] Ibid., 326–327.

authority'. Nations were to retain weaker national militaries to keep law and order in their own territories.[55]

Davies envisaged the air arm, the 'chariots of the air', to be the most powerful portion of his international police – it was to be the focus of investment and development, and would have its own factories and development facilities ensuring that advances in 'aeronautical science' would occur quickly (less wasteful competition) and accrue only to the international air force. National air forces, meanwhile, were to be gradually disbanded. Davies addressed liberal concerns over the cost of warfare by arguing that the force's monopoly of the most scientific (and so the most effective) weapons would ensure its cost-effectiveness. It would prevail over national armed forces which would only have recourse to the inferior less scientific weapons developed before the First World War. Scientists, meanwhile, would be freed from working on military weapons, and 'increase the productivity' of their national economies by 'lowering the costs of production and to discovering new designs of labour-saving machinery'.[56] Davies designed his proposal such that it could be part of an all new international organization, though could also be incorporated into the League with a radically redesigned Covenant. The international organization was to have two separate branches: an executive and a judicial branch, and the international police was to be the major tool, alongside international law, through which the organization would shape international relations.[57]

Davies was not the first to privilege aviation in his proposal for an international police force: the turn to aviation had in fact been underway from the early 1920s. As the belief in the desirability and viability of some form of international force strengthened in the 1920s, it was imagined that such a force would be predominantly aerial in nature. A League air force was suggested in the House of Commons in 1920 as an alternative to continued funding of British military aviation, and at the 1922 League Assembly by Robert Cecil. Although against a strongly armed League on the basis that it would further deter American participation, Cecil nevertheless suggested that making air forces the League's 'special weapon' would facilitate disarmament. Proposals ranged from a force formed when required from national contingents (and disbanded afterwards), to a permanent independently and internationally administered standing force with no national allegiance.[58]

---

[55] Davies, *The Problem of the Twentieth Century*, 380, 449–455, 466.

[56] Davies, 'More Armaments'; Davies, *The Problem of the Twentieth Century*, 297–339, 360–382, 475, 581.

[57] Davies, *The Problem of the Twentieth Century*, 360–382, 390, 425–502.

[58] Liberal Joseph Montague Kenworthy in the Commons: Hansard, HC, vol. 126, cols. 1639–1640, 11 March 1920. On Cecil, see Spaight, *An International Air Police*, 31; Letter,

Aviation, Empire, and Control

The internationalist turn to aviation was a manifestation of the growing prominence of aviation in Western European cultures. Since the days of ballooning aviation had been depicted as a potentially world-transforming invention, full of the possibility of widespread international transport and communication. By the beginning of the twentieth century aviation had come to be added to the increasing roll call of modern transport and communication machines, seemingly of civilian scientific origin, which promised to overcome time, distance, and national boundaries and bring the peoples of the world together. As early as the 1899 Hague Peace Conference one delegate had been astounded by the volume of 'aeronautical letter writers' who besieged the conference with calls for support for their inventions which would abolish national barriers and borders. The delegate had noted in her diary that, although 'every technical improvement, especially all means of easier communication, ultimately leads to universal peace', unfortunately '[a]ll new inventions are invariably employed by the war authorities'.[59]

This constellation of assumptions about aviation broadened and deepened after the First World War. The aeroplane came to be seen as the latest and greatest in a line of powerful civilian inventions which had transformed warfare, industrializing, mechanizing, and ultimately modernizing it. Aviators were often characterized as modern apolitical technical experts, untainted by nationalist or traditionalist impulses. The modernity, the transformative powers, and the inherently international nature of aviation and the aeroplane would work its magic on international relations, if only this invention were freed from the shackles of the nation-state. Aviation appeared to herald the arrival of a modern age, as architect and urban planner Le Corbusier put it in 1935: aviation was 'the symbol of a new age. At the apex of the immense pyramid of mechanical progress, it opens the NEW AGE; it wings its way into it'.[60]

The transformative powers of aviation were highlighted through its spectacular prevalence in American and European entertainment. Aerial entertainment reached an all-time high in the 1920s and 1930s through air races, transatlantic crossings, air shows, pilot celebrities, air

Robert Cecil to J. C. Smuts, 6 October 1922, 51076/112–114, Cecil of Chelwood Papers, British Library Manuscripts Collection, British Library.

[59] Quoted in: Sluga, *Internationalism in the Age of Nationalism*, 22. On pre-Second World War responses to aviation see: Robert Wohl, *A Dream of Wings: Aviation and the Western Imagination, 1908–1918* (New Haven, CT: Yale University Press, 1994).

[60] Le Corbusier, *Aircraft* (London: Studio Publications; Milan: Abitare Segesta, 1996), 13. Citations refer to the Studio Publications edition.

leagues and clubs, the press, books, radio, and cinema.[61] But to see the growing influence of aviation only through this spectacle would be to miss the prominence it gained through its use in the First World War and subsequently in the British Empire, for the aeroplane came widely to be seen as one of the great new scientific war-transforming weapons to emerge from the Great War. Supporters of a strengthened British military aviation, for example the official aerial war historian Walter Raleigh or the first Chief of the Air Staff (1918) Hugh Trenchard, emphasized its power and modernity.[62] By the late 1920s it was widely assumed that aviation would be the pre-eminent weapon in any future war, enabling the aggressor to inflict large amounts of bombing damage against which there was little or no defence. It was also often said that the use of poison gas, discharged from bombers over civilian populations, was inevitable. The notion that 'the bomber will always get through' to deal a 'knock-out blow' from the air came to be widely believed in the 1920s and 1930s.[63] The bombardment of Chinese cities by Japan in 1932, of Abyssinian cities by Italy in 1935, and of Spanish cities by Germany in 1936 and 1937 were highly publicized in the European press, and held up as a warning of the destruction that could be wrought on other European cities too. A British cinema newsreel of the Guernica bombing announced that 'these were homes like yours'; British pamphlets carried the headlines 'Children Like Ours! Mothers Like Ours! Wives Like Ours!'[64] H.G. Wells's best-selling science fiction was packed with destructive aerial warfare and heroic aviators.[65] Alexander Kodra's 1936 feature film *Things to Come*, written by H.G. Wells, showed in graphic detail what heavy aerial bombardment would do to a British 'Everytown'.[66] Public intellectuals, military men, and politicians warned that Britain, for centuries safe as an

---

[61] Wohl, *The Spectacle of Flight*, chapters 3–6; Brett Holman, 'The Militarisation of Aerial Theatre: Air Displays and Airmindedness in Britain and Australia between the World Wars', *Contemporary British History* 33, no. 4 (2019): 483–506.

[62] Walter Raleigh, *The War in the Air: Being the Story of the Part Played in the Great War by the Royal Air Force*, vol. 1 (Oxford: Clarendon Press, 1922); Tami Davis Biddle, *Rhetoric and Reality in Air Warfare: The Evolution of British and American Ideas About Strategic Bombing, 1914–1945* (Princeton, NJ: Princeton University Press, 2002), 69–88, 94–97.

[63] The most important works on the fear of the bomber are: Bialer, *The Shadow of the Bomber*; Brett Holman, *The Next War in the Air: Britain's Fear of the Bomber, 1908–1941* (Farnham: Ashgate, 2014).

[64] Ian Patterson, *Guernica and Total War* (London: Profile Books, 2007); Wohl, *The Spectacle of Flight*, 213–227; Richard Overy, *The Bombing War: Europe 1939–1945* (London: Allen Lane, 2013), 34.

[65] H. G. Wells, *The War in the Air* (London: George Bell and Sons, 1908); H. G. Wells, *The Outline of History: Being a Plain History of Life and Mankind*, 3rd ed. (New York: Macmillan, 1921); Wells, *The Shape of Things to Come*.

[66] *Things to Come*, directed by William Cameron Menzies (London Film Production and United Artists, 1936).

island protected by seas and the Royal Navy, was now threatened. A new type of warfare was on the horizon – one which would use aviation's long arm to reach Britain. Internationalists latched onto such pronounce-ments as evidence of the unique destructiveness of bombers, and of their unstoppability. For Noel Baker, writing in 1926, the threat of a knock-out blow from the air was a symptom of the new era of scientific warfare in which such a blow could arrive from a yet to be invented 'new scientific principle or device ... against which there is no known or discoverable defence', much in line with the 'the electric death-ray, the deadly bacteria, the atomic bomb' peddled by 'writers of fiction'.[67]

But if warfare in Europe allowed aviation's destructive power to be imagined, it was empire that allowed its speed, reach, and modernity to be projected onto a grander global stage. It was through empire that the aeroplane came to be seen as a tool for the military and bureaucratic control of vast territories. The aeroplane was put quickly to use in empire in the early 1920s, and by the end of the decade was considered, both by the British state and interested onlookers, to be indispensable for the functioning of its colonies. Its most important roles appeared to be in binding this far-flung and disparate polity together, and in policing and administrating its restless tribal provinces. The RAF's subjugation of insurrections in Somaliland (1920), Yemen (1928), Iraq (1922, 1923), and the North West Frontier and Waziristan (1919, 1920, 1925) were publicized as great successes for British military aviation, and held up as exemplars of aviation's usefulness not only as a military weapon but also as a tool of British imperial bureaucratic and political control. One 1924 Air Ministry memorandum noted that the value of 'air control' was not only military but also as a 'means to close co-ordination and co-operation of administrative effort'. As a form of 'administrative duty' aviation would provide 'an effective reminder to many of the existence and power of Government' and allow the colonial government to 'achieve a highly centralized yet widely understanding intelligence which is the essence of wise and economical control'.[68]

Aviation's colonial significance went beyond the narrowly military or administrative. When transposed globally over seemingly lawless and chaotic lands it appeared to stand for law, order, and civilization itself.

---

[67] Noel Baker, 'Disarmament', *Economica*. On the threat of aerial bombardment, see: Holman, *The Next War in the Air*.

[68] *Note on the Method of Employment of the Air Arm in Iraq*, Cmnd. 2271 (1924). See also: Omissi, *Air Power and Colonial Control*, chapter 2; and Satia, *Spies in Arabia*, chapter 7. In reality its usefulness was limited, see: James S. Corum, 'The Myth of Air Control: Reassessing the History', *Aerospace Power Journal* 14, no. 4 (2000): 61–77.

Aerial control, explained Christopher Birdwood Thomson (Secretary of State for Air in both the Labour governments of the 1920s) to readers of the *Chicago Daily News*, had brought 'law, order and prosperity' where there had been 'chaos'.[69] Speaking at the Royal United Services Institution in 1937, Air Commodore C.F.A. Portal pointed out that, although imperial aerial control was successful 'from the political point of view' because 'the tribesman regard the aeroplane as an impersonal agent of Government, it was our constant aim to get the native to think of a landing ground not only as a place from which he might get bombed, but also as a point of contact with civilization where he could obtain some of its benefits'. In Aden these benefits included medical attention, letters, and frequent Political Officer visits.[70]

Given these connections it is no surprise then that some of the most articulated thinking on aviation's impact on international relations emerged in connection with proposals for the development of commercial aviation in the British Empire. Imperial aviation began its consolidated expansion in 1924 with the foundation of Imperial Airways. This heavily subsidized 'chosen instrument' would carry over half a million passengers over its fifteen-year history, the vast majority in Europe, though by the mid-1930s passenger miles were roughly evenly split between Europe and Empire. Proponents of increased government funding for commercial aviation emphasized its importance for imperial development and for international relations within the Empire, and elucidated its beneficial impact in liberal internationalist terms. This rhetoric extended much beyond rejuvenated organizations such as the Air League of the British Empire – as well as in the press, aviation, and intellectual circles, it was also particularly apparent in relation to the government's Imperial Airship Scheme, a peculiarly liberal internationalist project for the development of a rigid airship route linking London to India to Australia. Although discussion on international air routes was widespread from 1919 onwards, the scheme generated a large discourse due to its contentious nature.[71]

---

[69] Christopher Birdwood Thomson, *Air Facts and Problems* (London: John Murray, 1927), 80–81.

[70] C. F. A. Portal, 'Air Force Co-operation in Policing the Empire', *Royal United Services Institution Journal* 82, no. 526 (1937): 343–358. Aerial policing of empire was in fact a long-standing imperial fantasy which predated mechanical flight – see Michael Paris, 'Air Power and Imperial Defence 1880–1919', *Journal of Contemporary History* 24, no. 2 (April 1989): 209–225. This thinking was reflected in interwar fiction as well, see: Dennis Butts, 'Imperialists of the Air-Flying Stories', in *Imperialism and Juvenile Literature*, ed. Jeffrey Richards (Manchester: Manchester University Press, 1989), 126–143.

[71] Gordon Pirie, *Air Empire: British Imperial Civil Aviation, 1919–39* (Manchester: Manchester University Press, 2009), 47–63; Teresa Crompton, 'British Imperial Policy and the Indian Air Route' (PhD diss., Sheffield Hallam University, 2014), 102–113.

An imperial airship route network was first suggested by the retired naval engineer and aerial entrepreneur Charles Dennistoun Burney, who proposed that he lead the armament firm Vickers to design the large airship which would ply imperial air routes. The incoming Labour government appointed aerial enthusiast Christopher Birdwood Thomson as Secretary for Air in 1924, who designed an Imperial Airship Scheme with provision for the development of two rigid airships: the R100 by Vickers's subsidiary the Airship Guarantee Company (headed by Burney), and the R101 by the state's Royal Airship Works. The scheme was contentious from the beginning in that it required the development of two airships, and remained so following numerous design problems, development delays, and cost overruns in the latter half of the 1920s. The scheme and its airships also captured the public imagination: *The Times* estimated that the final trials of the R101 in 1929 drew a crowd of up to a million.[72]

Thomson and Burney (who was elected as a Unionist MP 1922 to 1929) emerged as particularly public exponents of aviation's role in international relations and mobilized this aerial rhetoric in promotion of the ambitious airship programme. Their rhetoric included liberal internationalist themes such as a focus on international travel and communications and an erasing of international boundaries – an example of how it was possible to both support empire and hold liberal internationalist views.[73] Thomson's speeches and press articles, drawn together into his 1927 volume *Air Facts and Problems*, pressed home the notion (widespread in aviation circles at that time) that the future British geopolitical power would be built on 'air power', just as it had been on 'sea power' in the past. He called for the government to 'stimulate the air habit', and for the development of commercial aviation so as to 'knit together' the British Commonwealth. He envisaged a future twenty years hence in which 'giant airships circle the globe, transporting passengers at prodigious speed and in luxurious safety'. He emphasized the dual and mysterious nature of aviation, 'not comparable with any other human activity; it cannot yet be classified nor can its full political reactions be foreseen. It is a source of rivalry between nations, yet it unites them'.[74] Burney joined Thomson in claiming that the Empire's future depended on 'an efficient system of Imperial air communication'. Aviation, he claimed in his 1929 internationalist polemic *The World, the Air and the Future*, was 'the most fundamental and far-reaching of all factors that are working for unity and

---

[72] Pirie, *Air Empire*, 66–81, 130–138; John Duggan and Henry Cord Meyer, *Airships in International Affairs, 1890–1945* (Houndmills: Palgrave, 2001), 156.

[73] There was in fact a prominent strand of Empire-orientated thinking amongst liberal internationalists. These are explored in Chapter 3.

[74] Thomson, *Air Facts and Problems*, 57–58, 112, 145–146.

peace in the world'. It was 'breaking down the barriers of national separatism all over the world and preparing the way for a United States of Europe, and ultimately perhaps for a World State that shall embrace all civilised peoples'.[75]

Burney made the case for the development of British imperial aviation by stressing the 'air-centrality' of Britain. Britain, he argued, lay at the natural aerial centre of the world – a point at which crucial international aerial routes converged. Through a portrayal common amongst supporters of British aviation, and by the 1930s amongst internationalists too, he projected aviation as the natural heir to the navy in Britain's international relations – in its international connectivity, in its military strategies, and in its imperial affairs. Britain's centuries-long 'sea-centrality', he explained, was no longer relevant. The navy had played its role in the construction and maintenance of Britain's empire and international standing, and now it was to be replaced by aviation.[76]

Burney went further than Thomson, and indeed many others, by explicitly linking aviation to an internationalist outlook amongst populations. Most rhetoric on aviation and international communications characterized its transformative effects in political, economic, or social terms. Air power proponent Brig.-Gen. P.R.C. Groves, for example, addressing the internationalist think tank the Royal Institute of International Affairs (Chatham House) in 1927 and then again in 1929 on aviation and international relations, stressed aviation's natural tendency to 'draw the nations closer together', 'facilitate the exchange of knowledge and ideas', 'develop commercial discourse', and 'help emigration'. The benefits of aviation to the British Commonwealth were expressed in internationalist terms: it offered 'infinitely quicker communications, hence closer commercial intercourse and better political contact'.[77] These were common claims, as was Thomson's argument that a powerful national aviation base required the inculcation of an 'air habit' amongst the population – that is a familiarity with and affinity for aviation.[78] Burney also believed that a new 'air sense, an air psychology' was necessary, but went further:

[75] Charles Dennistoun Burney, *The World, the Air and the Future* (London: Alfred A. Knopf, 1929), 21, 156–157.

[76] Ibid., 82–140. Also: Davies, *The Problem of the Twentieth Century*, 647. On the bomber versus the battleship debate see: Peden, *Arms, Economics and British Strategy*, 117–122.

[77] P. R. C. Groves, 'The Influence of Aviation in International Relations', *Journal of the Royal Institute of International Affairs* 6, no. 3 (May 1927): 133–152; P. R. C. Groves, 'The Influence of Aviation in International Affairs', *Journal of the Royal Institute of International Affairs* 8, no. 4 (July 1929): 289–317.

[78] Thomson, *Air Facts and Problems*, 58. The idea that the English were naturally, or needed to be, 'air-minded' was widespread in England at that time, see for example Holman, *The Next War in the Air*, 11.

the new air age required the cultivation of an 'international outlook', which, along with aviation, would create a new 'internationalism'. This outlook and internationalism consisted of 'the clear conception of the civilised world as now, to all intents and purposes, an organic unity' and a realization that 'the close interdependence of States in the modern world, no nation can hope any longer to develop itself independently of, or at the expense of, its neighbours'.[79]

Although Burney and Thomson expressed support for civilian aviation, both had military backgrounds, and were interested in a militarily strong Britain rather than in disarmament or strengthening the League of Nations. Their rhetoric was, in fact, a testament to the strength of British military aviation in the 1920s. Following a few years of retrenchment after the end of the war, British aircraft manufacturers had repositioned themselves for growth by the mid-1920s. Between 1924 and 1930, the output, exports, and employment of the aircraft industry doubled. The air force, having also been cut following the war, stabilized by the mid-1920s and even enjoyed a period of modest expansion, buoyed by the success of its primary role: that of imperial policing.[80]

The collapse of the Imperial Airship Scheme in 1931 did little to dampen the rhetoric on aviation and international relations which it had helped inflate. Aviation continued to portend a new era of military power and international relations into the 1930s. It appeared to be, in the words of one historian, the 'Age of Air Control'.[81] For liberal internationalists it was no large leap to incorporate this supposedly world-changing technology into their proposals for disarmament, and announce that collective security could now be efficiently and effectively instituted thanks to the maturation of the flying machine. Empire and imperialist visions of colonial geographies and societies were sources of inspiration for proposals for international control. This is particularly apparent in the most detailed proposal of the period, David Davies's *Problem of the Twentieth Century*. Davies suggested that the international air force be positioned at strategic points that would be out of reach of the Great Powers but could reach trouble spots quickly and control global communication and supply lines. His scheme followed imperialist and status quo power cartographies, and envisaged a future world in which these would remain largely unchanged, save for a powerful world organization based in internationalized

---

[79] Burney, *The World, the Air and the Future*, 97, 143.

[80] Edgerton, *England and the Aeroplane*, 22; Malcolm Smith, *British Air Strategy between the Wars* (Oxford: Clarendon Press, 1984).

[81] Daniel R. Headrick, *Power over Peoples: Technology, Environments, and Western Imperialism, 1400 to the Present* (Princeton, NJ: Princeton University Press, 2010), 302–333.

territories. Davies identified four types of territories that would have to be handed over to the international authority for its international police bases: League of Nations mandated territories, neutral or demilitarized zones (such as the Rhine zone), and strategic points currently controlled either by the Great Powers as overseas territories (e.g. Suez, Gibraltar, Malta, Singapore, Hawaii, and Panama) or which are part of 'small states' (e.g. Monaco, Latvia, Estonia, Honduras, or Haiti). Crucially these 'police stations' would then become internationalized territories under direct control of the international authority. Davies highlighted Constantinople (which had in fact been present in his proposals since the First World War) as a special location which could serve as the main base for the Continent of Europe. States would willingly give up these locations, he reasoned, in return for monetary compensation, for the security offered by the international police, and the monetary saving by not having to retain military forces in these locations (e.g. the costly British naval base at Singapore). A second special location was the headquarters of the police force, which was to be mandated Palestine. Davies credited the British diplomat and expert on Ottoman and Middle Eastern matters Mark Sykes for first suggesting Palestine as a base, and reasoned that it would be ideal because of its symbolic and geographical value: a meeting place of East and West, of ancient civilizations, and of Christianity and Islam.[82]

Although relying on the experience and geography of empire, empire itself was often de-emphasized in visions of aerial collective security. Davies, for example, described his internationalized territories in federalist rather than imperial terms: Palestine was to be a 'District of Columbia around which the world confederation may group its States-members'.[83] Comparisons between aerial control of 'uncivilized' colonies and 'civilized' Europe may have been unpalatable, and anyhow many colonial administrators and policymakers who had implemented colonial air control did not see it as being applicable in Europe.[84] Moreover, for Davies and most others, an international air police was to be used primarily to deal with the problems of war and peace emanating from Europe, and only secondarily within Europe's colonial possessions and the rest of the world. *Problem of the Twentieth Century*, for example, was primarily concerned with European international relations. Some proposals called for the international force to be restricted to Europe only. When pushed to make their proposals more achievable, even those with a more global vision,

---

[82] Davies, *The Problem of the Twentieth Century*, 462–472.   [83] Ibid., 466–467.
[84] Satia, *Spies in Arabia*, 241. Some in the RAF nevertheless felt that it would be good training for a bombing war in Europe. Ibid., 253.

such as Davies, focused on Europe. The later abridged edition of Davies's book hoped that at the very least the 'four Great Powers of Western Europe' would pool their air forces. For one prominent supporter writing in the 1930s, Jonathan Griffin, Western Europe was to form a core, whilst 'the Dominions and the countries of the American Continent and East Asia join with Europe in the collective system as an outer ring of countries less rigidly bound, on the principle of "concentric circles of security"'. The United States, he (and others) hoped, would join once the scheme's commercial and security advantages became apparent.[85]

Nevertheless, in response to critics of aerial disarmament and an international air force (who feared that this regime would take away the very instrument used to pacify colonial possessions), internationalists emphasized that their proposals allowed for the continuation of aerial imperial policing.[86] Noel Baker acknowledged that 'air control can be more pacifying and more humane than military control' and warned that the abolition of bombing would 'compel a very large increase of expenditure on land garrisons to replace the relatively economical air forces now used to maintain order in Iraq, Trans-Jordan, Aden, and the N.W. Frontier'.[87] Davies's scheme allowed for the retention of small inferior air forces by imperial powers specifically for colonial policing – 'a 1918 type of aeroplane, which is obsolete in a modern war, but which is good enough for policing semi-civilised tribes', though his international air force could also cover Empire. Indeed it was to save money on imperial defence by making the costly British naval presence in empire redundant.[88] Even some of the earliest conceptions of an international police force had suggested that it would be useful at the edges of the civilized world. Dodd's *Captain Gardiner of the International Police* imagined a future in which the most serious threats faced by world government emerged from a coalition of Muslim countries and 'savage tribes of Africa and other dependencies of the Federation', led by a 'secret association which controls the government of China' which resented the civilizing work of the Federation.[89]

By the early 1930s an international air force was established in internationalist and international relations circles as a bona-fide solution to the problem of war and disarmament. Its growing acceptance was based on notions of scientific warfare and of aviation as the most modern

[85] David Davies, *An International Police Force: An Abridged Edition of 'The Problem of the Twentieth Century'* (London: Ernest Benn, 1932), v–vi; Griffin, *Britain's Air Policy*, 165.
[86] For a review of criticism see: Griffin, *Britain's Air Policy*, 63. See also Chapter 3.
[87] Philip Noel Baker to Maxwell Garnett, 27 April 1932, 5/140, Noel-Baker Papers.
[88] David Davies, 'An International Police Force?', *International Affairs* 11, no. 1 (January 1932): 76–99; Davies, *The Problem of the Twentieth Century*, 150, 360–382, 471, 625.
[89] Dodd, *Captain Gardiner of the International Police*, 19, 58.

manifestation of this application of science to war – a successor not only to the integrative communication and transportation inventions of earlier years but also to the powerful navies and the industrialized methods of warfare used in the First World War. Although not all those supporting an international force could be described as technocrats, the force itself was a technical and technocratic solution to the problem of war. Aviation as applied in the colonies already appeared to demonstrate its ability to govern large areas, both militarily and bureaucratically. Newly emergent experts on international relations were amongst those calling for technical solutions and rationalistic disarmament techniques as a solution to the problem of war. Lastly, internationally organized aviation was in keeping with the high hope internationalists held for international technically and bureaucratically manned organizations, especially the League and its organs. The radical trajectories of thinking on aviation and international relations did not stop there however – by the early 1930s aviation appeared to offer much richer possibilities for international relations.

3   The Shape of Things to Come
Aviation, the League of Nations, and the
Transformation of World Order

In 1933 H.G. Wells published his future history of the world, *The Shape of Things to Come*. In it he described how a global war, beginning in 1940, devastated the globe, killing millions, destroying national states, and isolating the remaining populations. Mankind's recovery and ultimate prosperity came from aviation and aviators. Commercial aviation recovered first from this destruction and gradually knitted societies back together into a larger global polity. Aviators formed a 'Central Board' which took direct control of international aerial and naval ports around the world, and regulated and increased international trade, commerce, and economic development. The board issued its own currency and used an international air police armed with a nonlethal poison gas, 'Pacificin', to crush nationalist uprisings, and eventually usher in a technocratically run world state. The book was made into a big-budget movie in the 1930s, the most expensive British production to date.[1]

As tempting as it might be to dismiss these ideas as science fiction, Wells was merely taking up proposals then commonplace in discussions on disarmament and international relations. *The Shape of Things to Come* was an engagement with the debates then raging about aviation's effect on warfare, society, and international relations. Wells stood alongside other British internationalists in understanding aviation to be a grand unifier – a bringer of law, order, and control. From the 1932 Geneva disarmament conference through the complex politics of the late 1930s the aeroplane became central to discussions on international relations. Earlier calls for an international air police transformed into much broader schemes for collective security and global transport through proposals such as the international control of commercial aviation. The call for internationalized aviation to bring order to a dangerous world became commonplace, profoundly affecting how not only international order but also aviation and science were thought about. Ultimately aviation became a central element of a significant anti-fascist ideology of the 1930s – a radical, armed, liberal

---

[1] Wells, *The Shape of Things to Come*, 284–305, 325–326, 330–333.

internationalism which looked to the aeroplane to defeat ultra-nationalism and militarism.

The Geneva Conference and the International Control of Aviation

On 2 February 1932, sixty nations gathered at the Palais des Nations in Geneva for the opening of the largest international conference since Versailles: the Conference for the Reduction and Limitation of Armaments. A Preparatory Disarmament Commission had produced a draft arms control agreement that lay before the assembled delegates at Geneva. Yet just three days in the French delegation astounded all by pushing the preparatory work aside, and putting forward a much more radical programme for collective security. The president of the French delegation, André Tardieu, called for the League to become a 'new and lasting power capable of organising peace, security and disarmament'. Aviation was to be the foundation of this new League-secured world order. His proposal called for all large military aircraft to be transferred to an International Air Force under the control of the League of Nations. Member states were to be forbidden from owning large national military aircraft. Other advanced weapons such as long-range artillery, heavy capital ships, and submarines were also to be placed under the control of the League. The international police force was to deter aggression, ensure disarmament, police the colonies, and oversee the development of aviation amongst the member nations. But most radically of all, moving far beyond any past official proposal, individual nations were to be prohibited from owning heavy civilian aircraft. These were to be transferred to an international civil transport service, under the League, which would then have a monopoly on establishing and operating all international airlines, including transport between nations and their colonies.[2]

Tardieu's proposals surprised the assembled delegates and observers, who did not expect such a quick departure from the work of the preparatory committee, and certainly not from Tardieu (a vocal supporter of French military might). 'Agitated' delegates feared that the proposal's radical nature would alienate Britain and the United States, and there was

---

[2] League of Nations, *Records of the Conference for the Reduction and Limitation of Armaments. Series A Verbatim Records of Plenary Meetings* vol. 1, 2 February to 23 July 1932, Disarmament 1932.IX.60 (Geneva: League of Nations, 1932), 46, 61; 'Proposals of the French Delegation, Conf. D.56, Geneva, 5 February 1932', in League of Nations, *Conference for the Reduction and Limitation of Armaments. Conference Documents* vol. 1, Disarmament 1932.IX.63 (Geneva: League of Nations, 1932), 113–116.

talk of 'the Conference having being torpedoed before it began'.[3] This did not happen, and instead the Tardieu Plan opened the floodgates to a series of competing proposals for the League control of aviation. This 'la bataille du désarmement', in the words of *Le Figaro*, helped propel aviation to the centre of internationalist thinking on international relations.[4]

We know now that Tardieu did not expect his plan to be accepted: it was drafted in order to occupy the higher moral ground at a conference through which he expected to achieve little, if anything at all.[5] In content it bore more than a passing resemblance to proposals increasingly emanating from internationalist conferences on the Continent, and from the centre-right of the political spectrum in France. Louise Weiss, editor and co-founder of the leading European-integrationist journal *L'Europe Nouvelle*, was quick to point out the similarity to a resolution calling for the international control of aviation adopted at a major internationalist conference on disarmament held only a few months earlier in Paris.[6] Participants included leading French Radical politicians and British and French internationalists, including Weiss.[7] The French Radical Party also claimed the aerial essence of Tardieu's plan as its own. The Party newspaper *La République* reprinted resolutions calling for the international control of aviation from the Party's November 1931 annual conference, and leading Radicals such as Pierre Cot singled out the diplomat and Radical party politician Henry de Jouvenel as the true originator of the plan.[8] Jouvenel had indeed proposed the internationalization of aviation in 1928, and authored, alongside other

[3] John Wheeler-Bennett, *Pipe Dream of Peace: The Story of the Collapse of Disarmament* (New York: William Morrow, 1935), 15.

[4] 'La contre-offensive Française a Genève', *Le Figaro*, 7 February 1932. On these schemes see: League of Nations, *Records of the Conference for the Reduction and Limitation of Armaments Series B: Minutes of the General Commission* vol. 1, 9 February to 23 July, Disarmament 1932.IX.64 (Geneva: League of Nations, 1932), 24, 35, 37; J. M. Spaight, *Air Power in the Next War* (London: Geoffrey Bles, 1938), 61–72; Carolyn Kitching, *Britain and the Problem of International Disarmament 1919 – 1934* (London: Routledge, 1999), 140–173.

[5] Maurice Vaïsse, *Sécurité D'abord: la Politique Française en Matière de Désarmement, 9 Décembre 1930 – 17 Avril 1934* (Paris: Pedone, 1981), 198. See also: Philip Charles Farwell Bankwitz, *Maxime Weygand and Civil-Military Relations in Modern France* (Cambridge, MA: Harvard University Press, 1967), 56–58.

[6] Louise Weiss, 'Bataille de discoures et de propositions', *L'Europe Nouvelle* 15, no. 731 (13 February 1932): 193–197.

[7] For more on La Conférence d'Etudes pour le Désarmement see: *L'Europe Nouvelle* 14, no. 721 (5 December 1931); and Pemberton, *The Story of International Relations*, part two, 12–18.

[8] Gabriel Cudenet, 'Un plan … ou un jeu ?', *La République*, 7 February 1932; 'Vingt-sixième Congres Radical: Ordre du jour sur la Conférence du désarmement', *La République*, 7 February 1932; Pierre Cot, 'M. Tardieu défendra-t-il a Genève tout le programme radical?', *La République*, 10 February 1932.

French internationalists, detailed proposals for a Union Aérienne Internationale in 1931.[9] The announcement of the Tardieu Plan led to a deluge of calls for international control from these constituencies. Louise Weiss's *L'Europe Nouvelle* reiterated its existing proposals, and a major peace conference held in Vienna in September 1932 also called for international control.[10]

The French proposal, rather than being dismissed, triggered an explosion of public debate on international control in Britain which lasted through to the end of the 1930s. British calls for the international control of civil aviation were present in the late 1920s, but not ubiquitous. However, by 1931 British internationalists had been heavily exposed to these ideas at Continental European gatherings. The November 1931 La Conférence d'Etudes pour le Désarmement (Paris) and the Congress of the International Federation of the League of Nations Societies (Budapest, May 1931), both attended by prominent British internationalists, produced resolutions calling for League control of aviation across Europe. The Geneva conference injected this radical thinking about aviation and disarmament into official diplomatic discussions and the mainstream press, emboldening British internationalists to incorporate civil aviation into their proposals for international control and disarmament.[11]

David Davies was quickest to capitalize on the Tardieu Plan and press speculation linking it to his pre-existing proposals.[12] He incorporated the

---

[9] Henry de Jouvenel, 'Internationalisez l'aviation', *La Revue des Vivants* 2, no. 10 (October 1928): 725–733; Pierre Brossolette, 'Sur un projet de réduction des armements aériens', *Notre Temps* 17, no. 118 (29 November 1931): 513–518. On Jouvenel's activism see also: Pemberton, *The Story of International Relations*, part two, 10–12.

[10] 'L'aviation, Problème-Clef du Désarmement', *L'Europe Nouvelle* 15, no. 763 (24 September 1932): 1123–1125; Laurent Eynac, 'L'internationalisation des transports aériens', *L'Europe Nouvelle* 15, no. 763 (24 September 1932): 1125–1127; Lucien le Foyer, 'La Question du désarmement', *La Paix* 7, no. 8 (July/August 1932): 150–165. Other proposals include: André Daniel Tolédano, *Ce qu'il faut savoir sur le désarmement (avec en annexe le projet de convention du désarmement)* (Paris: A. Pedone, 1932); and Louis Perillier, *De La Limitation des Armements par la méthode budgétaire et du contrôle de cette limitation* (Paris: Rousseau, 1932).

[11] Prominent British calls in the late twenties include: Hansard, HC, vol. 214, col. 214, 12 March 1928; Rennie Smith, *General Disarmament or War?* (London: George Allen & Unwin, 1927), 88–89. Quaker pacifist Hilda Clark also called for international control: Hilda Clark, letter to the editor, 'Budgetary Limitation of Air War Material', *Manchester Guardian*, 9 December 1930. On European internationalist activity 1931–1932 see: Davies, *The Possibilities of Transnational Activism*, 87–97; Pemberton, *The Story of International Relations*, part two, 12–18. See also: *L'Europe Nouvelle* 14, no. 721 (5 December 1931); Viscount Cecil of Chelwood, 'Disarm or Rearm!', *New York Herald Tribune*, 16 August 1931.

[12] Raymond Poincaré, 'La Paix Dans La Sécurité', *Excelsior*, 8 February 1932; '40,000 Women Plead for World Peace. Clash of Plans at Geneva: How Nations Received the French Proposals', *Reynolds News*, 7 February 1932.

internationalization of civil aviation into his existing scheme and quickly published an abridged version of his *The Problem of the Twentieth Century* which included an endorsement of the Plan. In October 1932 he founded the New Commonwealth Society to agitate for his proposals, as well as a New Commonwealth Institute which published more broadly on international relations.[13] The institute and the society would go on to push for international control through conferences, at the League of Nations Union and other internationalist organizations, and through publications throughout the 1930s.[14] Prominently, the institute organized lectures on an international air force and the internationalization of civil aviation for the eighth International Studies Conference of the Institute of Intellectual Cooperation held in London in June 1935.[15] As well as propagating David Davies's ideas, the society published calls for internationalization by others, including Henry de Jouvenel, imperial governor Edmund Allenby, and Professor of Military Studies Major-General Frederick Maurice.[16]

Supporters of the League of Nations now added international control of civil aviation to their existing proposals for disarmament and collective security. Noel Baker had called for aerial arms reductions and the formation of an international air force in the 1920s but had explicitly argued against international control of civil aviation. Post-Geneva he added the complete abolition of bombers and the international control of civil aviation to his proposals.[17] The Executive of the League of Nations

[13] Davies, *An International Police Force*.

[14] For example at a pacifist conference in Brussels in February 1934: New Commonwealth Society, *An International Air Force: its Functions and Organisation / being a Memorandum submitted by the Executive Committee of the New Commonwealth to the International Congress in Defence of Peace, Brussels, February, 1934* (London: New Commonwealth, 1934). It also expanded abroad, see for example: Ofer Ashkenazi, 'Transnational Anti-war Activity in the Third Reich: The Nazi Branch of the New Commonwealth Society', *German History* 36, no. 2 (June 2018): 207–228.

[15] These were subsequently published, and included: Rear Admiral Robert Neale Lawson, *A Plan for the Organisation of a European Air Service* (London: Constable, 1935); H. R. G. Greaves, *An International Police Board* (London: Allen & Unwin, 1935); and Pierre Cot, *Military Force or Air Police* (London: New Commonwealth Institute, 1935).

[16] Edmund Allenby, *Allenby's Last Message: World Police for World Peace* (London: New Commonwealth Society, 1936); Henry de Jouvenel, *Aviation for World Service* (London: New Commonwealth Society, 1933); Frederick Maurice, 'The International Police Force', *The New Commonwealth Research Bureau* Research Materials No. 1 (November 1934): 4–6. On the New Commonwealth see: Bauernfeind, 'Lord Davies and the New Commonwealth Society 1932–1944'; and Michael Pugh, 'Policing the World: Lord Davies and the Quest for Order in the 1930s', *International Relations* 16, no. 1 (2002): 97–115. See also: David Davies, *An International Air Force: Its Functions and Organisation* (London: New Commonwealth Society, 1934); David Davies, *Force and the Future* (London: New Commonwealth Society, 1934).

[17] Noel Baker, *Disarmament*, 240–243; Philip Noel Baker, 'The International Air Police Force' in Noel Baker, *Challenge to Death*, 206–239.

Union, including its President Robert Cecil, had for years resisted grass-roots and New Commonwealth pressure to back an international police force. It gradually succumbed from 1934 onwards – finally committing itself to complete international control in 1935, and holding a prominent conference on 'Aviation as an International Problem' the same year.[18] Other converts included the international relations theorist J.A. Hobson and the General-Secretary of the League of Nations Union Maxwell Garnett.[19] Eugenicist Julian Huxley welcomed international control as an example of the application of the scientific approach to war and disarmament.[20]

Powerful constituencies within the Labour and Liberal Parties, already proponents of disarmament and critics of government policy, threw their weight behind international control. The Labour Party had been committed to League 'pooled security' since 1928, and with Arthur Henderson's support the Labour Party Annual Conference passed its first resolution calling for an international police force in 1933. At the October 1934 Annual Conference in Southport international control of aviation was adopted as policy by the Labour Executive. Labourites attacked government opposition to international control, arguing that it was done in order to continue to increase military spending, support the arms and aviation industries, and continue bombing in the colonies. MP Clement Attlee called for the internationalization of aviation in the Commons in March and then again in November 1932, and went on to publish his views in a 1934 New Commonwealth Society pamphlet; and the Labour Party presented international control as a central foreign policy proposal in its 1935 General Election Manifesto, and reaffirmed its commitment in 1936. This was to be part of a wider commitment to international law, the 'ultimate object being', as Arthur Henderson explained: 'to abolish all national armed forces and entrust the defence of world law and order to an international police force, under the League of Nations'.[21] The Liberal MP Geoffrey Mander championed international control at the Liberal Summer Schools and in the House of

---

[18] League of Nations Union, *The Problem of the Air* (London: League of Nations Union, 1935).

[19] J. A. Hobson, 'Force Necessary to Government', *Hibbert Journal* 33, no. 3 (April 1935): 331–342; Maxwell Garnett, 'The World Crisis and the Disarmament Conference', *Contemporary Review* 141 (Jan/June 1932): 144–154; Maxwell Garnett, letter to the editor, 'Collective Defence. The Use of an International Air Force', *Manchester Guardian*, 11 June 1934; Maxwell Garnett, *Freedom of the Air* (Letchworth: Garden City Press, 1933).

[20] Huxley, *Scientific Research and Social Needs*, 175.

[21] Labour Party, *Labour and the Nation* (London: Labour Party, 1928); Labour Party, *Labour Party Annual Conference Report* (London: Labour Party, 1933), 192; Henry Ralph Winkler, *British Labour seeks a Foreign Policy, 1900–1940* (New Brunswick:

Commons. With his backing it remained on the agenda of the National Liberal Federation conferences between 1932 and 1935, when several resolutions were passed in support. A group of Liberals, seeking radical policies for their party, issued a manifesto in 1932 which included support for Davies's international air police.[22]

The liberal press, generally supportive of the League and disarmament, also took up the cause of international control. After initially condemning the Tardieu Plan as 'propagandist', the *Manchester Guardian* soon became a strong supporter. For the *News Chronicle* the French proposals were 'revolutionary' and 'more ambitious and far-reaching than any the League of Nations has ever conceived of'. The *Economist* (then headed by Walter Layton, an economist and leading Liberal intellectual) admitted that the French had the most to benefit from the formation of a League air force, but nevertheless welcomed aerial disarmament and declared that the case for internationalization was 'overwhelming' on the basis that 'Flying is supra-national in its very nature' and 'a safeguard against being bombed wholesale'.[23]

Transaction Publishers, 2005), 104; Labour Party, *For Socialism and Peace: The Labour Party's Programme of Action* (London: Labour Party, 1934); Vigilantes [Konni Zilliacus], *The Dying Peace* (London: New Statesmen and Nation, 1933), 13–14; Hansard, HC, vol. 263, cols. 228–231, 15 March 1932; Hansard, HC, vol. 270, cols. 533–534, 10 November 1932; Clement R. Attlee, *An International Police Force* (London: New Commonwealth Society, 1934); F. W. S. Craig, *British General Election Manifestos 1918–1966* (Chichester: Political Reference Publications, 1970), 81–83; National Executive Committee, *Labour's Immediate Programme* (London: Labour Party, 1937); Arthur Henderson, *Labour's Way to Peace* (London: Methuen, 1935), 45, 46. On Labour, the League, and collective security see: Lucian M. Ashworth, 'Democratic Socialism and International Thought in Interwar Britain', in *Radicals and Reactionaries in Twentieth-Century International Thought*, ed. Ian Hall (New York: Palgrave Macmillan, 2015), 75–100.

[22] Hansard, HC, vol. 263, col. 232, 15 March 1932; Hansard, HC, vol. 284, cols. 445–502, 13 December 1933; Geoffrey Le Mesurier Mander, 'Military Disarmament', *Forward View* 3, no. 32 (September 1929): 124; Richard S. Grayson, *Liberals, International Relations and Appeasement: The Liberal Party 1919–1939* (London: Routledge, 1990), 164–177. Also: 'Resolution on Peace and Security', *Liberal Magazine* 43 (June 1935): 56–60; Hubert Phillips et al., *Whither Britain? A Radical Answer* (London: Faber & Faber, 1932).

[23] 'Great Disarmament Sensation. France's Sweeping Proposals – League to Control War Forces', *News Chronicle*, 6 February 1932; 'Reasons for the French Move', *Manchester Guardian*, 6 February 1932; 'Civil Aviation and Air Warfare. Internationalisation', *Manchester Guardian*, 10 September 1932; 'Civil Aviation and Air Warfare. Internationalisation II', *Manchester Guardian*, 13 September 1932; 'Plan for the Air', *Manchester Guardian*, 21 May 1935; Editorial, 'The Disarmament Conference', *Economist*, 13 February 1932, 340–341; 'French Plan', *Economist*, 19 November 1932, 916–917; 'Disarmament in the Air', *Economist*, 5 February 1933, 399. The leftist *Daily Herald* greeted the Tardieu Plan by asking: 'Is France Sincere?', *Daily Herald*, 6 February 1932.

Beyond Collective Security: World Order Reimagined

Although international control was associated primarily with disarmament, it was now also conceived as a way of producing greater global governance and unification. The primary tool for this was to be the international air police, now envisaged to be a much more powerful force compared to pre-Geneva. For Noel Baker it was now to have significantly more functions than the one he had suggested in the 1920s. A 1934 essay divided its functions into three categories: peacetime, in times of crisis, and in times of conflict. During peace, the air police was to supervise national disarmament by acting as a neutral expert advisor on disarmament, and flying inspectors around the world. It was to stand guard at international airports and act as an international airline for the League of Nations. It would also manage the internationalized civil aviation network and provide meteorological services. Its research and development wing would develop better fighting machines, but also further the scientific and technological development of civil aviation. In response to the then growing concerns over international criminal activity, Noel Baker pointed out that the air force could police international criminal activity ('A race of air-bandits may grow up') which might seek to abuse 'freedom of the air'. In times of crisis the air police would allow the League to act swiftly by arriving at the scene of dispute quickly, transporting League representatives, setting up neutral zones, and patrolling the crises area. In times of war the police would stop the aggressor's air forces, and only in the last resort would it carry out bombing of military, communications, or even civilian targets.[24] At a European internationalist conference the same year Noel Baker suggested that 400 to 500 'fast-climbing, interceptor machines' would be sufficient, with a budget between five and ten million pounds.[25]

Bombardment emerged as a more central feature of these internationalist visions. The Tardieu Plan explicitly described its force as a 'repressive international force' armed with 'bombing aircraft'. Bombers were an integral component of David Davies's international air force, and emphasized in New Commonwealth propaganda. One pamphlet warned that in the case of 'La Guerre Totale ... the hands of the international policeman cannot be tied behind his back'. The international force should be free to bomb cities and civilians, even as a first

---

[24] Noel Baker, 'The International Air Police Force'. On growing concern over international criminality see: Paul Knepper, *International Crime in the 20th Century: The League of Nations Era, 1919–1939* (Houndmills: Palgrave Macmillan, 2011), 57–85; Michael D. Callahan, *The League of Nations, International Terrorism, and British Foreign Policy, 1934–1938* (Cham, Switzerland: Palgrave Macmillan, 2018), 1–8.
[25] 'A League Legion of the Air', *Manchester Guardian*, 17 February 1934.

strike.[26] Noel Baker's and Jonathan Griffin's forces, meanwhile, were to be composed of a permanent core of fighters, with civilian aircraft being drafted in for bombing purposes as a final resort. Maxwell Garnett dissented, calling for an international air force composed of fighters only as 'no good will come of attempting to punish the whole population of the offending nation'. One proponent of internationalized aviation, Henry de Jouvenel, had sanctioned actual heavy bombardment of colonies: the naval and land artillery attacks on Damascus during his time as High Commissioner of the French mandated territory of Syria.[27]

Long-standing platitudes about aviation's integrative effects were wheeled out in order to present international control as much more than collective security. Attlee, for example, presented internationalization as a 'great step forward in civilisation' during the March 1932 Air Estimates debate in the Commons. Aviation, he argued:

annihilates distance, and it ought to lead more towards internationalism than towards combative nationalism ... I am inclined to think that discussions on disarmament tend to range too much on a negative aspect, and not enough on a positive aspect. ... A constructive policy of peace would be a great deal more effective.[28]

In such rhetoric the military uses of aviation were sidelined, and instead aviation was envisioned as the latest in a line of integrative transport and communications inventions. Attacking the government's lacklustre record at disarmament during a Commons debate in November 1932, Attlee called once again for a 'constructive internationalism', but this time the internationalization of aviation was to be much more than a type of disarmament; it would be a 'prelude to a far greater degree of internationalisation', which would include 'all the great Continental railway groups' and 'an international mercantile marine, and the doing away with all navies save such naval forces as might be necessary to restrain piracy'.[29] By contrast, attempts at disarmament which simply limited weapons was disparaged by internationalists as 'negative disarmament' which would not end war but may actually 'prolong and intensify' it.[30]

---

[26] New Commonwealth Society, *The Functions of an International Air Police* (London: New Commonwealth Society, 1936), 8–9.

[27] 'Proposals of the French Delegation, Conf. D.56, Geneva, 5 February 1932', in League of Nations, *Conference for the Reduction and Limitation of Armaments: Conference Documents*, vol. 1, Disarmament 1932.IX.63 (Geneva: League of Nations, 1932), 113–116; Davies, *The Problem of the Twentieth Century*, 440; Noel Baker, 'The International Air Police Force'; Griffin, *Britain's Air Policy*, 116; Michael Provence, *The Great Syrian Revolt and the Rise of Arab Nationalism* (Austin: University of Texas Press, 2005), 109, 126, 127.

[28] Hansard, HC, vol. 263, cols. 228–231, 15 March 1932.

[29] Hansard, HC, vol. 270, cols. 533–534, 10 November 1932.

[30] David Davies, *Suicide or Sanity?* (London: Williams and Norgate, 1932), 25.

There were, in fact, widespread debates in Europe at that time on the transnational organization of railway, road, and electricity networks. Delegates at the International Economic Conference in Geneva in 1927 had discussed the organization of European resources and communications at a pan-European level. For some this was to be done through government cooperation, for others through the creation of specialist technical organizations, be they independent or allied to the League of Nations.[31] Prominent internationalists such as Albert Thomas, Director of the International Labour Office, and economist Francis Delaisi pushed proposals for the creation of integrated pan-European transport and electricity networks through various international organizations.[32] This 'technocratic internationalism' represented not only the urge to achieve economic and technical efficiency through centralized planning but also a desire to lay the foundations for further political integration.[33] These proposals intertwined with calls for the internationalization of aviation from 1932 onwards, with many proponents arguing that European aviation needed to be internationally organized in conjunction with European waterways, shipping, railways, and post.[34]

The popularity of international control was also bolstered by a growing sense that the national economy needed more technocratic centralized planning. Symptomatic of this was the formation of a number of radical liberal pressure groups in the 1920s and especially the 1930s. The Next Five Years Group, formed in 1934 by Baron Clifford Allen and others (Noel Baker was initially involved but dropped out along with other Labour supporters), was one of the more influential ones. Formed largely to call for more rationality in state decision making and state planning of the national economy (including an 'Economic General Staff'), the group was on the right of most of the Labour Party: it emphasized mass democracy, the safeguarding of political liberty, and lower tariffs. For the Group national prosperity required unfettered international trade and commerce, and so international peace. Science, rationality, and technical expertise threatened civilization through modern war but offered also the possibility of national and international technocratic control. Consequently, the Group pushed too for the internationalized planning and control of international aviation. Convinced that 'Aviation will either

---

[31] Vincent Lagendijk, *Electrifying Europe: The Power of Europe in the Construction of Electricity Networks* (Eindhoven: Aksant, 2009), 78. See also: Frank Schipper, *Driving Europe: Building Europe on Roads in the Twentieth Century* (Eindhoven: Aksant, 2009).

[32] Schipper, *Driving Europe*, 92–120.

[33] Schot and Lagendijk, 'Technocratic Internationalism in the Interwar Years'.

[34] For example: Lucien le Foyer, 'La Question du désarmement', *La Paix* 7, no. 8 (July/ August 1932): 150–165; G. D. H. Cole and Margaret Cole, *The Intelligent Man's Review of Europe To-Day* (London: Victor Gollanz, 1933), 776–777.

destroy or save our civilization', founder Clifford Allen ensured that the Group made international control central to its foreign policy demands.[35] The group emphasized the importance of the existing 'international civil service' in Geneva, which would be used for 'technical, statistical, and other kinds of research' into international questions of an economic or technical nature, including emigration, commodities, and civil aviation.[36] Allen was most publicly supported by fellow Group member Jonathan Griffin, editor of the weekly *Essential News*. Griffin published detailed proposals in 1934 and 1935 which were endorsed by Norman Angell and supported by the *Economist*.[37] Both Allen and Griffin took extreme positions on international control. Clifford Allen refused to sign a letter to *The Times* from fellow internationalists because it proposed an interim European international air force before the formation of a permanent one, and because national air forces were not to be immediately abolished, only withered down over time. He wanted instead complete and immediate international control. The problem, Allen explained in his own letter to the newspaper, was that 'The international control of civil aviation has so far been considered chiefly from the point of view of disarmament, whereas it ought surely to have been looked upon as the key to the world organization of security'.[38] Griffin went further than most proponents by also explicitly calling for the complete abolition of private aviation, which was, he maintained, simply a form of national military armament.[39]

---

[35] Next Five Years Group, *The Next Five Years: An Essay in Political Agreement* (London: Macmillan, 1935), 18; Clifford Allen, letter to the editor, *The Times*, 14 December 1933. See also: Clifford Allen et al., letter to the editor, 'A Disarmament Programme', *Manchester Guardian*, 22 December 1933; Clifford Allen et al., letter to the editor, 'Disarm or Rearm?', *Manchester Guardian*, 17 February 1934; Martin Gilbert, *Plough My Own Furrow: The Story of Lord Allen of Hurtwood as Told Through His Writings and Correspondence* (London: Longmans, 1965), 291–309, 331–349; Clifford Allen, 'Public Opinion and the Idea of International Government', *International Affairs* 13, no. 2 (March–April 1934): 186–207; Clifford Allen, *Britain's Political Future: A Plea For Liberty and Leadership* (London: Longman Green, 1934), 116–117. On the Next Five Years Group see: Next Five Years Group, *A Summary of the Book 'The Next Five Years': An Essay in Political Agreement* (London: Next Five Years Group, 1936); and Arthur Marwick, 'Middle opinion in the thirties: planning, progress and political "agreement"', *The English Historical Review* 79, no. 311 (1964): 285–298.

[36] Next Five Years Group, *The Next Five Years*, 249.

[37] Jonathan Griffin, *World Airways – Why Not? A Practical Scheme for the Safeguarding of Peace* (London: Gollancz, 1934); Griffin, *Britain's Air Policy*; Norman Angell, *The Menace to our National Defence* (London: Hamish Hamilton, 1934), 166–170; 'Collective Security and the Air', *Economist*, 12 October 1935, 708.

[38] See the correspondence between David Davies and Clifford Allen, B10/1, Lord Davies of Llandinam Papers, National Library of Wales (hereafter cited as Davies Papers); Gilbert, *Plough My Own Furrow*, 336.

[39] Griffin, *Britain's Air Policy*, 79. See also Griffin, *World Airways*; Allen, *Britain's Political Future*, 116–117.

Aviators, Experts, and the Brotherhood of the Air

The aviator stood alongside his machine as a central figure in these visions of aerial world security. Building on widespread understandings of the importance of technical experts, and the image of the modern heroic internationally minded flyer, internationalists imagined that aviators would guide the world to greater interconnection and ultimately peace. Aviators, it was generally thought, cared less for the nationalist allegiances and petty jealousies of land-dwellers, and felt a greater bond between themselves – a 'freemasonry', according to one-time Minister for Air Christopher Birdwood Thomson, which was 'an expression of a new international spirit evolved in the common struggle against Nature's forces for the conquest of the air'.[40] These images of aviators as not only chivalric warriors but technical experts built up and entered the public imagination during the First World War (though, as Michael Paris has noted, had earlier origins), and continued into the inter-war years. So, for example, in the wartime book *The Romance of Air-Fighting*, the 'noble' aviator was not only a 'duellist of the air' and an 'air warrior' but also a 'scientist' who 'must learn another dozen sciences' on top of aeronautics and meteorology.[41]

These characterizations were easily subsumed into internationalist discourses on international relations. An international air force would coalesce and be loyal to the League because, argued Noel Baker, 'we all know, that aviators are a class apart, bound by a common loyalty to each other, speaking already an "esperanto of the air"'.[42] In response to criticisms that his proposed internationalized commercial air fleet could still be misused for national military purposes, Griffin pointed to the 'international fraternity' of the 'world's most skilled pilots and airway administrators' which would render 'the misuse of civil aviation almost impossible'.[43] Salvador de Madariaga believed that 'if we developed a team of 40 to 50,000 pilots and mechanics of all nations, all working for the same line, their brotherhood would have been sufficient to ensure that not a single plane could be diverted to nefarious deeds'.[44] Pointing to

---

[40] Thomson, *Air Facts and Problems*, 188.
[41] R. Wherry Anderson, *The Romance of Air-Fighting* (New York: George H. Doran, 1917), 13, 16, 18; Michael Paris, 'The Rise of the Airmen: The Origins of Air Force Elitism, c. 1890–1918', *Journal of Contemporary History* 28, no. 1 (January 1993): 123–141; Michael Paris, *Warrior Nation: Images of War in British Popular Culture, 1850–2000* (London: Reaktion Books, 2000), 158–159; John H. Morrow Jr., 'Knights of the Sky: Rise of Military Aviation', in *Authority, Identity and the Social History of the Great War*, eds. Frans Coetzee and Marilyn Shevin-Coetzee (Oxford: Berghahn Books, 1995), 305–324.
[42] Noel Baker, 'The International Air Police Force', 239.
[43] Griffin, *Britain's Air Policy*, 48–49.
[44] Salvador de Madariaga, *Morning without Noon: Memoirs* (Farnborough: Saxon House, 1973), 251, 274.

the First World War, Attlee noted that 'there was probably greater cama-
raderie among the airmen of the different nations at war than in any of the
other Forces'. In internationalized civil aviation, airmen would 'become
truly international – trained together, working together, developing an
international esprit de corps'. This 'could cut off a large amount of
potential support for war by the development, not only of airmindedness, but of international mindedness', leading eventually to the
abolishment of national air forces.[45] Such sentiments easily blurred into
talk of world government – for Noel Baker, if the airman was brought into
the League of Nations, 'he would be helping to bring into being – and
airmen alone can do it – the beginning of international government
throughout the world'.[46]

Others placed greater emphasis on the technical nature of the aviator's
work as the source of his internationalism and effectiveness. Internationalists
built on a growing sense that technical experts were required to solve the
increasingly technical social and political problems bedevilling mankind –
expressed most vividly through H.G. Wells's call for an 'Open
Conspiracy'.[47] The League of Nations itself, as I have argued in
Chapter 1, was reconceptualized by its supporters as a successful technical
organization manned by apolitical experts who successfully advised on and
planned for international technical issues. David Davies hoped that the
apolitical nature of this technical expertise could be reproduced through
his international air police, which would double as a 'world technical college
of the highest standing'. Recruits would 'undergo an intensive educational
and scientific training' and emerge from service as 'graduates of a world
university superbly equipped with technical knowledge . . . and fully trained
to embark upon a civilian career in the allied professions and industries'. On
returning home they would spread the 'international spirit', reducing
'national outlook'.[48] The head of this air force would embody the best of
these technocratic values: for Noel Baker, as well as being an able leader and
loyal to the League, he would be 'highly qualified in both the technical and
the military science of aviation'.[49] This emphasis on technical expertise did
not replace earlier understandings of aviators as embodying aristocratic or
chivalric values. Still common in aerial literature, these surfaced too in
internationalist thought.[50] The Fabian internationalist and political scientist

[45] Hansard, HC, vol. 263, cols. 230–231, 15 March 1932.
[46] P. J. Noel Baker, 'Disarmament', International Affairs 13, no. 1 (January–February
1934): 3–25.
[47] H. G. Wells, The Open Conspiracy (London: Victor Gollancz, 1928).
[48] Davies, The Problem of the Twentieth Century, 448–449, 581, 700.
[49] Noel Baker, 'The International Air Police Force', 215.
[50] Edgerton, England and the Aeroplane, 50–53; Stefan Goebel, The Great War and Medieval
Memory: War, Remembrance and Medievalism in Britain and Germany, 1914–1940

G.E.G. Catlin declared in 1934 that in an international air force 'the chivalry of honour and the cavaliers of glory are able to serve a cause of lasting good to humanity'.[51] Rear Admiral R.N. Lawson, defending his proposal for a European Air Service in a letter to the *Spectator*, argued that 'surely the many stories we hear of the chivalry of the Air during the War are an earnest [*sic*] of the international comradeship of the air on which our hopes for a guardianship of peace must be based'.[52]

The internationalizing aviator also emerged in fiction, perhaps nowhere as prominent as Wells's *The Shape of Things to Come*, in which a world government evolved from the trading of apolitical commercially minded aviators. This vision of aviation's transformative power was far ahead of anything Wells had posited in the 1920s, when his writings placed aviation more broadly within a reconstructed world state, but not the driving force behind it. The third edition of his *Outline of History*, published in 1921, for example, talked of the need for the creation of a 'world state' which would require 'world control' of a number of social phenomenon, including not only 'air-ways' but also warfare, the mobility of people, labour, and a united world economic system.[53] Similarly, in the late 1920s he highlighted technically minded individuals as the driving force behind world unification and produced several works explaining the nature of this 'open conspiracy' to his readers. *The Shape of Things to Come*, by contrast, centred on the aviator.[54]

Although aviators viewed themselves as internationalist citizens of the world, only a few subscribed to these proposals and schemes. Leading aviation magazines such as *Flight* and *Aeroplane* were suspicious of League-orientated proposals for internationalization, seeing in them the loss of autonomy for British aviation and an unnecessary entanglement in European politics. 'No doubt we shall have international air lines, but they will come into being as ordinary commercial propositions, without any forcing by the League of Nations', noted Charles G. Grey, the influential editor of *Aeroplane*.[55] Air Commodore J.A. Chamier (one-time delegate to the Washington Conference on the Limitation of

---

(Cambridge: Cambridge University Press, 2007), 187–230. Examples: Bennett A. Molter, *Knights of the Air* (New York: D. Appleton, 1918); 'Chivalry in the Air', *Aeroplane*, 12 December 1934, 714.

[51] G. E. G. Catlin, 'The Roots of War', in Noel Baker, *Challenge to Death*, 21–39.

[52] R. N. Lawson, letter to the editor, 'A European Air Service', *Spectator*, 31 May 1935, 922.

[53] Wells, *The Outline of History*, 3rd ed., 1092.

[54] H. G. Wells, *The World of William Clissold* (London: Ernest Benn, 1926); Wells, *Open Conspiracy*; H. G. Wells, *What Are We To Do with Our Lives?* (London: William Heinemann, 1931); Wells, *The Shape of Things to Come*.

[55] C. G. Grey, 'On Disarmament', *Aeroplane*, 10 February 1932, 217–226; 'On Imperial Aviation', *Aeroplane*, 24 August 1932, 349–356.

Armaments and later secretary of the Air League of the British Empire) wrote in *Bookman* magazine that the internationalization of both civil and military aviation was unnecessary as the threat of bombardment had been exaggerated, and limitation of air armaments combined with international laws restricting bombing would be enough to reduce the threat from the air.[56] One prominent exception was serving RAF officer R.E.G. Fulljames, who suggested in 1935 that Britain, France, Italy, and Germany allocate aircraft to create a European force of fifty-two squadrons based in a neutral country (Switzerland, Hungary, or Greece), retaining fifty-two squadrons each of slower machines for their own air forces.[57]

Internationalists also placed limits on the internationalism of technical experts. Many were critical of disarmament negotiations which involved military officers: they were considered too wedded to warfare and to their militaries to be able to recommend disarmament.[58] This feeling was widespread enough for E.H. Carr to note 'utopian' disdain for disarmament 'experts' in his 1939 attack on internationalism, *The Twenty Years' Crisis*.[59] Attlee suggested a solution in the Commons: that these officers be replaced by civilians and 'simple soldiers' who represented the views of the masses and actually had faced the armaments under discussion.[60] One prominent exception to this attitude was Noel Baker, who was generally unwilling to criticize military delegates' pronouncements on their armaments, and instead referred to them, and treated them, as 'military experts'.[61] For military strategist B.H. Liddell Hart, the apolitical nature was deeply embedded within modern mechanical weapons themselves. In November 1932 at the Royal Institute of International Affairs lecture he argued that it was the mechanical sophistication of aeroplanes, machine guns, and tanks which made them peculiarly useful for an internationalized force. 'Modern weapons', especially long-range ones, he claimed, required less 'pugnacity' of their operators, and more 'pride of craftsmanship'. 'Both the gunner and the airman despatch their missiles in an essentially impersonal spirit', he suggested, and so 'the more mechanical weapons become, the less will the absence of national spirit be felt in an international force'.[62]

---

[56] J. A. Chamier, 'Air Bombing', *Bookman* 84, no. 503 (August 1933): 233–234.
[57] R. E. G. Fulljames, 'An International Air Police Force', *The Royal Air Force Quarterly* 6, no. 3 (July 1935): 245–250.
[58] See for example: Davies, *The Problem of the Twentieth Century*, 148, 149, 152.
[59] Carr, *The Twenty Years' Crisis*, 25–26.
[60] Hansard, HC, vol. 270, cols. 532–533, 10 November 1932.
[61] For example: Noel Baker, *Disarmament*, 77, 97, 100, 110, 145.
[62] B. H. Liddell Hart, 'An International Force', *Journal of the Royal Institute of International Affairs 1931–1939* 12, no. 2 (March 1932): 205–223.

Opposing the Pooled Frankenstein and His Monster

The air lobby and opponents of disarmament and the League disparaged the French proposals as soon as they were made. *Flight* dismissed them as an 'electioneering move', and Charles G. Grey of *Aeroplane* described them as a 'ridiculous and grotesque' attempt by the French to grab the limelight and to appear internationally minded. He held that, as Britain had a stronger air force, the French proposals would simply allow France to control British men and materials. *The Times* claimed that the French were only willing to reduce armaments that they themselves did not possess. A critic of the League, *The Times* warned that their proposals would transform it into an 'armed super-Power'. The *Daily Express* agreed; it ran the headline: 'France Staggers Geneva. Plan to Make the League Over-Lord of the World'. The *Spectator* agreed: although it was 'desirable to prevent the devilries of war advancing hand-in-hand with the benefits of science', the French proposal was a 'revolution' which would create an armed 'super-state'.[63] The government's response was negative, and was typified by Robert Vansittart, the Permanent Under-Secretary at the Foreign Office, for whom the French proposals were 'either a device for ensuring the military superiority of France in the immediate future' or 'deliberately inserted with the idea that they will be rejected by others, upon whom the blame for failure can be thrown'.[64]

The Air Ministry believed that the French proposals were a threat to British aviation and lobbied to convince government that they were unworkable and not in the British national interest. League control of civil aviation not only 'constituted an intolerable interference with purely commercial enterprise' but would anyway fail to prevent the conversion of commercial aircraft into military aircraft in times of war, which could be carried out quickly and secretly. Another Air Ministry charge was one of hypocrisy: their 'extreme measures', based on 'the fear of German re-armament through civil aviation' were 'in very marked contrast to the policy of intense nationalism which, in practice, informs their attitude to the operations over French territory of any non-French civil aviation

---

[63] 'Editorial Comment', *Flight*, 12 February 1932, 125–126; C. G. Grey, 'On Disarmament', *Aeroplane*, 10 February 1932, 217–226; 'A Discouraging Reception', *The Times*, 8 February 1932; 'Initiative Seized by France. Comment in Paris', *The Times*, 8 February 1932; 'Disarmament', *Spectator*, 13 February 1932, 204; 'France Staggers Geneva. Plan to Make the League Over-Lord of the World', *Daily Express*, 6 February 1932.

[64] Letter, R. Vansittart to the Foreign Secretary, 'The French Proposals', 24 November 1932, AIR 8/146, The National Archives, UK (hereafter cited as TNA). For more on the British response see: C. J. Kitching, *Britain and the Geneva Disarmament Conference: A Study in International History* (Houndmills: Palgrave Macmillan, 2003), 55; and Kitching, *Britain and the Problem of International Disarmament*, 140–173.

enterprises'.[65] The Air Staff had opposed any reduction in RAF numbers from before the Geneva Conference, a 1931 memo concluding that Britain was already weak in the air as 'we have carried unilateral disarmament too far'.[66] So they too lobbied against the French plan, arguing that it appeared so unfeasible that it had to be in reality nothing more than a 'political gesture'.[67]

Supporters of British aviation continued to use these arguments through the 1930s in their opposition to international control. In 1940 Grey would recall that in March 1933 he received a letter signed by forty-nine leading aviation organizations warning him that internationalization would 'hamper' British aviation's 'legitimate development'. Signatories included the Royal Aeronautical Society, the Royal Aero Club, the Society of British Aircraft Constructors, the Air League of the British Empire, and the Guild of Air Pilots and Navigators of the British Empire.[68] Opponents of international control also continued to support the British bomber. The most developed arguments were made by Air Ministry lawyer and air power advocate James M. Spaight. In works such as his 1932 book *An International Air Police*, he insisted that a strong national air force consisting of heavy bombers would best ensure peace through deterrence. An international air force, moreover, could not be restricted from waging aggressive war; bombers were too dangerous to be concentrated in the hands of one organization.[69] Spaight had already in the late 1920s suggested regional aerial defensive pacts as a practical form of collective security.[70] The Air League of the British Empire issued a pamphlet to mark its 1934 Great Empire Air Day emblazoned with a cover-page declaration announcing its opposition to international control and its support for 'logical and practical measures' to bring about world peace. These measures were described inside: a strong bomber force equipped with poison gas bombs (because 'bombers mean

---

[65] J. S. Ross, Air Ministry to the Under-Secretary of State for Foreign Affairs, 29 August 1934, AIR 2/665, TNA.

[66] J. Salmond, 'The Basis for the Limitation of Air Armaments – Metropolitan and Overseas Quota, DC (P) 36 and 37', 8 June 1931, AIR 8/128, TNA.

[67] Memo by the Air Staff, 'Note by the Air Staff on the French Proposals Contained in Document Conf.D.56 dated 5.2.32', ca. February 1932, AIR 8/132, TNA. For lobbying, see: J. M. Salmond to A. Eden, 10 March 1933, AIR 2/664, TNA; Telegram, J. M. Salmond to Air Marshal J. Steel, R. A. F. Simla, 26 July 1932, AIR 8/145, TNA; and the folder: 'India and the Proposal to Abolish Bombing', 9 May 1932 to 4 November 1933, AIR 8/145, TNA.

[68] C. G. Grey, *A History of the Air Ministry* (London: George Allen & Unwin, 1940), 230–231.

[69] Spaight, *An International Air Police*; J. M. Spaight, 'Self Defence and International Air Power', *Journal of Comparative Legislation and International Law*, 3rd series 14, no. 1 (1932): 20–29.

[70] J. M. Spaight, *Pseudo-Security* (London: Longmans, 1928).

peace').[71] Nigel Tangye, aviation correspondent for the *Evening News*, organized a book in support of British military and civil aviation which emphasized its importance for British defence and commercial prosperity. Contributors included editor of the *Aeroplane* C.G. Grey, First World War flying ace and Conservative MP Harold Balfour, Leonard Bairstow (holder of the Zaharoff Chair of Aviation at Imperial College), and J.A. Chamier, Secretary of the Air League.[72]

Opponents, nevertheless, shared important assumptions about the nature of aviation with proponents of international control. For them, too, aviation was a modern and modernizing international activity, a world-changing scientific invention in terms of both warfare and international communication. It was because of these characteristics, they argued, that more national aviation was needed. At a time when war was now more deadly, enemies more threatening, and traditional armaments more obsolete, military aviation could protect countries and even reduce international warfare through deterrence. Indeed, for Spaight, military aviation was largely defensive and its ability to exact a 'knockout blow' exaggerated.[73] For others the military aeroplane was simply too powerful to be halted by collective security; for Groves 'owing to the speed of the aerial arm, and the terrific nature and scope of its attack'.[74] Moreover, opponents argued, international control of civil aviation would stifle, not promote, the natural integrative qualities of aviation. Aerial entrepreneur and Conservative MP John Moore-Brabazon, for example, opposed internationalization in the Commons in March 1932 by characterizing it as a 'restriction' of an otherwise 'great new form of transport', which offered 'a great opportunity of linking one nation with another'. There should be 'free flight from one country to another' without interference from 'tin-pot countries all over the world' who 'thought the old law of the air belonging to themselves prohibited this free intercourse of aircraft from one country to another'. Only this 'freedom in the air' would lead to 'true internationalisation of communication in the air'.[75]

The Left, viewing the League as a conduit for Anglo-French imperialist policies, opposed proposals which strengthened the League militarily.

---

[71] Air League of the British Empire, *A Souvenir of the Great Empire Day of 1934* (London: Air League of the British Empire, 1934).

[72] Nigel Tangye, ed., *The Air is Our Concern* (London: Methuen, 1935). During the Second World War Tangye would go on to write a book and pen a three-episode BBC series, also titled *The Air is Our Concern*, celebrating British civil and military aviation; Nigel Tangye, *Britain in the Air* (London: William Collins, 1944).

[73] Spaight, *Air Power in the Next War*, 97–113. Also: Spaight, 'Self-Defence and International Air Power'; and the lectures by Groves noted subsequently.

[74] Groves, 'The Influence of Aviation in International Affairs'.

[75] Hansard, HC, vol. 262, cols. 2055–2058, 10 March 1932.

The British Communist Party theorist R. Palme Dutt warned that a League police would be a Great Power police. Similarly for the journalist and author H.N. Brailsford 'A League army to-day would be the picked police force of allied capitalism'. The Left wing of the Labour Party, set against League collective security, voiced its opposition to Labourite proposals for international control through the New Fabian Research Bureau and the Socialist League.[76] The Left was also suspicious of technical experts and their techno-utopianism. Communist intellectual Christopher Caudwell attacked the notion that salvation could only come through technocrats, scientists, or 'divine bureaucrats' arriving from Utopia 'in a glittering aeroplane to put things right from above'. The new world order was not to be moulded by them but rather from below, by the 'sole creative force of contemporary society', the proletariat.[77] Modern inventions, contended George Orwell in a 1944 *Tribune* column, had not in any way caused 'the abolition of distance' or 'the disappearance of frontiers'. Rather 'the effect of modern inventions has been to increase nationalism, to make travel enormously more difficult, to cut down the means of communication between one country and another, and to make the various parts of the world less, not more dependent on one another for food and manufactured goods'.[78] Yet even for Orwell scientific inventions inherently brought internationalism, if only they were controlled by socialists. Unfortunately, 'we haven't Socialism. As it is, the aeroplane is primarily a thing for dropping bombs and the radio primarily a thing for whipping up nationalism'.[79]

Pacifists were particularly conflicted in their response. Some saw in an international force the promise of collective security and peace in international affairs, whereas others continued to argue against any use of violence. The pacifist Helena Swanwick and the National Peace Council both supported the abolition of national air forces and the internationalization of civil aviation and yet also opposed the use of force, and so opposed an international air force.[80] A group of mostly British pacifist organizations sent a declaration to the Disarmament conference in July 1932 calling for the abolition of tanks and military aviation,

[76] Dutt, *World Politics 1918–1936*, 131; H. N. Brailsford, 'A Socialist Foreign Policy' in *Problems of a Socialist Government*, eds. C. Addison, et al. (London: Victor Gollancz, 1933), 252–286; Report by John Parker, New Fabian Research Bureau, 'Conference on Foreign Affairs', October 1934, 2/12, Noel-Baker Papers; Paul Corthorn, 'The Labour Party and the League of Nations: The Socialist League's Role in the Sanctions Crisis of 1935', *Twentieth Century British History* 13, no. 1 (2002): 62–85.

[77] Christopher Caudwell, *Studies in a Dying Culture* (London: Bodley Head, 1938), 73–95.

[78] George Orwell, 'As I Please', *Tribune*, 12 May 1944.

[79] George Orwell, 'As I Please', *Tribune*, 2 February 1945.

[80] H. M. Swanwick, *Collective Insecurity* (London: Jonathan Cape, 1937), 282.

but not the formation of an international air force or the internationalization of civil aviation. One member of the No More War Movement wrote
to the *Manchester Guardian* arguing that 'an international police force
would only serve to perpetuate the political and economic status quo in
Europe and to facilitate the repression of colonial populations'.[81]

Helena Swanwick emerged as a leading pacifist opponent of an international air force and collective security, particularly through books and
pamphlets such as her well-known 1937 tract *Collective Insecurity* and
Pooled Frankenstein and His Monster.[82] Her arguments included
a sustained criticism of the domestic analogy between national police
forces and international police, used by internationalists since the First
World War to rationalize the latter.[83] A 1933 letter to Clifford Allen
decried it as a 'false analogy' and listed seven vital differences between
a national and an international police, including the observation that
a policeman has a 'whistle and a truncheon', whereas an international
air force 'would have to be fully armed with the most modern and
destructive weapons' and so would be a force for 'terror and frightful-
ness'. *Collective Insecurity* similarly listed twelve differences.[84]

Swanwick recognized but scornfully dismissed public support for the
use of international force. In *Collective Insecurity* she branded the results of
the Peace Ballot as unconsidered and 'irresponsible' responses to 'a
highly complex and technical question'. She warned that even
a 'League War' for collective security would be destructive in which
'knights of the air' would not spread peace but bomb civilians and cities.
She was sceptical of the opinions of technical experts, for it was a 'Queer
wicked world, isn't it? that emerges from the labours of peace-professing
politicians, laboratory-minded, mild-mannered scientists, "nice-boy"
aviators, all driving the herd-like multitude!' Swanwick agreed with 'sanc-
tionists' that the Great War had inaugurated new scientific weapons, and

---

[81] Gerald Bailey, Directing Secretary of the National Peace Council, letter to the editor,
'Military Aviation and Public Opinion', *Manchester Guardian*, 31 May 1933; Percy
Bartlett, 'National Peace Council: A Constructive Peace Policy', 24 October 1933, box
89, Davies Papers; 'Socialists Warn M. Herriot. World Appeal to Conference',
Manchester Guardian, 19 July 1932; E. A. Williams, letter to the editor, 'An
International Police Force. Some of its Dangers', *Manchester Guardian*, 9 March 1934.

[82] H. M. Swanwick, *Pooled Security: What Does It Mean?* (London: Women's International
League, 1934); H. M. Swanwick, *Pooled Frankenstein and His Monster: Aviation for World
Service* (London: Women's International League, 1934); H. M. Swanwick, *New Wars for
Old* (London: Women's International League, 1934).

[83] For example: C. van Vollenhoven, 'The Opportunity of this Solution' and 'International
Police and Pacifism', in Vollenhoven, *War Obviated by an International Police*, 59–82.
Domestic analogies were widespread in internationalist rhetoric; see: Suganami, *The
Domestic Analogy and World Order Proposals*.

[84] Gilbert, *Plough My Own Furrow*, 337–338; Swanwick, *Collective Insecurity*, 230–232.

that the new air warfare could strike a knock-out blow, but was dismayed that they would use 'incendiary bombs, high explosives, poison gas, and – doubtless in time – a death-ray' in pursuit of their 'obsession of security'. The aviator, meanwhile, was a powerful figure in her thinking. *Collective Insecurity* warned that through rearmament 'a very dangerous military aristocracy is being built up by air forces in every country. Few, select, expert and young, the aviators may become the most dangerous tools of reaction, whether admittedly employed by a state or theoretically by a combination of states, such as the League'.[85]

Supporters of international control recognized the power of pacifist critiques and spent as much time combating them as they did supporters of British aviation. New Commonwealth propaganda emphasized the centrality of force for the achievement of peace through world organization and directly attacked arguments made by 'no-force pacifists', such as Swanwick.[86] The edited volume *Challenge to Death*, published in 1934 in support of internationalized aviation and League collective security, attacked what one author called 'extreme pacifism' as much as it did 'militarism', and was recognized by the *Manchester Guardian* as a 'manifesto' by the 'Right-Centre of the peace party', concerned 'as much with appealing to the pacifists as exposing the patriots'.[87]

There also remained a prominent strand of liberal internationalist activism which looked not to the League and aviation for a solution to the problems of international anarchy and nationalism, but instead to the British Empire and Commonwealth. This imperialist internationalism, although supportive of the use of aviation in Empire, opposed the centralization of aviation through international control. Scholar Alfred Zimmern called for the creation of a world order based foremost on an 'international mind', the leading South African statesman J.C. Smuts imagined a future 'holistic' world order based on the Commonwealth, and academic Lionel Curtis supported an 'organic' Commonwealth.[88]

---

[85] Swanwick, *Collective Insecurity*, 19, 26, 63, 66, 81, 246–247, 282. For more on Swanwick see: Lucian M. Ashworth, 'Feminism, War and the Prospects for Peace: Helena Swanwick (1864–1939) and the Lost Feminists of Inter-War International Relations', *International Feminist Journal of Politics* 13, no. 1 (2011): 25–43.

[86] W. Horsfall Carter, *Peace Through Police* (London: New Commonwealth Society, 1934), 9–11.

[87] 'Books of the Day: Challenge to Death', *Manchester Guardian*, 19 November 1934; Mary Agnes Hamilton, 'No Peace Apart from International Security: An Answer to Extreme Pacifists', in Noel Baker, *Challenge to Death*, 261–274; Vera Brittain, 'Peace and the Public Mind', in Noel Baker, *Challenge to Death*, 40–66.

[88] Alfred Zimmern, *Quo Vadimus?* (London: Oxford University Press, 1934); Zimmern, *The League of Nations and the Rule of law*, 493–496; Paul Rich, 'Reinventing Peace: David Davies, Alfred Zimmern and Liberal Internationalism in Interwar Britain', *International Relations* 16, no. 1 (2002): 117–133; Jeanne Morefield, *Covenants without Swords: Idealist*

These views, out of step with mainstream liberal internationalism by the mid-1930s and appearing (to many) outdated, were largely immune from the aerial internationalism of their time. At a time where aviation loomed large in the public's imagination, for example, the Marquess of Lothian, Philip Kerr, addressing Chatham House in June 1934, dismissed aviation as irrelevant for global power, war, and security. Sea power, he argued, remained central. Britain should stay out of any collective security arranged by Continental European states (in which aviation could, he conceded, play an important role), and Britain and her Commonwealth should join the United States in enforcing the Kellogg Pact using their navies. Zimmern, in a lecture series delivered at Oxford University in early 1939, reflected on the worsening international relations of the time to argue that policymakers had been 'misled by the hopes of short-cuts, of remedies that would require less effort, by the perpetual lure of something for nothing'; misled into believing that the 'tension between the things of Caesar and the things of God' could be 'relieved by some simple and mechanical means involving little or so effort'.[89]

H.G. Wells stood out as an internationalist proponent of aviation for world order who at the same time was a critic of mainstream proposals for internationalization. An opponent of the League of Nations, he argued that effective control would never be achieved through that 'ill-planned and ill-supported assembly' which was only of 'considerable usefulness' for the settlement of 'minor international difficulties . . . but as a guarantee

Liberalism and the Spirit of Empire (Princeton, NJ: Princeton University Press, 2005), 96–135; Peder Anker, *Imperial Ecology: Environmental Order in the British Empire, 1895–1945* (Cambridge, MA: Harvard University Press, 2001), 51–53; Mazower, *No Enchanted Palace*, chapter 1; Lionel Curtis, *Civitas Dei: The Commonwealth of God* (London: Macmillan, 1938), 910–912; Deborah Lavin, *From Empire to International Commonwealth: A Biography of Lionel Curtis* (Oxford: Clarendon Press, 1995), 114, 294, 314. On this type of internationalism see also: George Egerton, 'Conservative Internationalism: British Approaches to International Organization and the Creation of the League of Nations', *Diplomacy & Statecraft* 5, no. 1 (1994): 1–20; Long and Schmidt, *Imperialism and Internationalism in the Discipline of International Relations*; Jerónimo, 'A League of Empires', 87–126; Jeanne Morefield, *Empires without Imperialism: Anglo-American Decline and the Politics of Deflection* (Oxford: Oxford University Press, 2014), chapters 1, 3 and 5.

[89] Philip Kerr, 'The Place of Britain in the Collective System', *International Affairs* 13, no. 5 (September–October 1934): 622–650; Alfred Zimmern, *Spiritual Values and World Affairs* (Oxford: Clarendon Press, 1939), 45–46. Kerr was heavily criticized by other internationalists in the ensuing discussion. Political scientist C. Delisle Burns and journalist Wickham Steed attacked Kerr for failing to take into account the changing nature of modern war due to 'diabolical devices'. The next war would be launched, they pointed out, from the air. The 'conquest of the air', agreed New Commonwealth member Horsfall Carter, required that Britain involve herself in Continental European security. On the outdated nature of this internationalism by mid-1930s see: Paul Rich, 'Alfred Zimmern's Cautious Idealism: The League of Nations, International Education, and the Commonwealth', in Long and Wilson, *Thinkers of the Twenty Years' Crisis*, 79–99.

against graver quarrels it is beneath contempt'. In 1927 he criticized the League as a 'mischievous opiate' which lulled people into a 'false secur-ity', and in 1940 described it as the 'opiate of liberal thought in the world'. But Wells did not simply call for the replacement of the League with a political or collective security organization. He wanted a powerful organization with far-reaching powers to control monetary and fiscal policy, and international trade, communication, and transport (especially for commercial purposes). This led him to criticize internationalist proposals such as David Davies's, which relied largely on an international air police for empowerment. He particularly took exception to Davies's use of the domestic analogy, which Wells believed failed to understand the unique administrative and control potentialities of a state with global reach. Through global trade and communication, the aeroplane now allowed resources worldwide to be brought into a global planning process, allowing for far-reaching social and cultural transformations which would be impossible in the limited horizons of a nation-state.[90]

Convertibility and the Militaristic Perversion of Aviation

Deeply held presumptions about the fundamental nature of aviation and the convertibility of civilian to military aviation underpinned proposals for international control. The notion that existing civilian aircraft and associated facilities could easily be converted to military use, although not new, emerged as a strong aspect of air power rhetoric at the end of the First World War. Influential planning documents such as the 1918 Sykes Memorandum on the future of British aviation recommended high subsidies for the British commercial aviation so that it could, amongst other things, maintain an extensive network of civilian air routes and aerodromes, and develop airliners that could be converted into bombers at short notice.[91] The 'reserve' argument in support of highly funded civil aviation, which continued into the 1920s and 1930s, was buttressed by the assumption, common in both military and civilian writings on aviation, that military aviation was derived from civilian roots. Air Commodore R.H. Clark-Hall's 1924 lecture to the Royal United Services Institution on 'The Value of Civil Aviation as a Reserve to the Royal Air Force', for example, argued that civil and military aviation were 'two branches' of one aviation, which was in essence an invention of

---

[90] H. G. Wells, 'Foreword', in *Peace or War?*, ed. J. M. Kenworthy (New York: Boni & Liveright, 1927), xii, xiii; Wells, *The New World Order*, 16, 104–106.

[91] F. H. Sykes, 'Memorandum by the Chief of the Air Staff on Air-Power Requirements of the Empire', in *From Many Angles: An Autobiography* (London: Harrap, 1942), 558–574. Also: Sykes, *Aviation in Peace and War*, 102–104.

increased communication (like the railway, the steamship, the telephone, the submarine cable, and the motor car before it).[92] Walter Raleigh's 1922 official history of wartime aviation began with a history of the 'invention' of aviation and highlighted heroic civilian inventors as the progenitors of military aviation.[93]

The convertibility argument was also used to warn of the danger to Britain from German civil aviation, and so make the case for a strong British air force. The most prominent public proponent, Brigadier General (retired) P.R.C. Groves, warned through lectures and newspaper and journal articles that Germany was developing her civilian aviation such that her civilian aircraft could be used for military purposes, including bombing.[94] In a much publicized 1927 report to an Air Transport Co-operation Committee of the League, *The Relations between Civil and Military Aviation*, he located this potential for conversion within the broader natural technical development of large civilian aircraft: their 'constantly increased range of greater reliability, greater weight and carrying capacity and higher speeds' were 'precisely the requirements aimed at in the development of bombers. . . . The average air liner is a potentially far more efficient bomber than the air liner of seven years ago'.[95] One RAF officer, writing in the *Spectator* in 1935, estimated that Germany had 300 'modern' civilian aeroplanes with 'a real military value', along with a network of commercial aerodromes and facilities suitable for military use. The British state, contended these supporters of air power, needed to invest more in civilian and military aviation or risk quick defeat in a future war.[96]

Given these nationalist concerns about aerial power, it is no surprise that convertibility became an important component of debates about aerial disarmament. It cropped up as early as the post-First World War Paris peace conference, where it was readily accepted by both those arguing for and arguing against tight restrictions on German aviation. The Allied aeronautical commission charged with drawing up the aerial sections of the Treaty of Versailles concluded that German civil aviation needed to be restricted because 'aeroplanes and airships can be very easily

---

[92] R. H. Clark-Hall, 'The Value of Civil Aviation as a Reserve to the Royal Air Force in the Time of War', *Journal of the Royal United Services Institution* 69, no. 475 (1924): 415–432.

[93] Raleigh, *The War in the Air*, 122.

[94] P. R. C. Groves, *Our Future in the Air: A Survey of the Vital Question of British Air Power* (London: Hutchinson, 1922); P. R. C. Groves, *Behind the Smoke Screen* (London: Faber and Faber, 1934), 230, 237; Brett Holman, 'The Shadow of the Airliner: Commercial Bombers and the Rhetorical Destruction of Britain, 1917–35', *Twentieth Century British History* 24, no. 4 (2013): 495–517.

[95] Groves, 'The Relations between Civil and Military Aviation'.

[96] P. R. Burchall, 'Civil Aviation and Bombing', *Spectator*, 7 June 1935, 968.

and quickly transformed into weapons of war'.[97] Convertibility became central too for anti-disarmament arguments in the 1920s and 1930s. The Air Ministry and air power enthusiasts used it to demonstrate that the complete abolition of bombers would not prevent bombing, and consequently that any attempt at abolition would have to take civil aviation into account as well. This, the argument usually proceeded, was practically difficult and would retard the natural and beneficial growth of civil aviation.[98] Opponents suggested instead that states segregate military and civilian aviation in order to reduce convertibility and militaristic misuse of civilian aviation. The most detailed proposal was put forward in 1927 by a committee of military experts appointed by the League of Nations Preparatory Commission for the Disarmament Conference to consider the relationship between civilian and military aviation. Its concluding suggestions included that states separate their departments dealing with civilian and military aviation, that civilian pilots not be required to undergo military training, that governments not subsidize civil aviation for 'strategic' purposes (only for 'economic and social purposes'), greater international cooperation in civilian aviation, and that governments 'refrain from prescribing the embodiment of military features in the construction of civil aviation material'. These points were eventually incorporated into the draft agreement put before the 1932 Disarmament Conference – the agreement which was brushed aside by the Tardieu Plan and later disarmament proposals.[99]

From the early 1930s onwards internationalists relied heavily on convertibility arguments to make their case for the international control of aviation, but did so with the aid of a second claim – that the military was perverting the natural development of aviation. The military perversion argument had its roots in late nineteenth-century liberal thought, when intellectuals had come to posit an opposition between industry and trade on the one hand and war and militarism on the other. Growing industrialization and international trade, Richard Cobden and others had argued, led to international peace. Militaristic policies, on the other hand, were

[97] 'Minutes of the Meeting of the Supreme War Council Held at the Quai d'Orsay, Paris, Monday, 17 March 1919, at 3 p.m.', *Foreign Relations of the United States*, 1919, vol. 4, The Paris Peace Conference (Washington, DC: Government Printing Office, 1943), doc. 20; D. Carlton, 'The Problem of Civil Aviation in British Air Disarmament Policy, 1919–1934', *Royal United Services Institution Journal* 111, no. 664 (1966): 307–316.

[98] For example: 'Air Staff Memorandum on Limitation of Air Armament', December 1923, AIR 5/360, TNA.

[99] League of Nations, Air Commission of the Conference for the Reduction and Limitation of Armaments, *Objective Study on the Internationalisation of Civil Aviation*, Disarmament 1932.IX.43 (Geneva: League of Nations, 1932), 47–48.

detrimental to industrial and commercial development. Central to this understanding was the assumption that modern science and industry were essentially civilian and peaceful in character.[100] By the early 1920s this Cobdenite thinking had become incorporated into a liberal critique of militaristic perversion. For J.A. Hobson one of the *Problems of a New World* was that war had become 'the revolt of machinery against its maker – the malign perversion of the science of industry from production to destruction'.[101] Economist A.C. Pigou's well-known 1921 critique of militarism and the costs of war, *The Political Economy of War*, counted the militaristic perversion of civilian land and sea transport as one of the burdens of war, and foretold that aviation too would be 'twisted' from its 'normal development' as long as governments continued to 'exercise control over the design of commercial aircraft' and 'have a voice in preparing air routes and determining the situation of aerodromes'. This would interfere with the 'free play of economic forces' and lead to 'less efficient instruments of communications in normal times'.[102] This critique was thus an attack on the military's influence in scientific and industrial research and development, and a call for this development to be directed by civilian technical experts (or the free market), who, it was argued, were best placed to ensure that industries were developed for the social and economic wealth and well-being of the nation.[103]

By the late 1920s this critique had come to underpin internationalist calls for arms control and international regulation or 'control' of strategic national industries. Noel Baker's *Disarmament*, citing Pigou's work, extended its arguments to virtually all of heavy industry. Militaristic aims were 'diverting the normal channel of industrial development': the 'iron and steel industries, the engineering industry; some chemical industries, iron-mining and coal-mining' as well as 'big shipyards and aircraft factories'. These were modern technical and scientific enterprises, and as such then there was also the 'perversion of scientific and inventive genius': that is the diversion of scientists and engineers from commercial ('productive') to military ('unproductive') research and development. Complete disarmament, he concluded, could only be achieved when all modern industries and scientific research was shielded from the military.[104]

In the 1930s internationalists took up this argument in support of international control. It was a necessity, declared the Spanish diplomat

[100] A. Howe, 'Free Trade and Global Order: The Rise and Fall of a Victorian Vision', in Bell, *Victorian Visions of Global Order*, 26–46.
[101] J. A. Hobson, *Problems of a New World* (London: George Allen & Unwin, 1921), 176.
[102] Pigou, *The Political Economy of War*, 9, 10.
[103] For more on the militaristic critique see: Edgerton, *Warfare State*, 2, 12.
[104] Noel Baker, *Disarmament*, 11, 14.

and League official Salvador de Madariaga, because 'military and naval reasons warp, distort and even subvert economic laws at every turn'.[105] Only international control, it was argued, would protect aviation from the militaristic and nationalistic influences of nations, thus allowing it to develop along its natural commercial trajectory and fulfil its internationalist promise of global interconnection and peace. Some went further, and argued that the governing international organization could utilize convertibility itself by treating its own commercial aerial fleet as a military reserve, to be used if necessary for international policing.[106]

This militaristic perversion, internationalists argued, had begun in the First World War. The war had accelerated the military appropriation of civilian science whilst at the same time hindering its peaceful development. The central crisis of international relations itself arose, argued David Davies, from this time, whence science was forced to assume 'a dual personality – the one benevolent, the other malignant':

All the discoveries and inventions which ultimately were responsible for the introduction of modern weapons were primarily intended to increase the wealth of the world and to expand its peaceful industries. They were the products of civilian brains and occupations and were undertaken to promote the interests and the welfare of humanity. Transformed for the purposes of war, they became a menace to civilisation and a trap for the unwary. Instead of ministering to the economic needs of the peoples, they provided them with the most formidable engines of destruction with which they might lash and tear each other to pieces.[107]

The militaristic perversion argument fed off wider internationalist concerns about the political power of national militaries, the militaristic tendencies of government policymakers and the civil service, the growth in armaments ('rearmament'), and ultimately growing tensions in European international relations. It was also closely related to concerns about arms manufacturers ('merchants of death') and their pernicious effects on international stability.[108] Thus Noel Baker, turning to consider in 1934 why the Geneva Disarmament Conference was failing, blamed civil servants, politicians, and arms manufacturers. He faulted bureaucrats for their inertia, old-fashioned thinking, and aversion to change, and politicians for the hold General Staffs had on them. Military men, who

---

[105] de Madariaga, *Disarmament*, 7–8. Also Salvador de Madariaga, *The World's Design* (London: Allen and Unwin, 1938), 64–69.
[106] Zaidi, 'Aviation Will Either Destroy or Save Our Civilization'; Holman, 'The Shadow of the Airliner'.
[107] Davies, *The Problem of the Twentieth Century*, 318–319.
[108] David G. Anderson, 'British Rearmament and the "Merchants of Death": The 1935–36 Royal Commission on the Manufacture of and Trade in Armaments', *Journal of Contemporary History* 29, no. 1 (January 1994): 5–37.

'naturally' loathed to reduce their military forces, had in turn 'had a powerful ally in the vested interests of the armament firms'. These, he argued, 'by their manipulation of the Press, embroil the international situation, and obscure from hesitating politicians the strength of the popular demand for peace'.[109]

Fighting Fascism

Relations between European states significantly worsened through the 1930s, and faith in the League's political work evaporated. The Geneva disarmament conference adjourned without any substantial agreements in June 1934.[110] Japan and Germany withdrew from the League in 1933, and Italy in 1937. Italy invaded Abyssinia in 1935, Japan invaded China in the summer of 1937, and a civil war ravaged Spain between 1936 and 1939. Yet calls for international control remained strong in Britain until the mid-1930s, and did not entirely disappear in the later years of the decade. In the Liberal Party this move was gradual and was more or less complete by 1938 by which point the Party instead supported 'collective security through alliances' instead of League control of aviation. In 1937 the Labour Party dropped international control from its detailed foreign policy statement, *International Policy and Defence*, and it was not mentioned at all during the 1937 Annual Conference.[111]

In contrast to Britain, support in France for internationalization dissipated much more rapidly from 1934 onwards. The increasing polarization of French political life led to the thinning of the liberal Centre/Centre-Left, which abandoned international control as it turned to the Right or, as in the case of the bulk of the Radical Party in 1935, the Left. In 1934 Louise Weiss left *L'Europe Nouvelle* to campaign for women's rights, and Henry de Jouvenel passed away in 1935. *L'Europe Nouvelle*'s final call for internationalized aviation was in February 1934. The magazine subsequently became a supporter of national aerial rearmament. Cot (appointed Minister for Aviation in January 1933) continued to call for international control at the League and at internationalist meets, though ultimately forged ahead with French aerial rearmament.[112]

---

[109] Noel Baker, 'Peace and the Official Mind'.
[110] Kitching, *Britain and the Geneva Disarmament Conference*, 136–173.
[111] Grayson, *Liberals, International Relations and Appeasement*, 93–94; National Council of Labour, *International Policy and Defence* (London: National Council of Labour, 1937); Labour Party, *Report of the Annual Conference of the Labour Party* (London: Labour Party, 1937).
[112] Roger Beaucaire, 'Aviation Commerciale et Économie Fermée', *L'Europe Nouvelle* 17, no. 862 (10 February 1934): 141–143; 'L'Avenir de l'aviation française: Interview de

Support in Britain for international control of aviation was sustained into the late 1930s by several factors. First, internationalized aviation, and an international air force in particular, had moved beyond policy and intellectual discussions and entered popular culture. Wells's novel *Shape of Things to Come* (1933) and its movie adaptation (1936) were perhaps the best known, but there were also a number of other fiction novels, the best known of which were Victor MacClure's *Ultimatum: A Romance of the Air* (1924), Michael Arlen's *Man's Mortality* (1933), and Brian Tunstall's *Eagles Restrained* (1936).[113] Second, many internationalists hoped that international control could be made acceptable to at least some of the major European powers. Many delegations had, after all, agreed to it during the Geneva conference, and some had put forward their own proposals. Many internationalists believed that most of the blame for the failure of the conference lay with the British government, and consequently that disarmament negotiations, even if only limited to aviation, could be reconvened with a sincere British initiative.[114] Moreover, although for seasoned observers the League appeared to have failed on the diplomatic front, the idea of League-orientated collective security continued to carry significant rhetorical appeal. The 1934–35 Peace Ballot signalled popular support, leading the three main political parties to back collective security in the run up to the November 1935 General Election. Even the Left, critical of the League in the early 1930s, began to show some measure of support once the Soviet Union joined in September 1934, with the split in the Socialist League's ranks between those advocating collective action through the League and those still dismissing it becoming particularly pronounced by late 1935. Many supporters were undeterred by the failure of the League to act over the Italian invasion of Abyssinia (1935–1936) and the December 1935 Hoare-Laval Pact. Noel Baker, for example, stood on a strongly pro-League and pro-collective security platform at the July 1936 Derby by-election, and overturned a government majority of 12,000. Abyssinia and the Hoare-Laval Pact appeared to demonstrate Britain and France's

M. Pierre Cot', *L'Europe Nouvelle* 18, no. 882 (5 January 1935): 13–14; *L'Europe Nouvelle* Special Issue: *Le problème des armements devant l'effort militaire du Reich* 18, no. 901 (18 May 1935): 479–485. On the Radical Party's turn to the Popular Front see: Julian Jackson, *The Popular Front in France: Defending Democracy, 1934–1938* (Cambridge: Cambridge University Press, 1990), 36–51.

[113] Wells, *The Shape of Things to Come*; *Things to Come*, directed by William Cameron Menzies (London Film Production and United Artists, 1936); Victor MacClure, *Ultimatum: A Romance of the Air* (London: George G. Harrap, 1924); Michael Arlen, *Man's Mortality* (London: Heinemann, 1933); Brian Tunstall, *Eagles Restrained* (London: Allen & Unwin, 1936).

[114] For example: 'Wings Over Europe', *The Manchester Guardian*, 4 June 1934.

unwillingness to take disarmament and League collective security seriously rather than any intrinsic issues in the notion of League collective security itself. International control of aviation could still prevent war, if only it were given a chance.[115]

Many supporters of internationalized aviation were able to successfully reconcile their support for growing national rearmament with support for international control and collective security by arguing that British arms were needed to meet the obligations agreed to in the Locarno Treaty, and could in time be transferred to the League. Rearmament was a very obvious, and popular, public reality by 1936: in 1934 there existed forty-two Home Force air squadrons providing a first-line strength of some 800 aircraft, and by 1939 this would grow to 157 squadrons and 3,700 aircraft. The League of Nations Union reversed its long-standing opposition to rearmament in 1936, though continued to expound aviation-based collective security.[116] Proposals for international control even managed to attract critics of appeasement such as Winston Churchill, who assumed the Presidency of the British section of the New Commonwealth Society in 1936 declaring that he did not want the kind of collective security seen 'in a flock of sheep on the way to the butcher'.[117] Churchill had, in fact, been attracted to the notion of an international air force since the late 1920s. The *Aftermath* volume of his history of the Great War, published in 1929, had suggested that Wilson, Clemenceau, and Lloyd George could have created 'world peace' in Paris by abolishing national aviation and turning the 'power of the air' to the League of Nations for the collective security and mutual defence.[118]

Events such as the bombing of Guernica and the conquest of Abyssinia only reinforced the bomber's status as the pre-eminent weapon of war,

---

[115] Ceadel, 'The First British Referendum'; Corthorn, 'The Labour Party and the League of Nations'. For Stafford Cripps' continued criticism of League collective security see: Stafford Cripps, *The Struggle for Peace* (London: Victor Gollancz, 1936), 53–65. However, Cripps supported League collective security in parliament in 1933: Hansard, HC, vol. 281, cols. 677–690, 13 November 1933. For a collective security-orientated response to the Abyssinian crisis see: Liberal MP Dingle Foot's intervention in the Commons: Hansard, HC, vol. 304, cols. 2917–2925, 1 August 1935. 'Derby Election Result', *The Times*, 10 July 1936. Noel Baker brought some of these trends together in: Philip Noel Baker, 'The Future of the Collective System', in *Problems of Peace*, 10th series Anarchy of World Order, eds. R. B. Mowat, et al. (London: George Allen & Unwin, 1936), 178–198.

[116] H. Wickham Steed, 'The Search for Security against War', *The Journal of the Royal United Services Institution* 81, no. 524 (November 1936): 707–733; Pugh, *Liberal Internationalism*, 151.

[117] 'Our London Correspondent: Mr. Churchill among the Diplomats', *Manchester Guardian*, 26 November 1936; Pugh, 'Policing the World'.

[118] Winston S. Churchill, *The World Crisis*, vol. 4, *1918–1928: The Aftermath* (London: Thornton Butterworth, 1929), 27.

and worsened international relations in fact offered unusual opportunities for action. In 1938 David Davies gathered together a group of interested internationalists (including Pierre Cot, Noel Baker, and Ludwik Rajchman) whilst attending a peace conference in Paris and attempted to form a volunteer international air force to protect Chinese cities from Japanese aggression. In 1940 he even considered the formation of such a force to aid Finland during the Soviet–Finnish War, and accompanied Harold Macmillan on a visit there.[119]

Lastly proposals for internationalization were also bolstered by internationalists' ability to flexibly negotiate the issue of German rearmament. In the late 1920s and very early 1930s internationalization of aviation fitted in well with the wider internationalist approach to German arms, that of achieving 'equality' in armaments. As Germany was not allowed any military aviation, internationalists reasoned, so must other states forego their air forces.[120] Equality dropped away after Germany's withdrawal from the League of Nations, following which the threat of German rearmament began to loom larger in internationalist minds. 'Some maniac like Captain Göring', Noel Baker warned at a Royal Institute of International Affairs lecture in 1934, 'with an immense machine in his control – it was said that twenty or thirty thousand aircraft was the force Göring intended to create – would wipe out all the chief centres of civilisation in Europe before people knew where they were'.[121] Yet many proponents still hoped for Germany's participation in any League international control scheme, arguing that she could be coaxed to join through the promise of all-round armaments reductions and economic prosperity. In support they pointed to various continuing arms limitation discussions, most importantly those for an 'Air Locarno' between Germany, France, and Britain in 1935. For some, such as Geoffrey Mander and David Davies, the putative Western Air Defence Pact could become a basis for an international air force. Griffin, on the other hand, criticized the discussions for not going far enough.[122] Attlee, in response to pacifist and leftist criticisms that proposals for the

---

[119] Gwyn Jenkins, 'Lord Davies, Howard Hughes and the "The Quinine Proposition": The Plan to Set up an International Air Force to Defend Chinese Cities from Japanese Air Raids, 1938', *The National Library of Wales Journal* 21, no. 4 (Winter 1980): 414–422; Bauernfeind, 'Lord Davies and the New Commonwealth Society 1932–1944'.

[120] For example: Viscount Cecil of Chelwood, 'Disarm or Rearm!', *New York Herald Tribune*, 16 August 1931.

[121] Noel Baker, 'Disarmament', *International Affairs*.

[122] Hansard, HC, vol. 307, col. 1856, 18 December 1935; Hansard, HC, vol. 95, col. 735, 1 May 1935; Griffin, *Britain's Air Policy*, 163–183. See also: Noel Baker, 'Disarmament', *International Affairs*; Brett Holman, 'The Air Panic of 1935: British Press Opinion between Disarmament and Rearmament', *Journal of Contemporary History* 46, no. 2 (April 2011): 288–307.

international control of aviation served only to cement Anglo-French power in Europe argued (in 1937) that even limited internationalization would be sufficient to allow the League's economic policies to function, bringing prosperity to its members and enticing the fascist powers to join.[123] Griffin, noting in 1935 that 'the moral authority of the League will be transfigured if the League is no longer patently the instrument of keeping Germany and the other defeated Powers down', argued that international control in fact offered the most practical way of combining 'equality for Germany with security for France and other countries'.[124]

Yet in the latter half of the 1930s internationalists also tackled growing fears of German rearmament by pointing out that internationalized aviation's security umbrella would protect the contracting parties not only from each other but also from those, such as Germany, who might choose to remain outside. Attlee's 1937 *Labour Party in Perspective* warned: 'If the Fascist States prefer to remain outside, they must of course do so, and the League must maintain forces strong enough to resist any attempt at aggression'. This argument was most easily made by the New Commonwealth as its proposals focused on an International Tribunal which could operate independently of the League. An attack by a 'non-member aggressor' on any one of the contracting nations was an attack on the whole agglomeration, and was to be met with force. The most 'drastic and severe' measures, such as the bombing of cities, would only be carried out against non-member aggressors. The Commonwealth plan consequently appealed to hawks such as Winston Churchill who saw it as an anti-fascist measure. Untainted by any association with the failing League, Davies was able to respond to international crises by forcefully making the case for an international air force. Whereas Atlee's response to Italian aggression in Abyssinia was to suggest a collective response through the League, Davies called for Britain and France to join together with other powers to form an investigative tribunal backed by military force. His response to Japanese aggression in China and the Soviet invasion of Finland, as we have seen, was also to attempt to organize volunteer international air forces.[125]

---

[123] Clement R. Attlee, *The Labour Party in Perspective* (London: Victor Gollancz, 1937), 224–225. For criticism see: E. A. Williams, letter to the editor, 'An International Police Force. Some of its Dangers', *Manchester Guardian*, 9 March 1934.

[124] Griffin, *Britain's Air Policy*, 75, 76, 131.

[125] Attlee, *The Labour Party in Perspective*, 224–225; The New Commonwealth Society, *The Functions of an International Air Police*, 7, 10, 11; Memorandum, 'Lord Davies' Interview with the Rt. Hon. C. H. Attlee', 30 January 1936, folder: 'LC, NC, re. Italo-Abyssinian Crisis', box 90, Davies Papers; David Davies, *Nearing the Abyss: The Lesson of Ethiopia* (London: Constable, 1936).

Although the politics of the League and collective security vacillated through the 1930s, the power and reach of the aeroplane continued to loom large in the way international relations was understood through to the end of the decade. It remained the ultimate machine of war, and yet also the bringer of international interconnectedness, peace, and prosperity. Liberal internationalists remained committed to using the bomber to ensure peace – in the beginning of the decade through League collective security, and by 1939 through selective military alliances. These commitments would come together again during the Second World War once greater opportunities arose for the reconstruction of international relations.

4 Air Power for a United Nations
The International Air Force during the Second World War

In August 1944 the long-standing director of the US League of Nations Association, Clark Eichelberger, published his vision of post-war international relations, *Time Has Come for Action*. Backed by the prominent internationalist think tank, the Commission to Study the Organization of Peace, the book was released just before the all-important Dumbarton Oaks Conference on the formation of the post-war United Nations Organization (UNO). It was a time of intense public debate on the organization's final shape and powers, and Eichelberger called for it to be armed with the emergent new long-range weapon: the bomber. 'When one thinks of the tremendous possibilities of the new B-29 super-fortress', he asked, 'when one thinks of such planes being able to fly thousands of miles in a few hours, who can escape the conclusion that an international air force composed of a few such planes would be the most powerful police weapon ever dreamed of?'[1] And indeed who could doubt the power of the bomber. Newspapers had for the past three months been brimming with news of America's new super weapon, the B-29 Superfortress. The bombardment of Japanese cities by these technological marvels was front-page news, the B-29s introduced as the first 'global' bombers, the ultimate fulfilment of air power's long-standing promise.[2]

This chapter focuses on the role of military aviation in internationalist post-war planning and argues that it was a significant aspect of this planning.[3] As support for a post-war United Nations organization

---

[1] Clark M. Eichelberger, *Time Has Come for Action* (New York: Commission to Study the Organization of Peace, 1944), 16.

[2] See, for example: '20th Air Force Superfortresses Open New Era of Aerial Warfare', *Air Force* 27, no. 7 (July 1944): 3; 'B-29', *Washington Post*, 16 June 1944; Frank Sturdy, 'Super Fortress First Warplane for Global Use: Has 10-Ton Bomb Load; Range is 5,400 Miles', *Chicago Tribune*, 16 June 1944.

[3] There is some discussion of proposals for international policing in the historical literature, though there is little exploration of the aerial nature of such proposals: Beaumont, *Right Backed by Might*, chapter 4; Patrick J. Hearden, *Architects of Globalism: Building a New World Order during World War II* (Fayetteville: University of Arkansas Press, 2002), 152–153; Elliott V. Converse III, *Circling the Earth: United States Plans for a Postwar Overseas Military Base System, 1942–1948* (Maxwell Air Force Base, AL: Air University

gathered pace in Britain and the United States during the Second World War, it assimilated proposals for an international air force. British and US industrial and aerial ascendancy allowed internationalists to imagine an organization that would be supported by and dependent on Allied fire-power. The newly rejuvenated and transformed globalist thinking of the wartime years, turbo-charged by society's admiration of modern science and aviation, incorporated the bomber as a global machine of law and order. Interwar assumptions about science and international relations, as well as new arguments about total war and Allied cooperation, fed into these new visions of collective security. The idea of a United Nations air force became widely accepted in Britain and the United States, and, although it was never formed, the United Nations Charter carried its imprint into the post-war years.[4]

The United States and the International Air Force

Although the international police force had some support amongst US internationalists before the First World War, in the 1920s a broader turn away from foreign interventionism left it too radical a concept to be widely accepted in the United States. Proposals such as the 1932 Tardieu Plan were denigrated as an attempted 'superstate' by the Carnegie Endowment. Others thought the proposals simply ridiculous. 'Does any-one believe that Great Britain would agree to hand over to a group of men sitting in Geneva the right to send a fleet of aviators to drop bombs on London?' asked the American Peace Society's prominent journal *Advocate of Peace* in 1929.[5] That is not to say, however, that some did

Press, 2005), v, 1–49; Eric M. Kendall, 'Liberal Internationalism, the Peace Movement, and the Ambiguous Legacy of Woodrow Wilson' (PhD diss., Case Western Reserve University, 2012), 154–177. There is also a large literature on Roosevelt's notion of the four policemen; see for example: Warren F. Kimball, 'The Sheriffs: FDR's Postwar World', in *FDR's World: War, Peace, and Legacies*, eds. David B. Woolner, Warren F. Kimball, and David Reynolds (New York: Palgrave Macmillan, 2008), 91–122; Robert C. Hilderbrand, *Dumbarton Oaks: The Origins of the United Nations and the Search for Postwar Security* (Chapel Hill: University of North Carolina Press, 2001), 138–158.

[4] On wartime globalist thinking and culture see: Neil Smith, *American Empire: Roosevelt's Geographer and the Prelude to Globalization* (Berkeley: University of California Press, 2003), chapter 14; Van Vleck, *Empire of the Air*, chapter 3; Rosenboim, *The Emergence of Globalism*.

[5] Carnegie Endowment for International Peace, *The Challenge of Disarmament* 5 (10 June 1932). 'Editorials', *Advocate of Peace* 91, no. 2 (February 1929): 88. Also: 'Ten Objections to an International Police', *The Advocate of Peace* 77, no. 8 (August 1915): 188–189; 'Editorials: Coming Around', *Advocate of Peace* 89, no. 11 (November 1927): 587–588; Editorial, 'An International Force', *Advocate of Peace* 93, no. 4 (December 1931): 201–202.

not find the idea attractive. Pitman B. Potter's 1927 textbook *International Civics* introduced the international police force as the 'only entirely reliable solution' to the problem of peace but conceded that it was currently too radical and impractical an idea to be accepted. Social scientist Emily Greene Balch, a leader of the Geneva-based Women's International League for Peace and Freedom, travelled frequently to Europe and supported an international air force and the internationalization of aviation and waterways.[6] Engineer-businessman Oscar T. Crosby propagated David Davies's ideas in the United States and introduced the Welsh industrialist to a number of prominent businessmen and industrialists during tours to the country. They even reached into science fiction, with the short-lived speculative-fiction magazine *Air Wonder Stories* regaling its young readers in 1929 with a story about an 'air bandit' who robbed aeroplanes of the 'International Air Line' whilst evading the Anglo-American 'Atlantic Police Forces'.[7]

The international force re-emerged as part of a wider growing internationalism in the early years of the Second World War. International intervention and cooperation, it seemed, were required to defend democracy from fascist aggression. One manifestation of this 'second chance' for international organization was the spectacular success of Clarence Streit's federalist tract *Union Now!*, first published in 1938. Proposals such as Streit's, which would have sunk without trace only a few years before, struck a chord with public and intellectual opinion between 1939 and 1941 as anti-interventionist sentiment was eclipsed by a growing willingness to intervene in European affairs and act on a global scale. As a consequence internationalists began to receive a wider and deeper reception in government, and their proposals became bolder. In 1939 the Carnegie Endowment formed the Commission to Study the Organization of Peace to create and publicize proposals for post-war international organization, and the anti-neutrality Committee to Defend America by Aiding the Allies was formed in May 1940.[8]

---

[6] Potter and West, *International Civics*, 177–178; Mercedes M. Randall, *Improper Bostonian: Emily Greene Balch* (New York: Twayne, 1964), 286, 323, 330, 373–374; Kristen E. Gwinn, *Emily Greene Balch: The Long Road to Internationalism* (Urbana: University of Illinois Press, 2010), 84–85.

[7] Kuehl and Dunn, *Keeping the Covenant*, 105; Oscar T. Crosby, *International War: Its Causes and its Cure* (London: Macmillan, 1919), 149–161; Oscar T. Crosby, 'An International Justice of the Peace and His Constable', *Advocate of Peace through Justice* 93, no. 2 (May 1931): 110–119; Oscar T. Crosby, 'International Force', *Advocate of Peace through Justice* 94, no. 1 (March 1932): 44–48; Edward E. Chappelow, 'The Planet's Air Master', *Air Wonder Stories* 1, no. 2 (August 1929): 160–182.

[8] Divine, *Second Chance*, 29–46. Clarence K. Streit, *Union Now! A Proposal for a Federal Union of the Democracies of the North Atlantic* (New York: Harper & Brothers, 1939). On growing internationalism and internationalist activity 1939–1941, see: Andrew Johnstone,

Official government policy turned also in a markedly internationalist direction. The most visible manifestation of this was the Atlantic Charter, agreed in August 1941 by Franklin D. Roosevelt and Churchill. The Charter reflected many long-standing liberal internationalist desires and concerns, including a call for the 'abandonment of the use of force', international 'disarmament', and a 'wider and permanent system of general security'.[9] Although it did not commit to the formation of an overarching international political organization, the Charter nevertheless raised internationalist hopes that such an organization could be the linchpin for a new post-war world order. Just as importantly, it helped frame the war as one of internationalism, represented by the values and cooperative methods of the Allies, versus nationalism.[10]

Once the United States entered the war in December 1941 collective security quickly became a central feature of debates on post-war international relations. It was now easier for internationalists, policymakers, and the public at large to imagine that post-war international organization would need to defend itself and its members from the forces of nationalism or fascism. The Allies (designated the 'United Nations' in state and internationalist publicity) were already cooperating militarily and logistically, and it was a small step to visualize this continuing into the post-war years. International police forces emerged as the central aspect of this collective security. The widely publicized B2H2 Resolution presented to Congress in late 1943 calling for the establishment of a United Nations organization armed with a 'United Nations military force' helped popularize the concept.[11] By 1944 an international force was a standard component of most internationalist suggestions for post-war international relations.

These proposals were fuelled by state post-war planning and internationalist organizations committed to bringing post-war collective security and a liberal internationalist world order. Proposals by Europeans, including many dating back to the 1920s, circulated in elite circles and even amongst the public. Widely circulated magazines such as *Collier's*, *Atlantic Monthly*, *The Rotarian*, and many others published internationalist calls for an international air force, and the quality press

---

Against Immediate Evil: American Internationalists and the Four Freedoms on the Eve of World War II (Ithaca, NY: Cornell University Press, 2014).

[9] 'Atlantic Charter', The Avalon Project, accessed 2 August 2020, avalon.law.yale.edu/wwii/atlantic.asp.

[10] Borgwardt, *A New Deal for the World*, chapters 1, 2.

[11] Wilson D. Miscamble, *From Roosevelt to Truman: Potsdam, Hiroshima, and the Cold War* (Cambridge: Cambridge University Press, 2007), 21–22; Philip J. Briggs, 'Congress and Collective Security: The Resolutions of 1943', *World Affairs* 132, no. 4 (March 1970): 332–344; Hilderbrand, *Dumbarton Oaks*, 27–28.

gave coverage to proposals around the time of the Dumbarton Oaks Conference, when they were discussed in a diplomatic setting.[12] Proposals were not monolithic, and support was far from unanimous: this led, as this chapter shows, to a lively discourse which pervaded policymaking and internationalist settings, as well as the press and media.

A range of proposed arrangements abounded: some imagined the Great Powers taking separate responsibility for international security in their regions of influence, others anticipated national military forces combined together under an international organization for more centrally directed military action. Although separate Great Power policing was initially popular (Roosevelt was a proponent of the Four Policemen model for most of 1941 and 1942), by 1943 internationalist opinion had swung firmly behind police forces under the umbrella of an international political organization. Great Power policing independent of such an organization would leave it weakened and without teeth – a discredited League of Nations under a new name.[13] Although not always explicitly articulated, it was widely thought that the force should be largely or solely aerial in composition – an international air force with bombers for offensive capability and fighters for defence.

Think Tanks and the Organization of Peace

The New York-based liberal internationalist think tanks the Commission to Study the Organization of Peace (CSOP) and the Council on Foreign Relations (CFR) emerged as the most significant loci of articulated thinking about aviation and collective security in the United States. Alongside the American Association for the United Nations these organizations were by 1944 the most powerful sources of institutionally organized support for a post-war international organization.[14] Funded by the Carnegie Endowment, and headed by the prominent League of Nations Association internationalists James T. Shotwell and Clark Eichelberger, the CSOP was formed in 1939 to draw up and disseminate proposals for

---

[12] See publications listed in Julia E. Johnsen, ed., *International Police Force* (New York: H.W. Wilson, 1944).

[13] On Roosevelt's support for the Four Policemen idea see: Robert Dallek, *Franklin D. Roosevelt and American Foreign Policy 1932–1945*, 2nd ed. (New York: Oxford University Press, 1995), 342; Kimball, 'The Sheriffs: FDR's Postwar World'.

[14] See, for example: Glenn Tatsuya Mitoma, 'Civil Society and International Human Rights: The Commission to Study the Organization of Peace and the Origins of the UN Human Rights Regime', *Human Rights Quarterly* 30, no. 3 (August 2008): 607–630; Cornelia Navari, 'James T. Shotwell and the Organization of Peace', in *Progressivism and US Foreign Policy between the World Wars*, eds. Molly Cochran and Cornelia Navari (London: Palgrave Macmillan, 2017), 167–192.

international organization. Over the course of the war the Commission brought together leading internationalists and State Department officials, and produced or funded a barrage of internationalist publications, lectures, and radio programmes.[15] The intellectual weight of its thinking rested on a series of agenda-setting reports released at six- to twelve-month intervals, drafted in sections by committees of experts. The reports outlined the problems of post-war international relations and offered solutions, the most prominent of which was the call for the formation of a powerful post-war international organization.

The fifth and final wartime report, released in November 1943, reflected the wider internationalist turn to collective security, and made its case for an armed United Nations organization by appealing to the novel power of the aeroplane. In support it presented a cyclical history of armaments in which one newly invented offensive weapon dominated, followed by a defensive weapon, and so on. The current supreme offensive weapon was aviation, which because of its now global reach and destructive potential made possible powerful collective security through the formation of a United Nations air force. The force was to be created gradually: initially consisting of national contingents formed through voluntary enlistment, and finally becoming a fully internationalized force distinct from national air forces. Bombers were to be the primary instrument of policing; fighters were to be used to protect bombers and bases. Manufacturing was to be distributed amongst many nations, and research and development was to continue for the benefit of this force. National air forces were to be whittled down to a bare minimum over time. The report emphasized the need for a network of international air bases and the standardization of equipment and aeroplanes. It also stressed that the force was a police rather than a military force: 'a symbol and not the embodiment' of power, a 'reminder of law and order'.[16]

This commitment to aviation was reiterated later in more detailed reports and announcements, many prepared in anticipation of the Dumbarton Oaks Conference on the UNO due to begin in August 1944. The later reports did, however, pull back from radical proposals for an international force. The 'Framework' for a 'General

---

[15] Smith Simpson, 'The Commission to Study the Organization of Peace', *The American Political Science Review* 35, no. 2 (April 1941): 317–324; Andrew Johnstone, 'Shaping our Post-war Foreign Policy: The Carnegie Endowment for International Peace and the Promotion of the United Nations Organisation during World War II', *Global Society* 28, no. 1 (2014): 24–39.

[16] Commission to Study the Organization of Peace, 'Fourth Report of the Commission to Study the Organization of Peace', *International Conciliation* 22 (January 1944): 3–110, particularly 14, 51–67. The first report of the Commission was titled the 'Preliminary Report', the second report the 'First Report', and so on.

International Organization' published in September 1944 suggested a small fast air force instead of a large menacing one, a 'permanent international air patrol recruited by voluntary enlistment, ready to fly at once to a scene of trouble'.[17] The Commission's reports and publications were widely disseminated within internationalist circles and the State Department, and amongst the public – by one count 100,000 copies of the first report were produced and distributed.[18]

CSOP internationalists promoted their views on international relations in public whilst participating in the think tank discussions, with several placing emphasis on the need for an aerial international force. Prominent amongst them were Eichelberger and Shotwell, whose promotion of an international force was part of their attempts to build support for a successor to the League that would have teeth but still be acceptable to US public and political opinion.[19] In a June 1943 NBC network radio talk sponsored by the CSOP, Eichelberger explained that the UNO's collective security system needed to be built around an international air force. This system had five key requirements. The UNO needed to take over 'certain strategic spots' such as the Japanese mandate islands in the Pacific as a 'permanent symbol of the collective power of the world community to enforce peace', and also militarily police strategic places such as the Panama Canal and Gibraltar which would remain under national control. Third, commercial aviation would require regulation to prevent wasteful competition, and fourth 'large-scale disarmament' arrived at progressively. Lastly, the UNO would be equipped with inspectors who would have the right to inspect all national industries for secret weapons or national armies.[20]

Eichelberger used a range of arguments to assuage fears about the extent and impact of US support for this international air force and its apparatus. Aeroplanes were the basis of this system, he explained to the April 1943 meeting of the American Academy of Political and Social Science, because they 'make possible the policing of the world with the minimum number of men'. He reassured his audience that this would not mean European soldiers in the Americas: policing was to be regional, with US forces manning United Nations bases and strategic water passages in Latin America. This, he claimed, was not an imperialist position: rather

---

[17] Commission to Study the Organization of Peace, 'The General International Organization: Its Framework and Function', *International Conciliation* 22 (September 1944): 547–551.

[18] Divine, *Second Chance*, 36.   [19] Johnstone, *Dilemmas of Internationalism*, 63–85.

[20] Commission to Study the Organization of Peace, *For This We Fight: Program No. 5 Making the World Secure* (New York: Commission to Study the Organization of Peace, 1943), 5–6.

the actual imperialists were 'isolationists' who insisted on the United States policing the world by itself through occupied foreign bases.[21] In his 1944 *Time Has Come for Action* Eichelberger explained that his proposals were not utopian but a finely balanced compromise 'between what is most ideal and what seems most practicable'. He again reassured readers that an international police force was popular with the public, and that it was a US (and so not a foreign European) idea dating back to 1910. This force needed to be aerial as it could be quickly constituted, 'manned by volunteers wearing the international uniform', and 'arrive quickly at the scene of trouble'. Just like the English in the nineteenth century, who 'were able to develop their liberties with a minimum amount of militarism because they were protected by sea power', so too would the liberal values of the United States be safeguarded through aviation. The latest most destructive bombers made such a force a realistic possibility. He characterized the force's international air bases as 'police stations' and pointed out that these were already being used by US and British forces overseas. Their value was not just military: they were to represent the power and authority of the UNO, 'symbols of the international police'. Eichelberger also reassured his readers that the United States would not immediately give up its air force; instead the Great Powers would eventually negotiate a 'General Disarmament'.[22]

Shotwell, meanwhile, developed his CSOP memoranda into a detailed book on international organization, also published in 1944. *The Great Decision* maintained that the proposed post-war organization would need an international air force in order to fulfil its aims of international peace and justice. Shotwell, like Eichelberger, emphasized the differences between this new organization and the League, particularly its pragmatic nature. The new organization was emerging through the 'functional method of international cooperation', rather than being built up idealistically as a 'Temple of Concord set alongside the capitals of sovereign States'. As well as the swiftness, surprise, and reach of the aerial force, Shotwell emphasized that it would be economical and so protect the 'liberties of its users' as it would not require the diversion of national industry for the creation of a large war economy.[23]

For other CSOP participants the international air force was as much about building international law as it was about creating a new international organization or bolstering collective security. Legal scholars Quincy Wright and Clyde Eagleton had been at the forefront of US

[21] Eichelberger, 'Next Steps in the Organization of the United Nations'.
[22] Eichelberger, *Time Has Come for Action*, 3, 14–16.
[23] Shotwell, *The Great Decision*, 128–132, 222–223.

internationalist attempts to strengthen the emergent League-orientated international legal regime in the 1920s and 1930s. Their hopes for a reconstructed international legal order, incorporating new legal regimes such as individual 'human rights', now rested on the proposed United Nations organization, which could enforce both international peace and international law. Those breaking international law would need to be identified by the 'community of nations', which would also sanction the use of force. Great Power sanctions or military force could not represent this community, they believed. Only a broader coalition could, institutionalized in the shape of a general political organization.[24]

Wright published and lectured extensively on the need for an international force and why it had to be aerial in nature. At the 1944 meeting of the American Society for International Law he argued that the latest developments in military aviation made possible 'international policing action' which would be 'certain, rapid, and powerful'. A 'relatively small volunteer international air force' could 'provide an effective spearpoint for police action capable of immediate use'. Although this force may not be able to deal with Great Power aggression, it could, as demonstrated by imperial policing, deal with 'small disturbances'. Such an air force may have stopped the Japanese invasion of Manchuria in 1931, Hitler's rearmament in 1935, and his Rhineland invasion in 1936.[25] In other writings Wright highlighted additional reasons for aerial action, including cost-effectiveness and minimal reduction impact on national sovereignty (as it was a small force), because it was less likely to 'endanger national liberties' or be used to establish despotism. Once international security was established disarmament would naturally occur 'under taxpayer's pressure' as 'people do not like to spend money for armaments unless they think they have to'. In an article in the St Louis Star Times he added that it could be 'situated in relatively small contingents on islands placed under the jurisdiction of the international authority and strategically located in various parts of the world'.[26]

---

[24] Quincy Wright and Edward Warner, 'Enforcement of International Law', *Proceedings of the American Society of International Law at Its Annual Meeting* 38 (April 1944): 77–97. Also: Clyde Eagleton, 'Peace Enforcement', *International Conciliation* 20 (April 1941): 499–504; Quincy Wright, *Human Rights and the World Order* (New York: Commission to Study the Organization of Peace, 1943).

[25] Wright and Warner, 'Enforcement of International Law'.

[26] Quincy Wright, 'Peace Problems of Today and Yesterday', *American Political Science Review* 38, no. 3 (June 1944): 512–521; Quincy Wright, 'National Security and International Police', *The American Journal of International Law* 37, no. 3 (July 1943): 499–505; Wright and Warner, 'Enforcement of International Law'. See also: Quincy Wright, 'Fundamental Problems of International Organization', *International Conciliation* 20 (April 1941): 468–492; Quincy Wright, 'How an International Police

New York University-based Clyde Eagleton, a supporter of collective security since the 1930s, also called for an international air force outside of CSOP meetings, including in a series of lectures given at New York University (and subsequently published as *The Forces that Shape Our Future* in 1945) and in the 1948 revised edition of his long-standing textbook *International Government*. Eagleton took special care to warn against an international police overly reliant on the United States or Britain, which could be construed as 'imperialistic domination'. Instead his force was to be more broadly constituted, a 'universal system' based on the 'United Nations' which reflected his vision of an international organization in which smaller countries would have proportionately greater power.[27]

Notwithstanding the prominence of the CSOP, its influence on the State Department was ultimately less than that of a rival liberal internationalist think tank which was more closely attuned to the needs of the State. In September 1939, *Foreign Affairs* Editor Hamilton Fish Armstrong and Council on Foreign Relations Executive Director Walter H. Mallory offered the Council's services to the State Department for wartime and post-war planning. They were well received, and the Council initiated its Rockefeller Foundation-funded War and Peace Studies project under the leadership of Armstrong and CFR President Norman H. Davis in December. The project was initially formed of four study groups (later five) which eventually met over 250 times, usually over dinner in New York City, over the next five years, and produced 682 memoranda for the circulation to the State Department.[28] Discussions within the study groups reflected more conservative views on international organization. There were fewer internationalist radicals, and many members were more closely associated with the state, including State Department officials and consultants, and military officers.

Force Would Work', in *Winning the Peace*, eds. Benjamin Akzin, et al., 2nd ed. (St Louis, MS: St Louis Star Times, 1944), 14–17.

[27] Clyde Eagleton, *Analysis of the Problem of War* (New York: Ronald Press, 1937); Eagleton, 'Peace Enforcement'; Clyde Eagleton, *The Forces that Shape Our Future* (New York: New York University Press, 1945), 108–117; Eagleton, *International Government*, 2nd ed., 446–454.

[28] See: Council on Foreign Relations, *The War and Peace Studies of the Council on Foreign Relations* (New York: Council on Foreign Relations, 1946). On the liberal internationalism of the CFR see: Smith, *American Empire*, chapter 7. Also: Laurence H. Shoup and William Minter, *Imperial Brain Trust: The Council on Foreign Relations and United States Foreign Policy* (New York: Monthly Review Press, 1977), 117–187; Robert D. Schulzinger, *The Wise Men of Foreign Affairs: The History of the Council on Foreign Relations* (New York: Columbia University Press, 1984), chapter 4; Wertheim, 'Tomorrow the World', chapters 2, 3, and 4.

There was substantial discussion on an international police force, leading to a number of reports circulated to the State Department.[29] Most participants agreed that some form of force would be needed after the war and that it could be built around current Allied military co-operation. The force was envisaged in global terms, and discussants were careful not to portray it as explicitly anti-German or Japanese. Its feasibility was supported by reference to the Boxer Rebellion and the Battle of the Java Sea (February 1942), portrayed as examples of international forces assembled from national contingents.[30] For political scientist Grayson Kirk, who wrote several memoranda on it, the force was needed to supervise disarmament, enforce international law and international courts, and satisfy 'the general anticipation that the war settlements must provide some approach to the security problems than a return to the pre-1914 or even the pre-1939 situation'.[31] Once the Atlantic Charter was formulated, he argued that a police force would satisfy its security obligations and could develop through the Great Power policing that would be required in the immediate post-war period.[32]

The most significant debate was over whether the force was to be a permanent force under the control of an international organization, or whether it was to consist of a coalition of nationally contributed units,

[29] As well as those noted in the footnotes below, there were: Grayson Kirk, 'Emergency Policing at the End of the War', 22 October 1941, W-85-A-B32, fiche 376, Records of the Council on Foreign Relations, 1921–1951, British Library of Political and Economic Science, London School of Economics and Political Science (hereafter cited as CFR Records); The Armaments Group, 'Problems of International Policing', 15 July 1942, W115-A-B62, fiche 381, CFR Records; Walter R. Sharp, *National Sovereignty and the International Tasks of the Postwar World* (New York: Council on Foreign Relations, August 1942); E. P. Warner, 'Various Types of Problems Arising in the Operation of International Policing Arrangements', 29 March 1943, W-136-A-B83, fiche 384, CFR Records; The Armaments Group, 'Political Problems Involved in the Selection of Bases for Use in International Policing', 3 May 1943, W-140-A-B87, fiche 385, CFR Records; Payson S. Wild, *The Distinction between War and International Policing* (New York: Council on Foreign Relations, June 1943); The Armaments Group, 'Relations between an International Political Organization and an International Police Force', 21 June 1943, W-145-A-B92, fiche 385, CFR Records; H. W. Baldwin, G. F. Eliot, E. P. Warner, and G. Kirk, 'The Respective Roles of Land, Sea, and Air Power in International Policing', 10 July 1944, W-163-A-B110, fiche 389 and 390, CFR Records; H. W. Baldwin, G. F. Eliot, and E. P. Warner, 'Technical Problems of International Policing', December 1944, W-687, fiche 507, CFR Records. Also: Grayson Kirk et al., *Some Problems of International Policing* (New York: Council on Foreign Relations, January 1944).
[30] Baldwin et al., 'The Respective Roles of Land, Sea, and Air Power in International Policing', 8. A longer history was also prepared: Payson S. Wild, *American Participation in International Policing* (New York: Council on Foreign Relations, July 1943).
[31] Grayson Kirk, *International Policing (A Survey of Recent Proposals)* (New York: Council on Foreign Relations, October 1941), 7.
[32] Grayson Kirk, *The Atlantic Charter and Postwar Security* (New York: Council on Foreign Relations, December 1941), 4.

assembled when required. The former option would greatly empower the international organization, whereas the latter allowed states to retain greater control over the use of force in international affairs. In contrast to the reports issued by the CSOP, the weight of opinion at the CFR was against a fully internationalized and integrated force. Opposition was spearheaded by military (and ex-military) men such as Major George Fielding Eliot, Admiral Julius W. Pratt, and Colonel Thomas J. Betts, as well as the Executive Director of the CFR Walter H. Mallory. They wheeled out a powerful range of arguments against its practicality: the weakness of European states after the war (and thus the need to rely on Great Power policing); the current lack of integration between British and US forces; organizational difficulties, and domestic opposition to the placement of US forces under foreign command.[33] Eliot, a prominent writer on military affairs and long-standing critic of international governance, reiterated some of these arguments in reports produced for the CFR and in popular magazines. For him an integrated force could only function as a part of a powerful 'world government', which was unlikely to be formed at the end of the war. He called instead for the United States and the British Commonwealth to police the air and sea 'highways' and the USSR and China to police on land.[34] This suspicion of powerful internationally controlled forces reflected a broader opinion on collective security. A survey conducted by CFR regional committees in March 1944 showed support for a multi-service force consisting of separate national contingents, though perhaps under overall joint Allied command. The prevalent view was that a 'permanent international force' would 'not be practicable or advisable in the immediate future'.[35]

---

[33] See in particular: 'Armaments Series: Memorandum of Discussions. Thirtieth Meeting', Studies of American Interests in the War and the Peace: Council on Foreign Relations, 4 January 1943, W-30-A-A30, fiche 366, CFR Records; 'Armaments Series: Memorandum of Discussions. Thirty-Third Meeting', Studies of American Interests in the War and the Peace: Council on Foreign Relations, 29 March 1943, W-33-A-A33, fiche 367, CFR Records.

[34] Major G. F. Eliot, 'Some Thoughts on International Police Forces', 29 March 1943, W-138-A-B85, fiche 384, CFR Records; Major G. F. Eliot, 'The Security Functions of a United Nations Political Organization', 22 May 1944, W-474-P-B82, fiche 459, CFR Records. His views chimed with his earlier calls for the United States to secure itself by extending the Monroe Doctrine to cover larger parts of the world: George Fielding Eliot, *Bombs Bursting in Air: The Influence of Air Power on International Relations* (New York: Reynal & Hitchcock, 1939).

[35] W. P. Sharp and P. W. Bidwell, 'American Public Opinion and Postwar Security Commitments: Results of the Inquiry Addressed to Twenty Committees on Foreign Relations, Spring, 1944', October 1944, W-699, fiche 511, CFR Records; Council on Foreign Relations, *American Public Opinion and Postwar Security Commitments* (New York: Council on Foreign Relations, 1944).

The leading proponents of an integrated force made the case for a specifically aerial force. The aeronautical engineer-administrators T.P. Wright and E.P. Warner (Vice-Chairman of the Civil Aeronautics Board) saw in an international air force both a new role for US military aviation and a safeguarding of the nation's engorged aviation industry into the post-war years. For them any future UN international force would be largely composed of aeroplanes manufactured by US corporations and flown by US pilots. Wright (then in charge of planning US aircraft production) had already publicly called for aerial collective security in a lengthy 1941 article in *Aviation* magazine (subsequently released as a pamphlet). The article justified and explained plans for US aerial military expansion and concluded with the suggestion that this huge new industrial capacity would not be useless after the war but would be needed for post-war defence through 'the closer union of nations . . . backed by preponderant Air Power'.[36]

Wright and Warner suggested huge forces. Wright's March 1943 memorandum for the study group, *The Organization of an Airforce for International Policing*, recommended the formation of a permanent 'internationalized air force' consisting of 54,000 fighters, bombers, trainers, and transports based at fifty to sixty bases around the world (US military aircraft production in 1942, by contrast, was around 47,000). The force was to be supplemented by national naval and ground units, and called up as required by the international 'General Staff'. The aeroplane was 'ideally suited as the policing power' because of the 'mobility of the airplane and the fact that it effectively reduces the size of the world and makes it "of a manageable size"'. National air forces, though initially retained, were to be gradually phased out.[37] Warner cautioned against modest proposals such as Quincy Wright's and emphasized the force's non-bombing uses: it could 'attain almost instantaneous investigation and report' and transport ground military units to hot spots.[38]

Warner and Wright's calls for an international air force were part of wider support for such a force from supporters of an enlarged US aviation industry. Facing a potential crisis of excess capacity after the war, the

---

[36] T. P. Wright, *Aircraft Production and National Defense* (New York: Farrar & Rinehart, 1941), 3, 4, 32.

[37] T. P. Wright, 'The Organization of an Airforce for International Policing', 29 March 1943, W-137-A-B84, fiche 384, CFR Records; 'United States Aircraft Production during World War II', Wikipedia, last modified 6 February 2020, https://en .wikipedia.org/wiki/United_States_aircraft_production_during_World_War_II.

[38] T. P. Wright and E. P. Warner, 'Enforcement of International Law' in 'Armaments Series: Memorandum of Discussions. Thirtieth Meeting', Studies of American Interests in the War and the Peace: Council on Foreign Relations, 4 January 1943, W-30-A-A30, fiche 366, CFR Records.

industry sought government assistance and promoted the slogan 'Air Power is Peace Power'. Support for an international force was part of this public relations strategy. So, for example, in an article in the popular magazine *Flying* the President of Consolidated Vultee Company called for government support to maintain a large aviation industry as it would be required to equip an international force after the war to 'prevent a new Hitler from hatching plans'. Even supporters of independence (from the army) for the US air force referred to an international air force in order to make their case: only independence, argued one aviation consultant, would allow the air force to mature enough to provide the international force required for 'world federation' after the war.[39]

Some CFR discussants attempted to find middle ground in order to create an acceptable and yet effective international force. The Pulitzer Prize-winning war reporter Hanson W. Baldwin (and one-time navy service-man) argued that it made sense to include naval forces because they had been tried and tested as international police forces in the past, and because they provided unparalleled 'strategic mobility' which would be required even for the functioning of international air forces. Noting Anglo-American preponderance in naval arms, he observed that Anglo-American influence in an international police force would be even greater if navies were included. He suggested a 'semi-internationalized' force which would be part way between a 'completely internationalized' force (which would require a 'super-state') and 'ad hoc' forces assembled as required. The former required a 'super-state' and the latter were historically shown to have been inefficient or sometimes even ineffective. He suggested that large states 'permanently assign' a small proportion of their forces to an international police force ('perhaps wearing their national uniforms but with international identifying insignia') controlled by an enlarged Combined Chiefs of Staff, but be ready to temporarily contribute national forces if required.[40]

Many of the supporters of a United Nations police force within the CFR and CSOP were intimately familiar with interwar European pro-posals. Shotwell had played an active role in European diplomacy and attempts at disarmament in the 1920s and early 1930s, and had been involved in intellectual discussions (including as an active member and later head of the American National Committee on Intellectual

---

[39] Karen S. Miller, *The Voice of Business: Hill & Knowlton and Postwar Public Relations* (Chapel Hill: The University of North Carolina Press, 1999), 33–34; Harry Woodhead, 'A Fighting Plan for Peacetime Production . . . .', *Flying*, November 1943, 24–25, 85–96; Harold Evans Hartney, 'The Case for a Separate Air Force', *Flying and Popular Aviation*, September 1942, 38–42.

[40] H. W. Baldwin, 'II. Sea Power and International Policing', in Baldwin et al. 'The Respective Roles of Land, Sea, and Air Power in International Policing', 3–14.

Cooperation) on international relations in Europe. In the 1930s he had penned an approving introduction to a rare call for League control of aviation published in the United States.[41] Quincy Wright had formed close connections with European internationalists in the 1930s and was instrumental in forming the Chicago branch of Madariaga's World Foundation. This would eventually become the Chicago-based World Citizens Association, which held its first large wartime meeting in April 1941. Wright participated, and an international force was discussed at some length.[42] In 1932 Edward P. Warner, whilst editor of *Aviation* magazine, had brought the Geneva discussions on international control to the attention of *Foreign Affairs*'s readers, though had argued that they were impractical.[43] Norman H. Davis and Hamilton Fish Armstrong had attended the 1932 Geneva disarmament conference, and Davis, who attended as chief delegate of the United States, had been regarded as a leading expert on disarmament in the 1930s. Armstrong also attended the 1935 International Studies Conference in London where the New Commonwealth Society organized a barrage of lectures on an international air force and the internationalization of aviation.[44] More broadly, interwar and wartime British proposals for an international police (especially Davis's) were well known in internationalist circles in wartime America – many were included in the publisher H.W. Wilson's 1944 summary of international police force proposals and were cited in American proposals.[45]

---

[41] On interwar activities: Pemberton, *The Story of International Relations*, part two, 39–42, 66–75, 166–170; Elvira K. Fradkin, *The Air Menace and the Answer* (New York: Macmillan, 1934), 118, 188–193. On Fradkin see: Elizabeth Shepard and Mike Farrelly, *Legendary Locals of Montclair* (Mount Pleasant, SC: Arcadia, 2013), 102. Fradkin would remain an ardent supporter into the fifties, see: Elvira K. Fradkin, *A World Airlift: The United Nations Air Police Patrol* (New York: Funk & Wagnalls, 1950).

[42] Memorandum, Quincy Wright, 'Meeting with Salvador Madariaga', 20 February 1936, folder A1, box 4, World Citizens Association Central Committee Records 1939–1953, Special Collections Research Center, University of Chicago Library (hereafter cited as WCA Central Committee Records); Edwin M. Clough to Edgar Ansel Mowrer, 5 February 1940, folder A1, box 4, WCA Central Committee Records; World Citizens Association, *The World's Destiny and the United States: A Conference of Experts in International Relations* (Chicago: World Citizens Association, 1941), 49–52, 178. For more on the Chicago branch see: Kuehl and Dunn, *Keeping the Covenant*, 95–96.

[43] Edward P. Warner, 'Can Aircraft be Limited?', *Foreign Affairs* 10, no. 3 (April 1932): 431–443.

[44] Hamilton Fish Armstrong, *Peace and Counterpeace: From Wilson to Hitler* (New York: Harper and Row, 1971), 487–495; Maurice Bourquin, ed., *Collective Security: A Record of the Seventh and Eighth International Studies Conferences, Paris 1934–London 1935* (Paris: International Institute of Intellectual Cooperation, 1936).

[45] Johnsen, *International Police Force*. Also: Wright, 'National Security and International Police'; Zechariah Chafee Jr., 'International Utopias', *Proceedings of the American Academy of Arts and Sciences* 75, no. 1 (October 1942): 39–53.

The Science and Globalism of Aerial Control

In one memorandum produced for the CFR, T.P. Wright noted the 'popular appeal' of 'policing by aircraft':

People are 'sold' on airpower ... The position of the airplane in the public mind, and the gradual building-up of a general belief that in airpower is to be found the primary solution of the policing difficultly of any world organization, will make the concept of an internationalized airforce one which the public should rapidly assimilate.[46]

Wright's observation was part of a wider internationalist realization that the American public had embraced military aviation during the war, and that this embrace could lay the foundation for an international air force. Military aviation's status as a powerful weapon of war, as an artefact of modern science and industry, and as a global and globalizing force had indeed increased as the war wore on. The aircraft industry itself saw explosive growth during the war, and quickly emerged as an important component of the nation's economy: employment grew from 63,000 in 1939 to a wartime peak of more than 2 million in 1944. Over 96,000 aircraft were produced in the peak year of 1944. The US Army Air Force grew to over 2.4 million aircraft by 1944. US military aircraft now operated across the world, helped by the emergence of new military transport arms: the Naval Air Transport Command was formed in December 1941, the Troop Carrier Command and Air Transport Command in June 1942. The commercial airline industry contributed significantly in terms of staff, facilities, and material to the success of the military's transport operations, and by the end of the war the Air Transport Command was staffed with 209,200 military and 104,600 military personnel operating more than 3,700 aircraft.[47] Aviation industrialist Alexander P. de Seversky's calls for the further development of military aviation captured this sense of growing aerial industrial and military power and reach. His *Victory Through Air Power* argued that air power had transformed warfare and called for the massive development of offensive and defensive military aviation and its associated industries and institutions. He envisaged US 'air control of the world' through bombers and fighters flying long distance from the continental United States to bomb her enemies.[48]

As early as 1941 Edward P. Warner wrote that Douhet's vision of airpower was finally materializing, based on the 'invincibility of the

---

[46] Wright, *The Organization of an Airforce for International Policing.*

[47] Charles J. Gross, *Military Aviation: The Indispensable Arm* (College Station: Texas A&M University Press, 2002), 96–99.

[48] Alexander P. de Seversky, *Victory Through Air Power* (New York: Simon and Schuster, 1942), 315–316.

airplane and its inherent superiority in power and economy to all other weapons'.[49] As the war wore on Warner's view appeared to be borne out: new marvels such as the Boeing B-29 Superfortress, the Consolidated B-32 Dominator (collectively termed 'superplanes' or 'superbombers' by the press), and the V1 and V2 weapons appeared to justify the rhetoric of the air power theorists, and spurred the expectation that even more powerful weapons were around the corner. 'Superduperfortress Next!' exclaimed the *New York Times* following the first B-29 raids in June 1944, and a few days later: 'Army developing mightier planes; Wright Field Head Promises Bombers Far Ahead of B-29 and Improved Fighters'. In a series of six articles in *Newsweek* in November and December 1944, J.F.C. Fuller explained that this world war, like the first, was one of invention, and highlighted the three most important ones: the tank, the bomber, and now the V-1 and V2. These developments reinforced the sense that these new aerial weapons would play a central role in future international relations: one scientist explained the flying bomb to the press as 'a development as revolutionary as the airplane, with a tremendous future and a definite political problem'.[50]

Support for an international air force was also fuelled by the new geographies of the air age that gained widespread acceptance during the war. The new aerial cartographies that spread during the war were part of an aerial globalism which, as Chapter 5 shows, underlay both support and opposition to internationalization of civil aviation. But they were also used to justify the creation of an international air force.[51] And so, for example, Professor of Geography George T. Renner's 1942 *Human Geography in the Air Age* (prepared for an 'Air-Age Education' book series sponsored by the Institute of the Aeronautical Sciences) explained to students that geography was man-made, the world had shrunk, and that aviation had created wholly new economic, social, and political

---

[49] Edward P. Warner, 'Douhet, Mitchell, Seversky: Theories of Air Warfare', in *Makers of Modern Strategy: Military Thought from Machiavelli to Hitler*, ed. Edward Mead Earle (Princeton, NJ: Princeton University Press, 1941), 485–503.

[50] 'US Cuts Air Output, Stresses Superplanes', *Washington Post*, 11 August 1944; 'Superduperfortress Next', *New York Times*, 21 June 1944; 'Army developing mightier planes; Wright Field Head Promises Bombers Far Ahead of B-29 and Improved Fighters', *New York Times*, 9 July 1944; J. F. C. Fuller, *The Future of War* (New York: Newsweek, 1944); 'Sees Robot as a Problem for Peacemakers: Asks World Body to Control Weapon', *Chicago Tribune*, 28 June 1944. Also: 'Builds Giant Bomber; Consolidated-Vultee Produces "Stablemate" to B-29', *New York Times*, 9 August 1944. On the culture of air power see: Steve Call, *Selling Air Power: Military Aviation and American Popular Culture* (College Station: Texas A&M University Press, 2009), 25–33.

[51] On this air globalism see: Jenifer Van Vleck, 'The "Logic of the Air": Aviation and the Globalism of the "American Century"', *New Global Studies* 1, no. 1 (2007): 1–37.

geographies. The military implications were thought to be dramatic. Political scientist Malbone W. Graham explained that the aeroplane had created a new class of 'security states' which, although middling powers, now had special military-strategic significance in international affairs because of their aero-geographical location. As the international air force gained popularity it was easily connected to these ideas. In magazines such as *American Magazine* and *Collier's*, Renner explained that the 'new Air Age map of the world' dictated that the United States could not police the world without aviation, or without Britain. The four great powers needed to come together to police the world and 'patrol the earth's skyways' through aviation.[52]

Calls for a United Nations air force were also predicated on widespread praise for Allied air force cooperation and a sense that it could easily and fruitfully be continued after the war. One argument used by Warner for an aerial force was that the current 'intermingling' of air forces in the RAF could be maintained and copied by other air forces in preparation for the eventual formation of an integrated international air force.[53] Polls carried out in March to June 1944 by the Rockefeller Foundation and the Universities Committee on Post-War International Problems showed almost unanimous support for an international air force. It was the 'least difficult form of international policing' because 'the RAF, the USAAF, the RCAF, and other allied air forces are now pretty well scrambled together' and 'we now have in the cooperative action of the various United Nations air forces the nucleus of a world air force'.[54] This notion was reinforced by British propaganda in the United States. One senior RAF officer's article in *Flying* depicted the RAF as an 'international air force' in all but name, with 'a common ídeal and fighting spirit, a common tactical doctrine, standard equipment, a mutual sympathy and comradeship in arms which will never be forgotten, and a language common to the English speaking world'. The 'organized combination of the air strength of 40 to 50 of the great nations of the world' was a 'revolution' that would eventually allow for international

---

[52] George T. Renner, *Human Geography in the Air Age* (New York: Macmillan, 1942); Malbone W. Graham, 'Great Powers and Small States', in *Peace, Security and the United Nations*, ed. Hans J. Morgenthau (Chicago: University of Chicago Press, 1946), 57–82; George T. Renner, 'Peace by the Map', *Collier's*, 3 June 1944, 44, 47; George T. Renner, 'How the United Nations Must Police the World; Patrol of the Earth's Skyways', *American Magazine*, September 1943, 30–31. On Renner and his air age geography see: Susan Schulten, *The Geographical Imagination in America, 1880–1950* (Chicago: University of Chicago Press, 2001), 138–139; Smith, *American Empire*, 285–287.

[53] Wright and Warner, 'Enforcement of International Law'.

[54] Universities Committee on Post-War International Problems, 'Summaries of Reports of Cooperating Groups: Problem XIII-International Air Traffic After the War', *International Conciliation* 23 (April 1945): 249–260.

organization.[55] Allied air forces were sometimes portrayed as being part of a wider machinery of Allied cooperation. The January 1944 Foreign Policy Association Headline Series booklet *On the Threshold of World Order* included Joint Allied military operations as part of 'war machinery', alongside other cooperative efforts such as the Combined Raw Materials Board and the Combined Food Board, which had to 'fit the peace' in the post-war years. This fitting was to occur through the formation of an international force with 'relatively small highly mobile . . . air armadas' at its core.[56]

The dual nature and integrative effects of modern scientific machines played an important role in rhetoric on international organization and its air force by providing a context and rationale for these internationalist solutions. The CSOP's first two reports, published in 1940, stressed the role of science in creating the current crisis of international relations. They reiterated internationalist arguments dating back to the 1920s about the misappropriation of civilian science for warfare, and reasoned that the only solution to the 'ever-advancing machinery of modern scientific warfare' was the development of international organization.[57] These themes echoed through the early papers written for the Commission in 1940 and early 1941, particularly those by James T. Shotwell (who was a leading author of much of the first reports) and economist Albert Lauterbach on the nature of modern warfare. They were also disseminated amongst the public through, for example, one of the CSOP's NBC talks in which science journalist Waldemar Kaempffert suggested the formation of a World Scientific Commission to direct science towards more discoveries and useful inventions.[58]

The power of science was particularly prominent in Shotwell's rhetoric. One radio address, just a few days after the Pearl Harbor attack, framed the attack within the wider context of a shrinking world thanks to 'modern science' creating 'swift ships and swifter planes'.[59] His call for an armed

[55] A. S. W. Dore, 'Allied Air Forces', *Flying and Popular Aviation*, September 1942, 146–148, 273.

[56] Vera Micheles Dean, *On the Threshold of World Order* (New York: Foreign Policy Association, January 1944), 8, 55.

[57] Commission to Study the Organization of Peace, *Preliminary Report* (New York: Commission to Study the Organization of Peace, November 1940); Commission to Study the Organization of Peace, *Study Outline on the Organization of Peace* (New York: Commission to Study the Organization of Peace, January 1940).

[58] James T. Shotwell, 'War as an Instrument of Politics', *International Conciliation* 20 (April 1941): 205–213; Albert Lauterbach, 'The Changing Nature of War', *International Conciliation* 20 (April 1941): 214–221; Commission to Study the Organization of Peace, *For This We Fight: Program No. 2 Science and our Future* (New York: Commission to Study the Organization of Peace, 1943), 3–4.

[59] James T. Shotwell, 'After the War', *International Conciliation* 21 (January 1942): 31–35.

United Nations Organization in *The Great Decision* was supported with the argument that science, now a great danger in war and peace ('scientific militarism'), itself provided the means to prevent war in its manifestation as aviation:

... modern science, by the invention of the airplane, has at last provided the nations of the civilized world with the means for having just such a limited but efficient police force. Before its advent as a military weapon, it was impossible for a policing expedition to reach many a scene of possible disorders without providing for the transit of armies across the territories of unwilling states ... The airplane has changed all this, for there is now no spot anywhere on earth that cannot be reached within sixty hours from some central airport ... Here then is the instrument ready at hand for the enforcement of peace in true police technique.[60]

But these ideas could also be found more widely in internationalist rhetoric. As early as November 1941 historian Denna Fleming began his Presidential Address to the Southern Political Science Association with the sentence 'The controlling factor in world politics is the steady advance in the technology of destruction', and then went on to build his case for intervention in the war on the contention that scientific war and aviation made international unification inevitable. The only question was: would unification be carried out by the liberal democratic United States or the totalitarian powers?[61] By 1945 the integrative effect of modern war technologies was widely accepted. In one February 1945 speech Eichelberger pointed to the 'rocket plane', then prominent in the press, as 'the most revolutionary development in warfare since the invention of gun powder' and evidence of 'forces which have made our world so complex and destruction so easy'. Like the aeroplane this invention could 'produce a security and a prosperity, and a civilization of which we can scarcely dream'.[62] When ex-Governor Harold E. Stassen outlined his seven-point programme for international organization in a March 1945 address (including, of course, an international police force), it began with the statement that 'absolute national sovereignty' was dead – 'It died with the airplane, the radio, the rocket and the robomb'.[63] An American

---

[60] Shotwell, *The Great Decision*, 129.

[61] Denna Fleming, 'The Coming World Order, Closed or Free', *The Journal of Politics* 4, no. 2 (May 1942): 250–243.

[62] Press Release, American Association for the United Nations, 'Summary of Address by Clark M. Eichelberger, of New York', 24 February 1945, folder: 'January – February 1945', box 126, Clark M. Eichelberger Papers, Manuscripts and Archives Division, The New York Public Library (hereafter cited as Eichelberger Papers).

[63] 'Stassen Gives Plan for Unity in Peace', *The New York Times*, 9 March 1945. On Stassen and the United Nations see: Lawrence S. Kaplan, *Harold Stassen: Eisenhower, the Cold War, and the Pursuit of Nuclear Disarmament* (Lexington: University Press of Kentucky, 2017), 24–40.

Association for the United Nations leaflet, released to coincide with the April 1945 San Francisco Conference on the United Nations, included a photograph of a V-1 flying bomb in flight, with the warning: 'no hiding place … If there are wars in the future they will be total wars – and death and destruction will rain from the skies on all of us'.[64]

Capturing the Public Mind

Proposals for an international air force carried far and wide in the United States – much further than discussions amongst a narrow elite. By the end of 1944 they were so widespread that the publisher H.W. Wilson was able to release a short summary of contemporary debates and publications on the international police force as part of its Reference Shelf series on major contemporary issues.[65] In September 1943 *Time Magazine* conducted a stock-take and noted that 74 per cent in a recent Gallup poll had supported the idea;[66] one surveyor of wartime public opinion noted that the concept had 'captured the public mind'.[67]

Magazines and best-selling books played an important role in mixing and spreading these ideas in a melange of technological internationalism. The *Rotarian*'s Symposium of the Month for October 1942 presented three different arguments on post-war order for its readers. Internationalist political scientist Nicholas Doman, author of *The Coming Age of World Control*, argued that a future United Nations organization would require an international police force to function. With technology featuring heavily, and in language reminiscent not only of interwar British proposals but also of the imperial aerial control schemes that inspired them, he claimed that 'One of the most amazing results of our technological age is the ability of well-organized, well-equipped small force to control and police large areas. A rebellious population today cannot arm itself with weapons to match those of a mechanized army'.[68] The second article, by Eichelberger, presented a political lag argument in support of world organization (recent 'rapid spiritual and scientific progress' promised peace and prosperity but was unfulfilled because of the division of the world into sovereign nations) and called

---

[64] American Association for the United Nations, *That These Dead Shall Not Have Died in Vain* (New York: American Association for the United Nations, 1945).

[65] See Johnsen, *International Police Force*.

[66] 'International Police', *Time Magazine*, 13 September 1943.

[67] Jerome S. Bruner, 'Public Opinion and America's Foreign Policy', *American Sociological Review* 9, no. 1 (February 1944): 50–56.

[68] Nicholas Doman, 'Control of Force is Key to World Order', *Rotarian*, October 1942, 10–11; Nicholas Doman, *The Coming Age of World Control: The Transition to an Organized World Society* (New York: Harper & Bros., 1942).

for a new United Nations organization that would include 'technical bodies' to manage social, economic, and political activities, including disarmament and an international force.[69] The third opinion piece was by the Irish playwright and author George Bernard Shaw who did not refer to an international force, but instead called for disarmament and the formation of a new league composed of regional federations of countries.[70] Coincidently, these articles were followed by one on the new world of international air cargo transport and how it would engender international commerce and bring the world closer together.[71]

One-time Bridge player Ely Culbertson emerged as the single most widely read technological internationalist of the war. Culbertson, who placed the transformative powers of modern science and machines at the centre of his calls for world federation, reached large audiences through his best-selling books and *Reader's Digest* articles, and may also have had some influence on state post-war planners and atomic energy policymakers.[72] Science had already become central to Culbertson's thinking on international relations by the time he penned his autobiography in 1940, but it was in his 1943 *World Federation Plan* that he called for the creation of a 'Machine for Peace'. In the past disarmament was an 'idle dream', but now, 'as a result of the revolution in military weapons', most nations were 'virtually disarmed' because they did not have the capacity to 'build vast industrial plants for planes, ships and tanks'. A 'World Armament Trust', under the control of the World Federation, was to take possession of all modern scientific 'heavy' weapons and manufacture them for the international police force.[73]

[69] Clark M. Eichelberger, 'A Society of Nations as Wide as Possible', *Rotarian*, October 1942, 12, 57.

[70] George Bernard Shaw, 'Let Small Nations Join in Federations', *Rotarian*, October 1942, 11–12.

[71] John P. Knapp, 'Boxcars with Wings!', *Rotarian*, October 1942, 13–56.

[72] Prominently: Culbertson, *Total Peace*; Ely Culbertson, *Must We Fight Russia?* (Philadelphia: John C. Winston, 1946); Ely Culbertson, 'A System to Win This War – and the Peace to Come', *Reader's Digest*, February 1943, 135–142; Ely Culbertson, 'The ABC Plan for World Peace', *Reader's Digest*, July 1948, 82–83; James G. Hershberg, *James B. Conant: Harvard to Hiroshima and the Making of the Nuclear Age* (New York: Alfred A. Knopf, 1993), 246. On his possible influence: 'Donald C. Blaisdell Oral History Interview', Truman Library, accessed 2 August 2020, www.trumanlibrary.gov /library/oral-histories/blaisdel; Joseph Preston Baratta, *The Politics of World Federation: From World Federalism to Global Governance*, vol. 1 (Westport, CT: Praeger, 2004), 70–72.

[73] Ely Culbertson, *The Strange Lives of One Man: An Autobiography* (Chicago: John C. Winston, 1940), 692–693; Ely Culbertson, *Summary of the World Federation Plan: An Outline of a Practical and Detailed Plan for World Settlement* (New York: World Federation, 1943), 5–13, 28.

Culbertson's best-selling *Total Peace*, published later in 1943, developed these ideas along the lines of David Davies's 'principle of differentiation'. Culbertson allocated all 'world weapons' to a multiservice 'World Police'. These 'heavy, decisive weapons or fighting machines' included destroyers and heavier warships, bombers, heavy bombs and artillery, tanks, torpedoes, and chemical weapons. National forces were only to be armed with lesser 'state weapons'. Unlike Davies, however, he did not give aviation a privileged place in his scheme: Culbertson's police force was to consist of two million men, 50,000 aircraft, 100,000 tanks and artillery pieces, and 100 battleships or aircraft carriers.[74] This 'principle of the segregation of weapons' was central to Culbertson's proposals for world order – so much so in fact that his main criticism of the League of Nations and *Union Now!* was that they failed to take into full account the 'revolutionary changes' in armaments.[75] Culbertson would go on to endorse the growing World Federalist movement in the United States, many of whose members had by 1942 come to support an international police force.[76]

Comics helped the international air police to reach wider audiences: they proliferated as mass entertainment during the war, and were widely read by younger adults, especially servicemen.[77] Several comics featured teams of multinational airmen fighting international criminals; perhaps the most successful arrived in the August 1941 debut issue of *Military Comics* in which Polish airman Blackhawk fought injustice and criminality without care for national boundaries. This heroic aviator was popular enough to be given his own title, *Blackhawk*, in winter 1944 in which he led a group of multinational airmen, living together on Blackhawk island, to fight international criminals and Axis agents in their futuristic Grumman XF4 F Skyrockets and black leather uniforms.[78] Other international policemen were not aviators, but were nevertheless associated with high technology. 'Radar, the International Policeman' was created as a regular feature in *Master Comics* (first appearance May 1944) for the Office of War Information. Blessed with super vision powers, Radar joined a secretive international police formed by the wartime leaders of

---

[74] Culbertson, *Total Peace*, 274–276, 283.

[75] Culbertson, *Summary of the World Federation Plan*, 5–10.

[76] Tom Griessemer, 'An International Police Force', *World Affairs* 105, no. 4 (December 1942): 262–266.

[77] In 1942 it was estimated that roughly 15 million comic books were sold each month, with an average of five readers per comic. By early 1943 this was estimated to be 25 million copies. Bradford W. Wright, *Comic Book Nation: The Transformation of Youth Culture in America* (Baltimore, MD: Johns Hopkins Press, 2001), 31.

[78] Kurt Mitchell and Roy Thomas, *American Comic Book Chronicles: 1940–1944* (Raleigh, NC: TwoMorrows Publishing, 2019), 120, 272–273. Also: Mark Evanier, 'Point of View', Internet Archive Wayback Machine, accessed 2 August 2020, https://web .archive.org/web/20101217065650/http://povonline.com/cols/COL306.htm.

Britain, the United States, the Soviet Union, and China, and received its orders to fight fascists, international criminals, and industrial cartels from the police's mysterious headquarters at Mount Mysto.[79] Even the American Association for the United Nations paid for and distributed a comic book on how the UNO would ensure peace and security. *A Third World War Can Be Prevented Now!* (May 1944) used aviation to represent the scientific and destructive nature of modern warfare ('Bombers with a range of 6,000 miles! ... Automatic planes which need no pilots! ... Rocket bombs which travel faster than sounds!') as well as the swift arm of international justice (Figure 4.1). One panel depicted a swarm of United Nations bombers descending on a militant aggressor nation, and an organogram showed the location of the international police force within the UNO.[80]

State Department and the Military

Although an international force was suggested at the earliest State Department post-war planning meetings in 1940, interest only really emerged in 1942 as public support increased and internationalists became involved in the planning process.[81] There was widespread support for some type of international police force within the subcommittees of the Advisory Committee on Post-War Foreign Policy (formed February 1942). As well as State Department officials the Committee included several senior representatives of internationalist organizations, including Norman H. Davis and Hamilton Fish Armstrong of the CFR and Shotwell of the CSOP. The international force was thought important enough for Davis to request that the CFR prepare a specific study on it at the very first meeting of the Security

---

[79] Paul Hamerlinck, *Fawcett Companion: The Best of FCA* (Raleigh, NC: TwoMorrows Publishing, 2001), 94; Wright, *Comic Book Nation*, 65–68; Mitchell and Thomas, *American Comic Book Chronicles*, 282; Sean MacDonald, 'Radar the International Policeman', Writeups.org, accessed 2 August 2020, www.writeups.org/radar-the-international-policeman-fawcett-comics; Chris Sims, 'Bizarro Back Issues: Boxing Day With Captain Marvel! (1944)', Comics Alliance, accessed 2 August 2020, https://comicsalliance.com/bizarro-back-issues-captain-marvel-boxing-radar-the-international-policeman.

[80] *A Third World War Can be Prevented Now!* (New York: True Comics Magazine, 1945). Also: Press Release, American Association for the United Nations, 'Comics Enter International Field in Support of United Nations', 1945, folder: 'May–June 1945', box 126, Eichelberger Papers.

[81] 'Memorandum [by Hugh R. Wilson] Arising from Conversations in Mr. Welles' Office, April 19 and 26', in *Postwar Foreign Policy Preparation 1939–1945*, ed. Harley Notter (Washington, DC: Department of State, 1949), 458–460. On State Department post-war planning see: Hearden, *Architects of Globalism*, 147–228; Christopher D. O'Sullivan, *Sumner Welles, Postwar Planning, and the Quest for a New World Order, 1937–1943* (New York: Columbia University Press, 2008), chapters 3 and 4.

Subcommittee in April 1942.[82] Enthusiastic participants included the Under-Secretary of State Sumner Welles who went on to make a prominent call for an international force in his May 1942 Memorial Day service speech at Arlington Cemetery. For Welles, as indeed for other supporters, the international force and its attendant bases and aerial networks were to help extend US influence and power overseas.[83]

The topics covered in these discussions, and the arguments made, were strikingly similar to debates in the internationalist think tanks. Areas of deliberation included public reaction to US forces being under international command, the merits of a force composed of national contingents versus a permanent standing integrated force, the usefulness of commercial aviation for military purposes, and the extent of national disarmament that would be required for such a force to function.[84] But the most vigorous debates materialized over whether the force should be limited to air forces. Breckinridge Long (Assistant Secretary of State for Special Problems) argued that it was the 'global' nature of any future world organization which would necessitate an aerial force, which 'would provide an expeditious way to arrive at the site of the trouble'.[85] Against objections such as Kirk's that a multiservice force would be more effective, Shotwell retorted that an air force would maintain the 'element of surprise'. British air policing of its colonies had demonstrated not only the power of air police but also that it 'raised no such problems of international law' as compared to a force composed of 'all types of armaments'. As aviation was 'new', Shotwell reasoned, an international organization could enforce 'freedom of the air' making aerial movement easier than travel by land or sea. Moreover, 'in Germany a recollection of the devastation wrought from the air during the war would cause the Germans to fear air action more than any other type'. Shotwell's clinching

---

[82] 'Minutes of the First Meeting of the Working Subcommittee on Security Problems', 15 April 1942, folder 2.2, box 1, Disarmament Files 1942–1952, Lot 58D 133, RG 59, National Archives and Records Administration, National Archives at College Park, College Park, MD (hereafter cited as NARA).

[83] Sumner Welles, 'Under Secretary Of State Memorial Day Address at the Arlington National Amphitheater', Ibiblio, accessed 2 August 2020, www.ibiblio.org/pha/policy/1 942/420530a.html. For reaction: 'Memorial Day Brings Hope of a Better Post-War World', *Life*, 15 June 1942, 36–37; Johnstone, *Dilemmas of Internationalism*, 39. Wertheim, 'Tomorrow the World', 223.

[84] See meeting minutes in folder 2.2, box 1, Disarmament Files 1942–1952, Lot 58D 133, RG 59, NARA.

[85] 'Minutes of the 26th Meeting of the Working Subcommittee on Security Problems', 5 February 1943, folder 2.2, box 1, Disarmament Files 1942–1952, Lot 58 D133, RG 59, NARA.

Figure 4.1 Panels from *A Third World War Can Be Prevented Now!* (New York: True Comics Magazine, 1945).

argument was that, ultimately, the 'international air arm is the only one to which the nation are likely to give their continued support'.[86]

Representatives of the armed forces, and state department officials and consultants, presented the most formidable opposition to an international air force. The two leading military men on the Security Subcommittee, Chairman of the General Board of the Navy Admiral Arthur J. Hepburn and Army Intelligence Chief and War Plans Division head Major General George V. Strong, were both veterans of interwar arms limitation conferences. Strong emphasized the uncertainties surrounding the formation of

---

[86] Ibid.; 'Minutes of the 27th Meeting of the Working Subcommittee on Security Problems', 12 February 1943, folder 2.2, box 1, Disarmament Files 1942–1952, Lot 58D 133, RG 59, NARA.

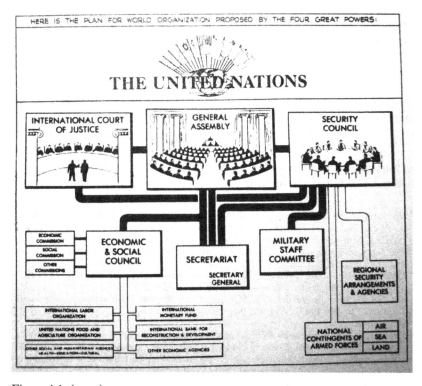

Figure 4.1 (cont.)

any future international organization and the possibility of disarmament, and so called for vague references to 'armed units' rather than an international air force in the subcommittee's reports. Instead of trying to second-guess the policing needs for an international organization which may take 'two or three or four generations' to form, he suggested that the subcommittee's proposals be restricted to 'principles' and should not be 'crystallized on any one conception or line of action'.[87]

State Department officials Grayson Kirk and Durward V. Sandifer argued against a specifically aerial force, as did Davis, who noted that, although the 'air force had caught the public imagination', it was 'wholly impracticable'. If there had to be an international police at all, it should be composed of national contingents which would include land and sea forces.[88] In a series of studies published under the aegis of the Yale Institute of International Studies, Kirk opposed a fully internationalized force, and instead suggested national contingents coming together as required.[89] Assistant Secretary of State Adolf A. Berle remained a powerful opponent of an international air force throughout this programme of post-war planning. He advised Vice-President Henry Wallace in March 1943 that, although the proposal was 'interesting . . . probably the nearest we can get to it is authorization to enter into agreements by which certain powers agree that they will use their air forces for international police work in certain specified contingencies'.[90] By July 1943 the Subcommittee on Security Problems had concluded that an international police force, whether aerial or not, was impractical.[91] Secretary of State Cordell Hull suspended the work of these subcommittees and

---

[87] 'Minutes of the 2nd Meeting of the Working Subcommittee on Security Problems', 29 April 1942, folder 2.2, box 1, Disarmament Files 1942–1952, Lot 58D 133, RG 59, NARA; 'Minutes of the 26th Meeting of the Working Subcommittee on Security Problems', 5 February 1943, folder 2.2, box 1, Disarmament Files 1942–1952, Lot 58D 133, RG 59, NARA.

[88] 'Minutes of the 29th Meeting of the Working Subcommittee on Security Problems', 26 February 1943, folder 2.2, box 1, Disarmament Files 1942–1952, Lot 58D 133, RG 59, NARA; 'Minutes of the 47th Meeting of the Working Subcommittee on Security Problems', 16 July 1943, folder 2.2, box 1, Disarmament Files 1942–1952, Lot 58D 133, RG 59, NARA.

[89] Grayson L. Kirk, *International Politics and International Policing* (New Haven, CT: Yale Institute of International Studies, 1944); Percy E. Corbett and Grayson L. Kirk, *The Outlook for a Security Organization* (New Haven, CT: Yale Institute of International Studies, 1944). On the Institute see: Paulo Ramos, 'The Role of the Yale Institute for International Studies in the Construction of United States National Security Ideology, 1935–1951' (PhD diss., University of Manchester, 2003).

[90] Memorandum, A. A. Berle Jr. to S. Welles, 2 March 1943, 800.796/258, Political Affairs – World aviation 1940–1944, RG 59, NARA.

[91] 'Minutes of the 47th Meeting of the Working Subcommittee on Security Problems', 16 July 1943, folder 2.2, box 1, Disarmament Files 1942–1952, Lot 58D 133, RG 59, NARA.

shifted responsibility for post-war planning to the newly formed Informal Agenda Group. This group also discussed the feasibility of an international police force, but eventually excluded such a force in the draft United Nations Charter it produced in August 1943.[92]

The military's initial planning for the acquisition of overseas military bases was also couched in terms of an international force. This framing was in response to a request by the President in December 1942 to the Joint Chiefs of Staff (JCS) for a wish list of foreign bases for US military use, which itself was framed in terms of requirements for a post-war international force.[93] Although Roosevelt had at times alluded to the possibility of separate Great Power policing (e.g. through an April 1943 article penned by Forrest Davis in the *Saturday Evening Post*), he was not generally in favour of a cohesive international force, aerial or otherwise. He brushed aside Eichelberger and the CSOP's suggestions for a force in late 1943 and early 1944; and a June 1944 statement assured sceptics of post-war international organization that the administration was opposed to 'a superstate with its own police forces and paraphernalia of coercive power'.[94] However, he had for many years been keen on securing foreign air bases for US military use. The fact that he chose to couch his request in terms of an international air force, as opposed to US air force deployment overseas, is probably as much a reflection of the popularity of such a force at that point as of a desire not to appear imperialistic.[95]

Roosevelt provided little guidance with his request, and the Joint Strategic Survey Committee (JSSC), to whom JCS assigned this project, was forced to make its own assumptions and rely on recently drafted proposals, such as that for a 'World Security Force' by a War Department General Staff Officer.[96] The various reports produced by the JSSC eventually formed the basis of JCS advice to the President, given in November 1943, on the location of such bases. This advice mapped Roosevelt's preference for post-war regional policing by the Great Powers onto an aerial police force. The military

---

[92] From Gary B. Ostrower, *The United Nations and the United States* (New York: Twayne, 1998), 8–10.

[93] Memorandum, Capt. J. L. McCrea to Admiral W. D. Leahy, 28 December 1942, JCS 183, CCS 360 (12-9-42) Sec.1, Central Decimal Files 1942–1945, RG 218, NARA.

[94] Forrest Davis, 'Roosevelt's World Blueprint', *Saturday Evening Post*, 10 April 1943, 20–21, 109–111; Clark M. Eichelberger, *Organizing for Peace: A Personal History of the Founding of the United Nations* (New York: Harper & Row, 1977), 242–244; Cordell Hull, *The Memoirs of Cordell Hull*, 2 vols. (London: Hodder & Stoughton, 1948), 1688–1689.

[95] The catalyst for Roosevelt's eventual request may have been an Australian memorandum, received just a few days before, suggesting the development of an alternative air route across the Pacific in case a Japanese advance cut off existing routes. This new route would have required the construction of airbases at numerous locations across the Pacific – construction which the Australians suggested could be shared between themselves and the United States. Converse III, *Circling the Earth*, 2–3.

[96] Ibid., 3–4.

planners visualized the eventual formation of a 'world-wide organization for collective security' through three distinct post-war periods: an initial period of 'enforced peace in Europe and war in the Pacific', a second period of 'World-wide peace enforced under the Four Power Agreement, pending establishment of a world-wide organization for collective security', and a third period when peace would be 'maintained by formally established world-wide machinery'. During the first two periods the globe was to be divided into three areas of responsibility: the Western Hemisphere for the United States; Europe, Africa, and the Middle East for Britain and Russia; and the Far East for all three as well as China. Seventy-two international airbases were proposed for US air force use in the first and second periods. In the third period all airbases were to be turned over to the United Nations and manned by an international air force, which, crucially, would be regional in character – so the United States would remain primarily responsible for the bases she already occupied.[97] In a letter to the State Department, JCS dismissed a 'denationalized' international police due to 'serious technical difficulties' with its implementation.[98] These JCS reports, and in particular their thinking on international air bases, were approved by Roosevelt through a series of meetings and memos in November 1943.[99]

The US Army Air Force produced its own reports on post-war bases for an international air force through the Air Transport Command (ATC)'s Plans Division and the Air Staff Plans Division. The ATC's views on post-war international aviation were largely in line with that of JCS – they argued against the internationalization of commercial aviation and against a unified transnational international air police. This was not surprising: many in the ATC Plans Division were aviation industry insiders, including at least two who had served with Pan American. They included the international law expert Oliver Lissitzyn, author of the well-regarded 1942 *International Air Transport and National Policy* which had argued against both an international

---

[97] Memo, W. D. Leahy on behalf of JCS, to the President, 'Memo for the President: U.S. Requirements for Post-War Air Bases', 15 November 1943, CCS 360 (12–9-42) Sec. 2, box 270, Central Decimal Files 1942–1945, RG 218, NARA; Joint Strategic Survey Committee, 'Air Routes Across the Pacific and Air Facilities for International Police Force', 25 March 1943, CCS 360 (3–27-43) Sec. 1, box 269, Central Decimal Files 1942–1945, RG 218, NARA.

[98] W. D. Leahy on behalf of JCS, to the Secretary of State, 'International Regulations of Armaments and Armed Forces; Supply and Use of International Armed Force and Facilities', 28 March 1944, CCS 388.3 (3–20-44), box 385, Central Decimal Files 1942–1945, RG 218, NARA.

[99] Memorandum, F. D. Roosevelt to JCS, 'U.S. Requirements for Post-War Air Bases', 23 November 1943, CCS 360 (12–9-42) Sec. 2, Central Decimal Files 1942–1945, RG 218, Central Decimal Files 1942–1945, NARA. See also: Mark A. Stoler, *Allies and Adversaries: The Joint Chiefs of Staff, the Grand Alliance, and U.S. Strategies in World War II* (Chapel Hill: University of North Carolina Press, 2000), 158–160.

air force and the internationalization of commercial aviation. The Air Staff Plans Division's reports, more widely circulated outside the Air Force than the ATC's, agreed with the latter over internationalization. The Plans Division warned that not only would a fully internationalized air force not be 'a complete substitute for a purely nationalistic plan of defense' but such a force and national self-defence might be 'mutually exclusive'.[100]

Britain and the International Air Force

Proposals for an international air force arose in Britain as well during the war and had as much popular support (if not more) as proposals in the United States. There was certainly more political support, with many in government and the Labour Party in particular calling for an international force. The more subdued economy, the greater (relative) demands of the war, and shortages of paper meant that proposals were generally less visible in popular printed culture than in the United States, though they were more visible in the quality press. Proposals in Britain also had a somewhat different initial impetus: they not only built on a rich heritage of support stretching back to the 1920s but were also rooted in hopes for growing Anglo-American cooperation and for European unification.

When the war began internationalist aspirations were boosted by hopes that growing Anglo-American cooperation would continue and broaden into a wider alliance later in the war and into the post-war years. Although the Atlantic Charter was the most overt official manifestation of this growing alliance, the popularity, just as in the United States, of Streit's *Union Now!* suggested a more widespread internationalism. Streit called for a political union of the 'North Atlantic democracies' (consisting of countries from Western Europe, and Canada, the United States, Australia, New Zealand, and South Africa) with a large military force for peace-keeping within the Union and external defence. International aviation within the Union would be merged into an 'Air Union', later to be expanded to include other forms of international transport and communications.[101] A stream of internationalist proposals soon flowed, many calling for an international force. Many took inspiration from Streit, such as Stephen King-Hall's 1941 call for the formation of a 'joint British-American Fleet and Air Force'. The force, required in order to ensure 'total victory', was to police the world for ten to fifteen years after the war. It was to be 80 per cent US–British, and its

---

[100] Converse III, *Circling the Earth*, 16–32. The Navy's General Board also produced reports on bases for an international force, see Ibid., 12–16.

[101] Streit, *Union Now!*.

naval component was to be three times bigger than the largest national fleet, and air force four times bigger than the largest national air force.[102]

The second current driving proposals was the sense that European unification was a realistic possibility. Nazi aggression appeared to demonstrate the need for closer political and economic union after the war, both for defence and for reconstruction.[103] In September 1941 one Member of Parliament noted that 'Hitler has done by force one good thing':

> He has brought into practical politics the possibility of a reunion of Europe ... We now have an opportunity, I believe, of restoring the essential unity of civilised Europe .... Of course, there would be international control of transport and postal facilities ... The private arms industry would be abolished and military forces limited and controlled .... If this plan emerged, Europe and the world would have all the advantages of the Nazis' new order without its evils.[104]

As early as February 1940 Robert Cecil issued a memorandum through the League of Nations Union calling for a post-war international air force alongside international control of civil aviation by a European General Staff.[105] In December 1940 Home Secretary Herbert Morrison called for the abolition of European national air forces (that 'tommy-gun of the international gangster') and their replacement by a pan-European air force.[106] Most proposals for European unification included a proposal for a European international force and the abolition of national air forces.[107]

David Davies's rejuvenated proposals combined visions of Anglo-American power with the possibilities of a post-war politically integrated Europe. At the beginning of the war the New Commonwealth Society issued a revamped set of objectives which emphasized the creation of governing institutions, including a 'common defence force'.[108] Roosevelt's May 1940 address to Congress (which dwelled on the power of military aviation and

---

[102] Stephen King-Hall, *Total Victory* (London: Faber and Faber, 1941), 214–222. Also: Philip Kerr, 'Address of the Marquess of Lothian', *International Conciliation* 20 (January 1941): 9–19. Andrea Bosco, 'Lothian, Curtis, Kimber and the Federal Union Movement (1938–40)', *Journal of Contemporary History* 23, no. 2 (July 1988): 465–502.

[103] Peter Wilson, 'The New Europe Debate in Wartime Britain', in *Visions of European Unity*, eds. Philomena Murray and Paul Rich (Boulder, CO: Westview, 1996), 39–62; Andrea Bosco, *June 1940, Great Britain and the First Attempt to Build a European Union* (Cambridge: Cambridge Scholars Publishing, 2016).

[104] Hansard, HC, vol. 374, cols. 135–137, 9 September 1941.

[105] Robert Cecil, 'Memorandum on Post-War Organisation against Aggression', 20 February 1940, folder: 'NC Memoranda, 1940', box 92, Davies Papers.

[106] Victor Altman, *The International Police Force and World Security* (London: Alliance Press, 1945), 9.

[107] For example: W. Ivor Jennings, *A Federation for Western Europe* (Cambridge: Cambridge University Press, 1940), 101–104.

[108] New Commonwealth Society, *The New Commonwealth Declaration* (London: New Commonwealth Society, February 1940).

suggested a production target of 50,000 military aircraft a year) led Davies to renew his push for an international air force. In June 1940 he called for an Anglo-American police force with a powerful aerial arm to 'impose disarmament upon every country' and to police a newly federated Europe. In January 1941 he wrote to US internationalist Oscar Newfang informing him of the 'modification' to his earlier proposals. Now 'our aim can best be reached by the creation after the war of a British-American Commonwealth which would undertake temporarily the policing of the European Continent until a federal structure had been created there'.[109]

The opportunity for British aerial control over post-war Europe was strong enough to convert aviation journalist C.G. Grey, who had opposed an international air force in the 1930s fearing the diminution of British aerial power. In a 1941 book celebrating the bomber, Grey conceded that the success of colonial air policing demonstrated the viability of 'Control without Occupation ... which may – I say it with reserve – be used in the future to assure the peace of the world'.[110] His 1940 history of the Air Ministry gave 1930s internationalization proposals more credence than the sharp dismissal he accorded them in 1932, and conceded that an international air force could work.[111]

These early ideas were not uncontested: some internationalists feared this post-war joining together of British and US military power. For Harold Laski it had the makings of a 'new Roman imperium'. He called for a larger more inclusive international organization with a force composed of units from many countries.[112] Continental internationalists were also not enthused by proposals for Anglo-American domination. Czech émigré Joseph Roucek grouped Streit's and King-Hall's proposals together as 'utopian' and pointed out that Continental allies 'may not be enticed by parochial insistence on the unique importance of the United States and Great Britain'.[113]

Just as in the United States, by 1942 allied aerial cooperation was increasingly portrayed as laying the groundwork for a future international force. The US volunteer air force operating in China came to be commonly referred to as an international air force in Britain.[114] The RAF was also an important

---

[109] David Davies, 'Aeroplanes May Become the Sky Police of the World', 19 July 1940, folder: 'NC Memoranda, 1940', box 92, Davies Papers. David Davies to Oscar Newfang, ca. January 1941, folder 12, box 5, Series I, Quincy Wright Papers, Special Collections Research Center, University of Chicago Library.

[110] C. G. Grey, *Bombers* (London: Faber and Faber, 1941), 70–75.

[111] Grey, *A History of the Air Ministry*, 229; C. G. Grey, *The Civil Air War* (London: Harborough, 1945), 135–138.

[112] Laski, *Reflections on the Revolution of Our Time*, 244.

[113] Joseph S. Roucek, 'King-Hall, Stephen, Total Victory', *Annals of the American Academy of Political and Social Science* 222, no. 1 (July 1942): 199–200.

[114] Stephen G. Craft, *V.K. Wellington Koo and the Emergence of Modern China* (Lexington: University Press of Kentucky, 2004), 141.

reference point. For an anonymous letter writer to the *Manchester Guardian* the RAF was already a 'nucleus' of an international force; for Labour Party intellectual Harold Laski an international air force was a realistic possibility because already 'the Royal Air Force is an international air force in composition'.[115] The 1942 Ottawa Air Training Conference attracted significant internationalist rhetoric. The *Economist* saw the Combined Committee on Air Training in North America, created by the Conference, as 'the latest device of United Nations collaboration' which would bring 'efficiency in air operations'. 'If an international police force of the air emerges in some future settlement of the world', it announced, 'its genesis may be traced to the Ottawa conference of 1942'. Roosevelt declared Canada the 'Airdrome of Democracy' as it sent 'thousands upon thousands of her own men and men of the other United Nations to fight in the cause of liberty'.[116]

The Labour Party reaffirmed its commitment to an international air force at the beginning of the war. Clement Attlee's *Labour's Peace Aims* (1939) declared the formation of an international air force, 'of such overwhelming strength that no would-be aggressor would dare to challenge it', a central post-war objective.[117] This support remained through the war. The National Executive Committee reiterated its commitment in a report circulated to the 1942 Annual Conference, and the Party agreed, at the 1944 Annual Conference, that future international authority must have an internationalized force.[118] Others calling for an international police force at that time included H.R.G. Greaves and Leonard Woolf (writing with the blessing of the Fabian Society), as well as the Conference of Labour Parties of the British Commonwealth of Nations, which met in London in September.[119]

---

[115] Polyglot, letter to the editor, 'International Policing', *Manchester Guardian*, 8 May 1942; Laski, *Reflections on the Revolution of Our Time*, 239–244. There are similarities to the 1930s when some internationalists had argued that the French Foreign Legion was an international force already in existence and proof that an international police force was feasible. Arthur L. Martin, 'An International Force in Being', *Spectator*, 5 January 1934, 13–14; Arthur L. Martin, 'An International Force of Tomorrow', *Spectator*, 12 January 1934, 8–9.

[116] 'Air Training', *Economist*, 20 June 1942, 861–862; F. J. Hatch, *The Aerodrome of Democracy: Canada and the British Commonwealth Air Training Plan, 1939–1945* (Ottawa, ON: Department of National Defence, 1983), 99–111.

[117] Clement R. Atlee, *Labour's Peace Aims* (London: Labour Party, 1939), 11, 13.

[118] National Executive Committee of the Labour Party, *The Old World and the New Society: A Report on the Problems of War and Peace Reconstruction* (London: Labour Party, 1942); William Arnold-Forster, *The United Nations Charter Examined* (London: Labour Party, 1946), 33.

[119] H. R. G. Greaves, Memorandum, 'The Minimal Requirements of an International System', n.d., ca. July 1944, folder: 'Section 1E2', Section E, Leonard Woolf Papers, Special Collections, University of Sussex Library; Leonard Woolf, *The International Post-War Settlement* (London: Fabian Publications, September 1944); Fabian Society, *Dumbarton Oaks: A Fabian Commentary* (London: Fabian Publications, December 1944), 17–20.

The war had rejuvenated existing conceptualizations of the dual nature of scientific inventions and the civilian scientific origins of modern weapons. These were once again incorporated into internationalist rhetoric. Churchill's address read before the 1944 Royal Society meeting in India, for example, dwelled on the destructive and constructive capabilities of aviation, radio, and modern chemistry, and how these would need to be used by the United Nations to fulfil the promises made in the Atlantic Charter.[120] These truisms were also linked directly to an international air force by its supporters. For Attlee in *Labour's Peace Aims* an international air force was required 'because men have so far been unable to control the results of their own inventions', and for Stephen King-Hall an Anglo-American force was to be a continuation of Britain's historical reliance on the latest scientific developments for international policing. This force was not simply for collective security, it was to control the 'results of our scientific progress', which otherwise would 'become more and more harnessed to destructive purposes . . . mankind may slide back into barbarism and primitive savagery, but in a mechanized, chemicalized, and organized society'.[121]

The air force and its supporters alleged that an international force was impractical, reflected a dogmatic approach to politics and aviation, and would lead to authoritarian rule. A *Flight* magazine editorial characterized it as an 'ideal' which was 'unobtainable' due to 'practical difficulties' and an 'unwillingness to intervene in other people's quarrels'.[122] The military remained dead set against: Lt. General Gordon MacReady, a member of the Combined Chiefs of Staff of the United States and Britain, told the American Academy of Political Sciences in May 1945 that a combined international air force was not 'within the realm of practical politics'.[123] In August 1943 the Foreign Office formed a Post-Hostilities Planning Committee which considered and then rejected the possibility of the formation of an international air force.[124] As part of its deliberations it asked Spaight to produce a report laying out the arguments against a fully internationalized international police force.[125] The report focused on a generic, and not an aerial, force, and gave several reasons for why such a force would

---

[120] 'Science in the New World Order', *Nature* 153, no. 3872 (15 January 1944): 63–64.
[121] Atlee, *Labour's Peace Aims*, 11, 13. Stephen King-Hall, *Britain's Third Chance: A Book about Post-War Problems and the Individual* (London: Faber and Faber, 1943), 15, 20–22.
[122] Editorial, 'An International Air Force', *Flight*, 30 July 1942, 110.
[123] Gordon N. MacReady, 'The Provision of Military Forces for Use by the World Organisation', *Proceedings of the American Academy of Political Sciences* 21, no. 3 (May 1945): 72–78.
[124] See: Julian Lewis, *Changing Direction: British Military Planning for Post-war Strategic Defence, 1942–1947* (London: Sherwood, 1988), 65–69.
[125] A. T. Cornwall-Jones, Secretary, PHP subcommittee, 'An International Police Force. Note by the Secretary', 11 May 1944, FO 371/40795, TNA.

not be practical, including the need to develop international law (particularly in relation to defining aggression), the need to create a workable and acceptable international organization, nationalist bias in command and staff, and operational and financial logistics for what would be a vast force. The report also pointed out the problems inherent in modern scientific armaments. The continuing advance in 'scientific development' may lead to the force's armaments becoming obsolete. It would consequently need its own research and development wing, which would be problematic because this new research would naturally spread and could not be kept away from nations. Finally, as 'modern war is total' the 'international authority would in fact have to control all the economic resources of the world'. The force's realization, warned the report, would ultimately require a 'World State'.[126]

Dumbarton Oaks

The international air force was also discussed at the August to October 1944 Dumbarton Oaks Conference on post-war international organization, where the Soviet delegation first raised the possibility of its formation.[127] The British delegation were initially against the suggestion (though Alex Cadogan would later record that the idea had 'merits'), but Churchill expressed some enthusiasm and directed the British delegation to accept the Soviet proposals.[128] The US delegation (save for General Muir S. Fairchild) was also opposed, though they finally agreed a compromise solution: the formation of an international police force composed of national land, naval, and air contingents, provided as and when requested by the international organization.[129] It was this suggestion which was finally incorporated into the Dumbarton Oaks plan, though a special clause was inserted for nations to also have ready 'national air force contingents for combined international enforcement action' for particularly 'urgent military measures'.[130] The Chinese, supporters of collective security throughout the 1930s, were also keen on

---

[126] Brigadier van Cutsem and J. M. Spaight, 'An International Police Force Study by the PHP Sub-Committee', 11 May 1944, FO 371/40795, TNA.

[127] Soviet interest in the force was signalled to the Americans just before the Conference, see: 'Soviet Proposal Made Debate Topic; World Air Corps to Keep Peace Gets Split Reception in Washington', *New York Times*, 16 August 1944; 'No Army Included; Soviets Call "International Air Force" Impractical', *New York Times*, 15 August 1944. See also: Hilderbrand, *Dumbarton Oaks*, 45, 141.

[128] David Dilks, ed., *Diaries of Sir Alexander Cadogan, 1938–1945* (London: Cassell, 1972), 660–661. On Churchill's direction see: Zaidi, 'Aviation Will Either Destroy or Save Our Civilization'.

[129] Stoler, *Allies and Adversaries*, 204.

[130] Hilderbrand, *Dumbarton Oaks*, 138–158. For the proposals see: Clark M. Eichelberger, *Proposals for the United Nations Charter: What was Done at Dumbarton Oaks* (New York: Commission to Study the Organization of Peace, 1944), 25–32.

a United Nations international air force but had little influence at Dumbarton Oaks.[131]

Although there was a resurgence of interest following the news that an international force had been discussed at the conference (there were even calls in Congress for aircraft to be set aside for the proposed force), the collective security provisions of the Dumbarton Oaks proposals received a mixed reception from internationalists in the United States.[132] They were condemned by world federalists such as Emery Reves and Ely Culbertson as part of a wider rejection of the structure of the proposed United Nations organization itself. Reves included a critique of the Dumbarton Oaks proposals for an international air force in his best-selling 1945 internationalist work *The Anatomy of Peace*. His argument was that, as the organization was to be only 'an association of sovereign states' with no law regulating relations between them, its police force would simply represent the will of the Great Powers. Culbertson felt that with the veto the organization would be ineffective, a 'revived League of Nations' which would fail to stand up to aggression. He informed the Senate Committee on Foreign Relations that the 'Charter is a system of collective security which is neither collective nor secure'.[133] Quincy Wright struck a more positive note, hoping that the 'compromise' air force envisaged by the Dumbarton Oaks proposals would eventually evolve into a permanent internationalized one.[134]

Supporters of a United Nations organization threw their weight behind Dumbarton Oaks. Shotwell and Eichelberger headed an educational campaign, sponsored by the CSOP and the American Association for the United Nations, which publicized the benefits of the international organization and emphasized that its air force could be further developed. 'World Air Force Held Key to Peace; Study Commission Wants it Set Up as Instrument of New International Body' announced the *New York Times* headline following a CSOP press release.[135] The Universities Committee

---

[131] See: Craft, *V.K. Wellington Koo and the Emergence of Modern China*, 170, 178, 180.

[132] 'Mundt Proposes World Air Patrol; Speech in House Calls for a Force of 5,000 Planes to be Used to Keep Peace', *New York Times*, 31 January 1945; William C. Johnstone, 'The Hot Springs Conference', *Far Eastern Survey* 14, no. 2 (31 January 1945): 16–22.

[133] Reves, *The Anatomy of Peace*, 215–224. This criticism was cut from the abridged version serialized in *Reader's Digest* that same year. Ely Culbertson, 'Russia and World Security', *Russian Review* 4, no. 1 (Autumn 1944): 10–17; US Congress, Senate Committee on Foreign Relations, *The Charter of the United Nations: Hearings before the Committee on Foreign Relations, United States Senate, Seventy-Ninth Congress, First Session, on the Charter of the United Nations for the Maintenance of International Peace and Security, Submitted by the President of the United States on July 2, 1945: revised July 9, 10, 11, 12, and 13, 1945* (Washington, DC: Government Printing Office, 1945), 416.

[134] Wright, 'The International Regulation of the Air'.

[135] 'World Air Force Held Key to Peace; Study Commission Wants it Set Up as Instrument of New International Body', *New York Times*, 4 September 1944; John W. Davis et al., 'The President and Peace Forces: Letter to the New York Times', *International Conciliation* 22

on Post-War International Problems polled its university groups in December 1944. The 'overwhelming majority' saw the Dumbarton Oaks proposals on collective security as a step in the right direction. Seventeen out of the forty-six polled would have liked a permanent international police force formed alongside the national air contingents stipulated in the proposals. The remainder appeared satisfied with national contingents only.[136] T.P. Wright also continued to hope for more: he called for the formation of a United Nations international air force at a Royal Aeronautical Society speech during a May 1945 visit to London.[137]

State propaganda did the opposite: in an effort to minimize potential opposition to the proposed organization, it downplayed the international force provisions. Statements by senior officials explained that, although the new organization would be more empowered than the failed League, the Security Council did not have its own forces, and so would only call national forces into action. It, moreover, couldn't force states into war. Secretary of State Edward Stettinius Jr reassured readers of *Reader's Digest* that 'The Security Council does not in any way become the arbitrary master of the world's military resources'.[138] The US Office of Education's *Education for Victory* magazine described the proposals as 'not an international police force' but rather separate 'national contingents'.[139]

The final Charter agreed at the April–June 1945 San Francisco Conference on the United Nations followed the Dumbarton Oaks proposals in most respects, including with regard to an international air force. Article 43 called on members to make available 'armed forces' at the request of the Security Council, and Article 45 stipulated that members hold ready air force contingents – the extent of which was to be determined by the Security Council in conjunction with a United Nations Military Staff Committee (MSC).[140] There was some concern during the Senate hearings on Charter

---

(December 1944): 795–800; American Association for the United Nations, *Dumbarton Oaks Proposals* (New York: American Association for the United Nations, 1944); Eichelberger, *Proposals for the United Nations Charter*, 9, 10, 23; Josephson, *James T. Shotwell and the Rise of Internationalism in America*, 255–256. On other support: 'Dumbarton Plans win Endorsement; League of Nations Association Announces Full Agreement on Proposed World Body', *New York Times*, 11 October 1944.

[136] Universities Committee on Post-war International Problems, 'Summaries of Reports of Cooperating Groups: XVIII-The Dumbarton Oaks Proposals: The Enforcement of Peace', *International Conciliation* 23 (October 1945): 671–682.

[137] T. P. Wright, 'Aviation's Place in Civilization', *The Journal of the Royal Aeronautical Society* 49, no. 414 (June 1945): 299–342.

[138] Edward R. Stettinius Jr., 'What the Dumbarton Oaks Peace Plan Means', *Reader's Digest*, February 1945, 1–7. Also: Leo Pasvolsky, *Dumbarton Oaks Proposals* (Washington, DC: Department of State, 1944), 9.

[139] 'Building the Peace', *Education for Victory* 3, no. 19 (3 April 1945): 4.

[140] Leland M. Goodrich, Edvard Hambro, and Anne Patricia Simons, *Charter of the United Nations: Commentary and Documents* (New York: Columbia University Press, 1969). On

ratification that it would bind the United States to sending men to fight abroad, to which the internationalist response was that the US contribution to the force would be 'mechanized weapons', such as bombers. Political scientist William T. Fox would note in late 1945 that this clause was an outcome of 'a general belief in the efficacy of the airplane as an instrument of coercion', and that the precise and prescriptive collective security clauses of the UN Charter were 'a vindication of the French position during most of the inter-war period'.[141]

Internationalist organizations continued to support the formation of a UN air force past the San Francisco Conference, but support soon whittled away, leaving behind some residual support for a contingent-based multi-service force. Perhaps the most prominent call for an international force was by Churchill in his March 1946 Fulton Missouri speech. Remembered today for phrases such as the 'iron curtain' and the 'special relationship', the address also suggested that states assign air squadrons which would 'wear the uniform of their own countries but with different badges', and rotate from one country to another, but not be expected to attack their own country.[142] The burgeoning world federalism movement did not focus on aviation in its manifestos and conferences, even though it continued to emphasize the transformative effects of science and technology on international relations.[143] The CSOP reiterated its wartime calls for international air force in its first post-war report, issued June 1947 and widely known, thanks to an article in *Newsweek*.[144] By the time of the Korean War, however, its reports no longer talked of an air force, rather more vaguely of a 'United Nations Legion' backed by 'National Contingents'.[145] Internationalist legal scholars, meanwhile,

the MSC: Jonathan Soffer, 'All for One or All for All: The UN Military Staff Committee and the Contradictions within American Internationalism', *Diplomatic History* 21, no. 1 (Winter 1997): 45–69.

[141] US Congress, Senate Committee on Foreign Relations, *The Charter of the United Nations*, 527–529, 539; William T. R. Fox, 'IV. Collective Enforcement of Peace and Security', *The American Political Science Review* 39, no. 5 (October 1945): 970–981.

[142] Winston Churchill, 'The Soviet Danger: The Iron Curtain', in *The Speeches of Winston Churchill*, ed. David Cannadine (London: Houghton Mifflin, 1989), 295–308.

[143] See: 'Appendix F: World Federalist Declarations' in Baratta, *The Politics of World Federation*, vol. 1, 539–555.

[144] 'Security Report: A Plea for Peace Through Public Opinion', *Newsweek*, 16 June 1947, 35–36.

[145] See: 'Fifth Report: Security and Disarmament under the United Nations (June 1947)' and 'Seventh Report: Collective Security under the United Nations (July 1951)' in: Commission to Study the Organization of Peace, *Building Peace: Reports of the Commission to Study the Organization of Peace, 1939–1972*, vol. 1 (Metuchen, NJ: Scarecrow Press, 1973), 185–215, 230–263.

continued to hope for UN collective security, but in effect supported Great Power law enforcement, such as the Nuremberg Trials.[146]

Growing hostility between the United States and the Soviet Union was the major factor in this decline, leading in particular to the failure of negotiations for the formation of a UN air force. In February 1946, the Security Council directed the MSC to implement Article 43. The US and Soviet delegations quickly found themselves in disagreement – the Americans proposed a larger force composed largely of their contingents, whereas the Soviets called for a smaller force with equal national contributions. At one point a force consisting of 1,250 bombers, 2,250 fighters and fighter-bombers, and 300 reconnaissance planes was being mooted by the US Delegation.[147] Discussions within the MSC continued until April 1947, at which point its divergent views were presented to the Security Council.[148] No agreement was reached, and no UN air force was ever formed.[149] Supporters of US military aviation now added the failure of an international force and collective security to their list of reasons for an expanded national air force, with the Air Policy Commission's 1948 report *Survival in the Air Age* beginning with the assumption that 'the United Nations can never develop as a permanent instrument of universal peace'.[150] A second reason for the reduced interest in an international air force was the advent of atomic weapons, which offered a new locus for internationalist activity from August 1945 onwards. This turn is explored in Chapters 6 and 7.

[146] Quincy Wright, 'The Law of the Nuremberg Trial', *The American Journal of International Law* 41, no. 1 (January 1947): 38–72.

[147] 'U.S. Proposes 3,800 Planes for the World Police Force; U.S for Big Force to Police World', *New York Times*, 1 July 1947.

[148] 'Report by the Military Staff Committee to the Security Council on the General Principles Governing the Organization of the Armed Forces Made Available to the Security Council by Member Nations of the United Nations, April 30, 1947', *International Organization* 1, no. 3 (September 1947): 561–574.

[149] Goodrich, Hambro, and Simons, *Charter of the United Nations*, 317–326; Soffer, 'All for One or All for All'; 'Donald C. Blaisdell Oral History Interview'; Edward Johnson, 'British Proposals for a United Nations Force 1946–48', in *Britain and the First Cold War*, ed. Ann Deighton (New York: Palgrave Macmillan, 1990), 109–129.

[150] Air Policy Commission, *Survival in the Air Age* (Washington, DC: Government Printing Office, 1948), 5.

5 Wings for Peace: Planning for the Post-War Internationalization of Civil Aviation

'We look forward to the day when each transport plane, arriving at an airfield, will carry on its wings not the mark of one nation but a symbol of the world's practical acceptance of its new unity'.
– *Wings for Peace: Labour's Post-War Policy for Civil Flying* (1944).[1]

During the Second World War international civil aviation took its place amongst the plethora of post-war issues that required planning and organization. Industry and state planners in Britain and the United States were keen to restore international services as soon as possible, whilst also making use of the opportunities afforded by the war and fending off threats to their respective industries. An internationalized post-war civil aviation regime, controlled or even directly owned and operated by an international organization, emerged as a serious policy option in Britain. Internationalists hoped that this framework would safeguard Britain's aviation industry and at the same time contribute to peace by limiting the militaristic use of civil aviation and spreading prosperity and international understanding. As in previous years there was a spectrum of proposals: from radical schemes for global internationalization of both domestic and foreign aviation to calls for internationalization by region, in which aviation in each region would be dominated by its leading aerial power. Opponents, especially in the United States, called for 'freedom of the skies' with minimal international regulation and a regime dominated by multilateral agreements. Although the sense of a new beginning and new opportunities drove these proposals and counterproposals, they nevertheless fell back on long-standing arguments about the nature of aviation and its impact on international relations.

Internationalization or Americanization

In June 1939, Parliament passed the British Overseas Airways Bill, authorizing the amalgamation of the country's international airlines into

---

[1] Labour Party, *Wings for Peace*.

a single nationalized entity, the British Overseas Airways Corporation (BOAC). It was to be a new beginning for British civil aviation. Plagued for years by charges of inefficiency and ineptitude, British airlines were at last to be rationalized and taken in a more commercially and financially efficient direction by the state. This vision never materialized. The outbreak of war interrupted the formation of BOAC, whose aircraft and operations were requisitioned by the Royal Air Force for the war effort.[2] By 1941 concerns over the ability of civilian aviation to resume effective post-war commercial operations grew. For whilst British civil aviation was held in abeyance, US civil aviation expanded both nationally and internationally, with new air routes being added and US manufacturers producing a new generation of transport aircraft. It appeared that the government would need to rethink civil aviation for the post-war period. How could British aviation rebuild after the war whilst meeting these challenges?

For state planners, politicians and interested observers, civil aviation was part of a bigger problem of post-war economic and social reconstruction. From 1940 onwards there was a growing consensus in liberal and leftist circles that, in return for the national war effort, the nation expected and deserved stable incomes and a buoyant economy after the war. This, it was increasingly believed, could only occur through state planning of the economy and commercial and industrial enterprises, as well as state intervention in social matters.[3] There was a significant international dimension to this concern, aptly captured in the Atlantic Charter, issued August 1941, and Roosevelt's Four Freedoms speech (January 1941). For progressives international peace and prosperity could not be implemented without the eradication of poverty and unemployment on a global scale. This required the creation of international technical organizations for the planning and the pooling of scarce resources – for Labourite socialist intellectual Harold Laski the 'organization of an economics of expansion in the postwar world'.[4] There were other reasons too for post-war cooperation. An awareness of Britain's reduced international influence further impelled internationalists to push for cooperation with the United States, the Dominions, and other allies. Intellectuals and policymakers were also cognizant of Britain's growing political, economic, and

[2] Robin Higham, *Britain's Imperial Air Routes 1918–1939* (London: G.T. Foulis, 1960), 1813–1818; Robin Higham, *Speedbird: The Complete History of BOAC* (London: IB Tauris, 2013), 5–8.

[3] Richard Toye, *The Labour Party and the Planned Economy, 1931–1951* (London: The Royal Historical Society, 2003), 87–155.

[4] Laski, *Reflections on the Revolution of Our Time*, 249.

military entanglement and influence in mainland Europe, which appeared to portend a leading role in post-war European affairs.[5]

These concerns were shared by Labour ministers, none more so than the senior Cabinet Minister Arthur Greenwood, who in early 1941 was in charge of post-war reconstruction. Writing to Archibald Sinclair (Secretary of State for Air) in March that he was 'deeply impressed' with the part civil aviation would play in the 'new world order', he set in motion planning for post-war civil aviation by forming a committee to make policy recommendations for the 'reconstruction, organisation and development' of post-war international civil aviation.[6] The committee's final report, produced in January 1942, reflected contemporary concerns about the future viability of the British aviation industry as well as the turn to international planning. It was also shaped by the internationalist bent of its chairman, Director-General of Civil Aviation Francis Shelmerdine, formed during his time on a League of Nations committee in 1932 charged with producing proposals for the integration of European commercial aviation.[7]

The Shelmerdine Report presented long-standing liberal internationalist arguments for why 'complete international control' was now required. Interwar civil aviation was depicted as inefficient and unprofitable, a 'manifestation of exaggerated nationalism' which was 'hampered in its development by nationalistic aims and by international jealousies', and surviving only because of its presumed 'war potential'. Post-war internationalization would remedy this whilst also complementing aerial disarmament. The report sought support in the Atlantic Charter, and argued that as a 'service to humanity' civil aviation fell under its remit. Civilian aviation should consequently be organized under the cooperative lines envisaged in Article V, contribute to peace as per Article VI, and contribute to a system of international security as per Article VIII.[8] The report reviewed, in some depth, British reasons for opposing the 1932 French proposals for internationalization, and contended that these reasons no longer held. It warned that Canada, the United States, and Australia might balk at complete internationalization, and that Britain should be wary of entering into a watered-down scheme. Acknowledging that, as global

[5] Carr, *Conditions of Peace*, 163–209.
[6] A. Greenwood to A. Sinclair, 3 March 1943, CAB 117/187, TNA; C.V. Davidge, Joint Secretary, Interdepartmental Committee on Civil Aviation, Memorandum, 'Composition and Terms of Reference', 7 August 1941, CAB 117/187, TNA.
[7] On the work of the Special Sub-Committee on the Constitution and Operation of a Main Network of Permanent Air Routes see: Air Commission, *Objective Study on the Internationalisation of Civil Aviation*, 51–52.
[8] Interdepartmental Committee on Civil Aviation, 'Interim Report to the Minister without Portfolio', 5 January 1942, CAB 117/187, TNA.

internationalization was unlikely to occur soon after the war, a start needed to be made with a 'nucleus' of European states and their dependencies.[9]

Ministerial response was divided, largely along party lines. Liberal and Labour Party members (including the senior Labour MP and Cabinet member Arthur Greenwood and Foreign Secretary Anthony Eden) mostly supported the report's recommendation.[10] Conservatives and those representing Commonwealth and Empire were either lukewarm or entirely negative in their response.[11] Secretary of State for India Leo Amery described the report as a 'half-baked, ill thought out blend of bad economics and bad politics'. A 'Civil Aviation League of Nations' would 'kill all progress in civil aviation', and anyhow would not bring peace. That would require the 'internationalisation of the air forces, armies and navies of the world under the control of the supreme world government'. These, he noted, were not envisaged by the Atlantic Charter.[12] Military advisors, including Spaight, also opposed the Shelmerdine Report, characterizing it as a backward reversion to the 1932 French proposals. The proposals were, they argued, part of a larger system of universal disarmament which would not be required after the Second World War as Germany would be 'disarmed to the bone'.[13]

Conservative opponents also responded by presenting alternative proposals for limited regionalized internationalization. In a scheme presented to the Cabinet a few months later, Amery suggested the pooling of aviation in separate regions of the world, of which the British Empire would be one. 'Free Air' would then only apply between these aerial groupings, and Britain would control aviation in her Empire and Commonwealth.[14] The Conservative Minister of Aircraft Production Moore-Brabazon made a similar suggestion: Britain and her Commonwealth, the USSR, and the United States should regionally divide all international aviation amongst themselves, whilst also acting as a 'world police force'.[15] He made this

[9] Ibid.
[10] A. Greenwood, Memorandum, 'Observations by the Minister without Portfolio', 23 January 1942, CAB 117/187, TNA; H. Johnstone to A. Greenwood, 27 January 1942, CAB 117/187, TNA; 'Summary of Comments by Ministers', 24 March 1942, CAB 117/187, TNA; Hansard, HL, vol. 127, col. 272, 15 April 1943.
[11] Viscount Cranborne to A. Greenwood, 5 February 1942, CAB 117/187, TNA; Memorandum attached to: Moyne to A. Greenwood, 6 February 1942, CAB 117/187, TNA; Morrison to A. Greenwood, 6 February 1942, CAB 117/187, TNA.
[12] L. Amery, Memorandum, 'Comments on the Interim Report of the Interdepartmental Committee on Civil Aviation', 26 January 42, CAB 117/187, TNA.
[13] Rear-Admiral R. M. Bellairs, Brigadier W. E. van Cutsem, and J. M. Spaight, Memorandum, 'Note by the Military Sub-Committee', 20 July 1942, CAB 117/187, TNA.
[14] L. Amery, 'Air Transport: An Empire Policy', War Cabinet Report RP(A)(43) 8, 3 March 1943, FO 371/36433, TNA.
[15] J. Moore-Brabazon, Memorandum, 'Observations by the Minister for Aircraft Production', 12 February 1942, CAB 117/187, TNA.

suggestion public later that year through a lecture at the Royal Aeronautical Society and an article in *Nature*, the latter of which also claimed that internationalization would free civil aviation from military influence. He even listed technical developments which had been held back by militaristic influences: 'assisted take-off, high wing loading, non-inflammable fuel, better landings, pressurized cabins, jet propulsion'.[16]

There was sufficient support for the Shelmerdine Report for a second committee to produce a more detailed proposal for internationalization. The Finlay Committee's *Internationalisation of Civil Aviation After the War* was submitted as a Cabinet Paper in December 1942.[17] By this time America's aerial and industrial might were making even deeper impressions in Britain: that very month the *Aeroplane* noted with concern that a House of Commons discussion had warned that 'The Americans had a monopoly in the Pacific; they had been given an entree in Africa; at least two American aeroplanes crossed the North Atlantic for one British .... American air lines would shortly be operating from Aden and from India'.[18] This concern replaced the Shelmerdine Committee's earlier focus on the Atlantic Charter. Judge William Finlay's covering letter summarized the choice facing the British Government as one between 'Americanisation and internationalisation', and warned of Pan-American Airways being used by the US state for 'peaceful penetration in those regions of the world where the local government are too weak, too poor, or too inefficient' to provide their own aviation services. It was under internationalization 'that British interests will best be served': the British aviation industry would flourish through contracts they would receive from the proposed 'Air Board of Control'.[19]

The report advocated radical 'comprehensive internationalisation' which was sketched out in an 'Outline of a Draft Convention on Internationalisation of Air Services'. This envisaged the complete global internationalization of both internal and external air services, with the exception of internal services in the Soviet Union and the United States (which, the argument went, were able to run services efficiently because of their size). The report attacked more conservative frameworks for internationalization such as that suggested by Moore-Brabazon and Amery. 'Regionalisation' would 'tend to perpetuate between regions the

---

[16] J. Moore-Brabazon, 'Ad Astra', *Aeronautical Journal* 46, no. 382 (October 1942): 247–260; J. Moore-Brabazon, 'Past, Present and Future of Aviation', *Nature* 150, no. 3796 (1 August 1942): 132–133.

[17] Lord Finlay, 'Internationalisation of Civil Aviation After the War', War Cabinet Report RP(42) 48, 18 December 1942, AIR 2/5491, TNA.

[18] 'The Future of Air Transport', *Aeroplane*, 25 December 1942, 724. Also: 'Aviation and Empire', *Spectator*, 1 July 1943, 3–4.

[19] Finlay, 'Internationalisation of Civil Aviation After the War'.

competition that it is desired to eliminate between states' and 'act as a centrifugal force', pulling the British Commonwealth and Empire apart. Internationalization limited to 'certain specified air routes of world importance' was dismissed as a 'meagre' copy of interwar aerial 'pooling' arrangements.[20]

Finlay had rejected the Air Ministry's suggestion that he conservatively approach internationalization, and the Ministry responded by launching strong objections to his report. For the Under-Secretary of State for Air, H.H. Balfour, it was putting the 'cart before the horse'. Instead of using civil aviation as the 'spearhead of internationalisation', Britain should wait until the United Nations was prepared to accept internationalization.[21] The Air Ministry's new Director-General of Civil Aviation, William P. Hildred, described the report as 'confused'. Whilst he acknowledged growing American aerial influence in the Commonwealth, responding with the 'creed' of internationalization went too far. He instead suggested the rationalization of British international air routes.[22] One dissenting voice within the Air Ministry was that of Hugh Michael Seely (Lord Sherwood), Parliamentary Joint Under-Secretary of State and prominent Liberal MP. His response to critics was that under internationalization air services would be more efficiently run, replacing 'cut-throat national competition'. He warned that 'freedom of the air' would lead to American penetration of global air routes now serviced by British airlines. He emphasized the Foreign Office's support for internationalization, and himself advocated regional internationalization based on the 1932 French proposals.[23]

The contentious nature of Finlay's report led the War Cabinet to create a further Ministerial subcommittee to consider post-war civil aviation. This committee, chaired by Chairman of the Committee on Reconstruction Problems William Jowitt, took a more cautious approach, recognizing in particular American opposition. The first meeting, held January 1943, decided that, although there was a 'strong probability of full internationalisation being rejected' by the United States, internationalization should still be proposed as it would place 'the onus of rejection' on the United States and perhaps also get 'full internationalization recognized as the ultimate objective'. The committee suggested that the

[20] Ibid.
[21] H. H. Balfour, Memorandum, 'Note on the Finlay Report', 29 December 1942, AIR 2/5491, TNA.
[22] W. P. Hildred, DGCA, Memorandum, 'Summary of Lord Finlay's Report (RP (42) 48) with Comments', 16 January 1943, AIR 2/5491, TNA.
[23] From: Untitled Memorandum, H. M. Seely to A. Sinclair and W. P. Hildred, 5 January 1943, AIR 2/5491, TNA. See also: Untitled Memorandum, H. M. Seely to A. Sinclair and H. H. Balfour, 19 January 1943, AIR 2/5491, TNA.

Government be prepared to enact a limited form of internationalization across Europe and its dependencies only.[24] There was some debate over whether imperial aviation should be combined with European aviation, during which supporters such as Noel-Baker (then Joint Parliamentary Secretary at the Ministry of War Transport) and Chancellor of the Exchequer Kingsley Wood warned that not including imperial aviation would 'divide the world into more or less watertight compartments' and not provide 'as efficient a barrier to American infiltration'.[25]

Orme Sargent, whose input had been crucial to the internationalist flavour of the Finlay Report, was an important factor in the pro-internationalization stance of the Foreign Office. The Permanent Under-Secretary Alexander Cadogan left much of the day-to-day running of the Ministry to Sargent, who was consequently able to place his stamp on the Foreign Office's attitude to internationalization. He outlined his views in a February 1943 internal Foreign Office memo. The government's objective, being to 'obtain an adequate and assured market for the British aircraft industry', could only be attained 'within the framework of internationalisation ... without involving ourselves in a head-on conflict, not only with the American air lines, but with the US Government itself'. Regional internationalization limited to Europe or the Empire would raise the ire of the Americans and the suspicions of the Dominions and the Europeans.[26] He opposed the conclusions of a later report (the Barlow Report), which suggested internationalization in Europe and her empires only, as being 'unfairly weighted' such that 'the difficulties in the way of universal internationalization are magnified and stressed'.[27]

British planners also began detailed discussions on post-war civil aviation with the Dominions and India in February 1943. Although there was considerable sympathy for internationalization (initial Canadian planners had also advocated the fullest possible internationalization), discussants were unable to agree on a common policy.[28] Meeting in June 1943, the Cabinet consequently abandoned global or imperial

---

[24] 'Draft Minutes of the 1st Meeting of the Subcommittee on Civil Aviation', 26 January 1943, AIR 2/5491, TNA.

[25] From: R. Law to W. Jowitt, 3 February 1943, AIR 2/5491, TNA. For other expressions of support see: K. Wood to W. Jowitt, 8 February 1943, AIR 2/5491, TNA; and P. Noel-Baker to W. Jowitt, February 1943, AIR 2/5491, TNA. Philip Noel Baker hyphenated his surname to Noel-Baker in 1943.

[26] O. Sargent, Memorandum, 'How to Obtain an Adequate and Assured Market for the British Aircraft Industry', 21 February 1943, FO 371/36431, TNA.

[27] Note, O. Sargent, 26 March 1943, FO 371/36433, TNA. Gladwyn Jebb also supported internationalization: Note, G. Jebb, 11 April 1943, FO 371/36434, TNA.

[28] David MacKenzie, *Canada and International Civil Aviation 1932–1948* (Toronto: University of Toronto Press, 1989), 121–125.

internationalization as a policy option. Herbert Morrison regretted that no agreement had been reached as it could have at the very least put the onus on the US government to 'wreck' the proposals during the upcoming international conference.[29]

Nevertheless the opportunity to dominate aviation in Europe seemed too good to pass up. Noting continuing calls for international control in the Commons, Foreign Secretary Anthony Eden hoped that aviation in Europe could never the less be internationalized. Churchill spoke against global internationalization in June 1943, especially 'a kind of Volapuk Esperanto cosmopolitan organization managed and staffed by committees of all peoples great and small, with pilots of every country from Peru to China (especially China), flying every kind of machine in every direction'. European internationalization, on the other hand, was agreeable, so long as it was not simply an 'amalgamation' of different countries' aviation interests, but rather genuinely controlled and organized through a pan-European organization. He even suggested that Britain agree a quid pro quo with the United States: their domination of aviation in North America in return for British domination over Europe.[30] Churchill's confidant and advisor Lord Beaverbrook agreed. Later that year he would write to Churchill that 'Our geographical position and our vital commercial interests entitle us to expect that Britain can become and should become a dominating factor in the air transport system of the Continent'.[31]

In July 1943 a committee under the Secretary of State for Dominion Affairs, the Conservative Peer Viscount Cranborne, produced a statement of principles to be used as a basis for discussion with the United States. These accepted 'freedom of the air', but suggested that certain Commonwealth trunk routes be internationalized, and left open the possibility of internationalization of European air transport. A White Paper issued in October 1944 laid out the policy for the upcoming November Chicago conference on international aviation. It proposed an 'International Air Authority' which would license international air operators and 'provide for the elimination of uneconomic competition by the determination of frequencies ... the distribution of those frequencies between the countries concerned and the fixing of rates of carriage'. It was recognized that the United States would oppose even this suggestion,

[29] W.M.(43), 88th Meeting, 24 June 1943, Cabinet Secretaries' Notebooks on War Cabinet Meetings, CAB 195/2, TNA.

[30] Ibid. Christopher Brewin, 'British Plans for International Operating Agencies for Civil Aviation, 1941–1945', *International History Review* 4, no. 1 (1982): 91–110.

[31] Memorandum, Lord Beaverbrook to W. Churchill, 14 October 1943, FO 371/36444, TNA.

and the recently appointed Minister for Aviation was given authority to make concessions in order to arrive at some form of multilateral agreement.[32]

Chatham House and the Future of Air Transport

State planning spread interest in internationalization amongst the Labour and Liberal Parties and internationalist organizations, with particularly detailed discussions taking place at the Royal Institute of International Affairs (Chatham House). Although the Institute advised the state on wartime and post-war matters through its Oxford-based Foreign Research and Press Service, it also hosted a number of discussion meetings at its London headquarters. This included one on civil aviation, which between March and October 1943 brought aviation industrialists, policymakers, and other interested parties together to debate the merits of internationalization.[33] Participants included aviation industry representatives such as Frederick Handley Page, leading aviation journalist Peter Masefield (of the *Aeroplane* and the *Sunday Times*), government representatives (including Noel-Baker for the Ministry of War Transport), and interested internationalists (most prominently Air Chief-Marshal Sir Arthur Longmore and Jonathan Griffin, author of *World Airways*).[34] The discussion group took as its starting point a December 1942 article by Chatham House transportation expert Harry Osborne Mance on 'Air Transport and the Future'. Director of Canals at the Ministry of Transport, Mance was a long-standing believer in the constructive effects of modern transport and communications. His paper proposed the 'complete internationalisation of air transport either in Europe alone or throughout the world' and suggested the 1932 French proposals as a blueprint. Internationalization would ensure 'freedom of the air' without nationalistic interference, and would anyhow be a necessary addendum to the international control of military aviation if it were instituted after the current war.[35]

---

[32] *International Air Transport Policy*, Cmnd. 6561 (October 1944).

[33] On the liberal internationalism of Chatham House see: Inderjeet Parmar, *Think Tanks and Foreign Policy: A Comparative Study of the Role and Influence of the Council on Foreign Relations and the Royal Institute of International Affairs* (Houndmills: Palgrave Macmillan, 2004), 70–74.

[34] From industry, also present were: L. M. J. Balfour (Portsmouth Aviation), Major K. M. Beaumont (BOAC), Thomas G. John (Alvis), and Captain L. Hope (Philip & Powis).

[35] Quotes from: H. O. Mance, 'Air Transport and the Future: Some International Problems', *Bulletin of International News* 19 (26 December 1942): 1173–1178. Also: H. O. Mance, 'The Influence of Communications on the Regulation of International

Participating internationalists reeled out a series of long-standing arguments for internationalization. Noel-Baker called for this 'most marvellous invention of modern times', which had so far been 'a colossal evil and an almost unmitigated evil', to be used to bring 'adventure, romance, happiness, health, understanding to the human race'.[36] James E. Wheeler, co-author of Mance's paper, advanced the efficiency argument: 'it was a great waste of money to have twenty-five different companies with twenty-five or more types of equipment'.[37] But the most powerful arguments emphasized the opportunities offered through the post-war reconstruction of mainland Europe. Reconstruction made internationalization a realistic possibility because European aviation would be in a 'transitional period' after the war (with aerial networks, airlines, and facilities being recreated or built up for the first time) when the Allies would need to heavily regulate military and civil aviation. During this period civil aviation could be run by a United Nations air force in which international personnel already within the RAF could pilot the air forces of smaller countries which would be organized into 'regional international air police forces'. In the longer term European international aviation would be internationalized (and monopolized) through a company which would own property and equipment at national airports throughout Europe.[38] Longmore suggested the international travel company Wagons-Lits (which owned its own railcars but used national rail tracks) as a model.[39]

Aviation industry representatives opposed internationalization and instead emphasized the need for light regulation. Handley Page believed that it would be best to 'dissociate governments from the direct running of air lines and also from directly subsidising air lines'.[40] They criticized proposals for internationalization as being born, in the words of R.H.

Affairs', *Journal of the British Institute of International Affairs* 1, no. 3 (May 1922): 78–89; H. O. Mance, *International Air Transport* (London: Oxford University Press, 1943).

[36] Informal discussion meeting, 'The Future of Aviation', 24 March 1943, 8/986, Records of the Royal Institute of International Affairs, Chatham House, London (hereafter cited as Chatham House Papers).

[37] 'Private Meeting, British Air Policy in the Pacific', 19 May 1943, 8/987, Chatham House Papers.

[38] 'Private Meeting: British Air Policy in the Pacific', 19 May 1943, 8/987, Chatham House Papers. Also: 'Informal discussion meeting: The Future of Aviation', 24 March 1943, 8/986, Chatham House Papers.

[39] On the use of Wagons-Lits in aerial discourse see: Waqar Zaidi, '"Wings for Peace" versus "Airopia": Contested Visions of Postwar European Aviation in World War Two Britain', in *Linking Networks: The Formation of Common Standards and Visions for Infrastructure Development*, eds. Martin Schiefelbusch and Hans-Liudger Dienel (London: Ashgate, 2014), 151–168.

[40] 'Informal Discussion Meeting: The Future of Aviation', 24 March 1943, 8/986, Chatham House Papers.

Thornton, Chairman of the Aviation Committee of the General Council of British Shipping, 'entirely out of fear' that civilian aircraft would be used for bombing.[41] In public Handley Page accepted that civil aviation may be misused for military purposes after the war, particularly by the defeated Axis, and suggested a multilateral agreement amongst aviation firms and their governments in which they laid open their financial statements to ensure that aviation was not being used as an 'instrument of state policy'.[42] Their views received some support from E.P. Warner, the Vice-Chairman of the US Civil Aeronautics Board, who attended one meeting and pointed out that, although there was some support in the United States for an international air force, there was no appetite for internationalized civil aviation.[43]

Although the Chatham House discussions did not lead to any coherent recommendations, there were important linkages to the wider discourse on internationalization and to governmental planning on aviation. Prompted by these discussions *The Times*, which had already come out in support of complete internationalization, produced two feature articles on civil aviation in late 1942 and early 1943.[44] In August 1943 Longmore was asked to join the Foreign Office's Post-Hostilities Planning Committee as the RAF representative, where he called for a post-war international air force. The committee, which included opponents of internationalization such as Spaight, decided against his suggestion.[45] Mance's Chatham House paper was circulated to the Foreign Office, where it was greeted with significant interest, with one official noting that it had much in common with the Ministry's own deliberations.[46]

World Airways and World Air Authorities

The Labour Party, committed as it was to post-war international organization, became a major source of support for internationalization.[47]

[41] 'Private Discussion Meeting: The Future of Aviation', 21 October 1943, 8/992, Chatham House Papers.

[42] Frederick Handley Page, 'Commercial Air Transport: its Past History and Future Prospects', *Flight*, 3 June 1943, 586–587.

[43] 'Private Meeting: The Future of Civil Aviation: An American View', 24 May 1943, 8/988, Chatham House Papers. On Warner see the next chapter.

[44] 'The New Air Age', *The Times*, 18 December 1942; 'Freedom of the Air', *The Times*, 6 February 1943; 'Problems of Civil Aviation: II – Danger of International Rivalries, The Special Case of Europe', *The Times*, 30 December 1943.

[45] See: Lewis, *Changing Direction*, 65–69; Arthur Longmore, *From Sea to Sky: 1910–1945* (London: Geoffrey Bles, 1946), 301–302.

[46] Filenote, 'Future of International Civil Aviation', 5 January 1943, FO 371/36430, TNA.

[47] Labour Party, *The Old World and the New Society: A Report on the Problems of War and Peace Reconstruction* (London: Labour Party, 1942).

Unlike in the 1930s, its support was now articulated separately and distinctly from its calls for an international air force, even though an important rationale for the former was still international security. As we have already seen, Labour Party members supported internationalization within government and other internationalist organizations. The most senior supporter through the war (though mostly in private discussions) was Noel-Baker, by then an MP and a member of the Labour Party's National Executive Committee. In a letter to fellow Labourite William Arnold-Forster in June 1943 he argued that a growing distrust of for-profit business provided an opportunity for centralized international planning and control of aviation. 'I think that civil aviation has become more urgent than the Atlantic Charter pamphlet ... I believe the forces in favour of an all-in policy can be overwhelmingly strong if we try to use them, and we have the advantage that the opposite side is being openly run by business interests, whose credit after the war will not stand very high with anyone'.[48] Noel-Baker's most public intervention was a presentation to the International Bureau of the Fabian Society in August 1941 (before he joined the wartime coalition government) which emphasized the importance of internationalization for the 'future of democracy'.[49]

The most vocal public supporters, although not without influence, were on the fringe of the Party. The ex-Liberal Labour MP Frank Bowles was particularly active within Parliament, and headed a World Airways Joint Committee which propagandized for a 'world communications authority' which would start with aviation but eventually include 'all international communications – railways, shipping, telegraphs, posts etc.' This Committee was funded by David Caradog's World Unity Movement, a prominent wartime internationalist organization whose members included Vera Brittain, G.E.G. Catlin, Leonard Hill, L. Hogben, Madariaga, John Boyd Orr, and Olaf Stapledon.[50] Michael Dunlop Young, Director of the think tank Political and Economic Planning (PEP) and (from 1945) head of Labour research, wrote press articles and pamphlets calling for internationalization. A 1943 study by

[48] Noel-Baker to Arnold-Forster, 1 June 1943, 3/15, Noel-Baker Papers.
[49] Meeting Minutes, International Bureau of the Fabian Society, 7 August 1941, 2/57, Noel-Baker Papers.
[50] Frank Bowles, 'M.P. Wants World Authority to Control Civil Aviation', *Evening Standard*, 16 March 1944. Also: World Airways Joint Committee, *Civil Aviation and World Unity: A Manifesto* (London: World Airways Joint Committee, 1944); F. G. Bowles and Major W. F. Vernon, *World Airways: What We Want* (London: World Airways Joint Committee, 1946). Prominent interventions in parliament were:
   Hansard, HC, vol. 390, cols. 56–63, 1 June 1943 (where he cited the Americans Ely Culbertson and E. P. Warner in support); Hansard, HC, vol. 397, cols. 1321–1383, 29 February 1944; Hansard, HC, vol. 398, cols. 164–168, 14 March 1944; Hansard, HC, vol. 403, cols. 2731–2770, 20 October 1944.

PEP suggested internationalization with a separation of roles: independent regional International Air Corporations could own and operate aircraft on major routes, whereas regulation and supervision would be carried out by an International Board for Air Navigation modelled on the League's International Commission for Air Navigation. In 1947 Young's exposition of Labour policies, *Labour's Plan for Plenty,* emphasized Labour's continued commitment to world government alongside internationalized civil aviation.[51]

The Labour Party's official policy appeared in April 1944, timed no doubt to influence planning for the upcoming air conference in Chicago. *Wings for Peace: Labour's Post-War Policy for Civil Flying* was produced by the Party's Committee on Transport, and drafted largely by William Arnold-Forster. The report suggested a radical form of internationalization as the ideal long-term solution. It demanded that a 'World Air Authority' own and operate a 'World Airways' with a monopoly on all international trunk routes. The Authority would own all civil aircraft (even those flying on domestic routes), which would then be leased back to national airlines if required. As an interim and more realistic short-term aim the committee suggested regional internationalization, including the formation of a pan-European 'Europa Airways' and a 'British Commonwealth Airways'. Unlike the more liberal PEP scheme, which envisaged states of each region owning shares in the region's independent aviation corporation, the Labour Party's Committee on Transport envisaged airlines owned and directed by regional organizations.[52]

Wings for Peace was the most prominent detailed public proposal made during the war, and press reaction to it split along similar lines to reaction in the 1930s. The *Manchester Guardian* declared it a 'bold framework' and supported regionalized internationalization, which it felt would be more acceptable to the Americans. Opponents such as the aeronautical correspondent for *The Times* retorted that it reflected only the Labour Party's 'political and economic principles'. 'Regulated competition' was better for the development of aviation, and that anyhow the proposals were unrealistic as they did not take American opposition into account. The

---

[51] Michael Dunlop Young, *Civil Aviation* (London: Pilot Press, 1944); James Avery Joyce and Michael Young, *Chicago Commentary: The Truth about the International Air Conference* (London: n.p., 1945); 'International Air Transport', in Political and Economic Planning, *Building Peace out of War: Studies in International Reconstruction* (London: Political and Economic Planning, 1944), 90–110; Michael Dunlop Young, *Labour's Plan for Plenty* (London: Victor Gollancz, 1947), 125–129, 131–135.

[52] Labour Party, *Wings for Peace.* For earlier discussions see: Report, Civil Aviation Subcommittee of the Committee on Transport, Labour Party, December 1942, 3/15, Noel-Baker Papers.

Economist's response fell in between. It commended the Labour Party for putting forward 'the ideal solution' of 'complete international control', but criticized the proposal for not taking British, American, or Soviet government objections into account. The magazine suggested an imperial approach as the most realistic: 'Freedom of the International Air, with the Commonwealth treated as a unit, and with a mutual sharing of imperial cabotage rights with the French and Dutch Empires'.[53]

Proposals which joined together international control of military and civil aviation were much less detailed than those of the Labour Party or PEP, and were pushed primarily by David Davies, and through the League of Nations Union, the Federal Union, and the Liberal Party. Davies continued to push his proposals in Liberal circles, often with some success. Prominently, at the September 1942 Liberal party conference, Davies's amendment to the Party programme calling for an international police force, internationalization of civil aviation, and a federation of English-speaking peoples was voted down by the Executive, but was carried by the Conference.[54] Liberal support continued through the war but by the time of the party's 1945 election manifesto was reduced to a call for an international force under the aegis of the UN ('to crush ruthlessly and immediately any attempt by an aggressive nation to go to war').[55] New Commonwealth support for an international air force continued through to the late 1940s, even though its own research showed declining British public interest – a Mass Observation report commissioned in 1946 showed a divided public.[56] The Society was dissolved in 1950; however, the associated think-tank,

---

[53] 'World Aviation', *Manchester Guardian*, 18 October 1944; 'Air Transport', *Manchester Guardian*, 13 March 1943; 'A World Air Authority', *The Times*, 27 April 1944; Ronald Tree et al., *Air Transport Policy: By Four Conservative Members of Parliament* (London: n.p., 1943); 'Air Transport', *Economist*, 6 March 1943, 286–288; 'Civil Aviation', *Economist*, 5 June 1943, 713–714; 'Civil Aviation', *Economist*, 1 April 1944, 424–426; 'Wings for Peace', *Economist*, 29 April 1944, 583.

[54] 'Liberal Party Programme. Reconstruction and a Social Policy', *Manchester Guardian*, 5 September 1942; Liberal Party Organization, *Final Agenda for the Meeting of the Assembly of the Liberal Party* (London: Liberal Party Organization, 1942), 6–8; Archibald Sinclair, *The People's Party* (London: Liberal Party, 1942).

[55] Liberal Party Organization, *20 Point Manifesto of the Liberal Party* (London: Liberal Party Organization, 1945).

[56] Mass Observation, 'World Organisation and the Future: An Interim Report by Mass-Observation describing peoples' hopes and expectations for the future of the world, as they were during the UNO sessions at Westminster', March 1946, folder 2370, Mass Observation Archives, Special Collections, University of Sussex Library (hereafter cited as Mass Observation Archives). Prominent post-war propaganda included: New Commonwealth Society, *World Security Force* (London: New Commonwealth Society, 1949); G. A. Rowan-Robinson, *An International General Staff* (London: New Commonwealth Society, 1948).

the David Davies Institute, continued to foster interest in a United Nations force, including a highly publicized 1964 study.[57]

The newly formed Federal Union, although not well disposed towards internationalization, considered embracing it in an attempt to build bridges with the New Commonwealth.[58] A 1942 approach to the New Commonwealth announced that 'inter-state transport, particularly shipping and civil aviation, should remain under international control at the end of the war and not be allowed to return to the anarchy of competitive private enterprise'. The 'super-abundance of long-range bombers with great carrying capacity' would 'take care of the first steps to reconstruction. This might well be the foundation of "World Airways"'.[59] Cecil and the League of Nations Union also continued to advocate the formation of an international air force, the abolition of national forces, and the internationalization of civil aviation. The Union, a shell of its former self, built these notions into its proposals for the post-war reconstruction and retained its commitment to an international air force after its reformation into the United Nations Association in early 1945.[60] United Nations Associations abroad also remained interested in the formation of an international police force – Pierre Cot for example introduced a resolution calling for one at the 1947 congress of the World Federation of United Nations Associations.[61]

Critics were not as unequivocal as one might expect, with many supporting some form of regional agglomeration. The threat of internationalization was taken seriously enough for four Conservative MPs to band together in 1943 and form a committee to lobby against it in Parliament. Their manifesto reasoned that internationalization was not required for

---

[57] D. W. Bowett, *United Nations Forces: A Legal Study of United Nations Practice* (London: Stevens and Sons, 1964).

[58] 'Point and Counterpoint', *Federal News* 111 (May 1944): 4. On the Federal Union see: Richard Mayne, John Pinder, and John C. de V. Roberts, *Federal Union: The Pioneers. A History of Federal Union* (Houndmills: Palgrave Macmillan, 1990); Bosco, 'Lothian, Curtis, Kimber and the Federal Union Movement (1938–40)'.

[59] Konni Zilliacus and N.V (for Federal Union) to New Commonwealth Society, 23 June 1942, folder: 'NC Memoranda 1943', box 92, Davies Papers. Federal Union continued to call for an international force well into the 1950s, e.g. during the Suez Crisis: *Proposals for a Permanent United Nations Force* (London: Federal Union, 1957; 1960).

[60] 'League of Nations Union. International Air Force', *Manchester Guardian*, 28 June 1941; League of Nations Union, *Annual Report for 1944* (London: League of Nations Union, 1945); Robert Cecil, *A Real Peace* (London: Hamish Hamilton, 1941); Cecil, 'Memorandum on Post-War Organisation Against Aggression', LNU paper S.707, 20 February 1940, folder: 'NC Memoranda, 1940', box 92, Davies Papers; Executive Committee, League of Nations Union, 'Draft Pact for the Future International Authority', *International Conciliation* 22 (February 1944): 131–139.

[61] 'An International Police Force. Federation's Appeal to U.N. Members', *The Times*, 22 August 1947.

'air security', and was not 'practical politics' as it would not be accepted by the House of Commons, let alone the United States or the USSR.[62] The head of this group (and Vice-President of the British Airline Pilots Association), Ronald Tree, called instead for a scheme similar to Amery's, with global aviation carved up into regions, each under the direct control of a separate Great Power.[63]

The airframe industry trod a fine line between its distrust of internationalization and its fear of US competition. Handley Page's critical review of Mance's proposals in *International Affairs* maintained that strict international licensing of airlines would be sufficient to prevent misuse of civilian aviation after the war. As 'it is impossible to separate the manufacture of civil aircraft from that of military aircraft', Axis' aerial industry would need to be 'under drastic control' after the war. However, this did not portend internationalization for the future: convertibility was 'difficult to estimate' and so it was 'unwise to be dogmatic'. Airliners 'may follow the ship and become more and more distinct; or, conceivably, the invention of some new kind of weapon which did not need special gun-turrets and bomb-bays might make them interchangeable merely by the installation or removal of special war equipment'.[64] The Society of British Aircraft Constructors, recognizing no doubt the benefits for British-made aircraft, did not oppose internationalization, but demanded as much 'freedom of operation' as possible. This included the demand that Britain continue research and development in transport aircraft, that ground facilities remain under national control, that Britain abandon its 'single chosen instrument' policy, and that 'British-designed and British-built transport aeroplanes must be used on World air routes'.[65] Others in manufacturing called for immediate bilateral negotiations with the United States as an alternative way of staving off post-war competition.[66]

Globe-Girdling Airways

The future of international commercial aviation was considered to be important also in the United States – in April 1945 newly appointed President Harry S. Truman would tell Henry Wallace (then Secretary of Commerce) that, along with reparations, international civil aviation

---

[62] Tree, *Air Transport Policy*.   [63] Hansard, HC, vol. 390, cols. 51–52, 1 June 1943.

[64] F. Handley Page, 'International Air Transport. By Sir Osborne Mance, assisted by J.E. Wheeler', *International Affairs* 20, no. 2 (April 1944): 281–282.

[65] 'The Aircraft Industry on Air Transport', *Aeroplane*, 2 July 1943, 4–5.

[66] 'Air Transport: A Statement by the Joint Air Transport Committee', *Flight*, 27 May 1943, 560.

was 'the most important postwar international problem'.[67] International American airlines, most importantly Pan American, were already positioning themselves for a massive expansion overseas, including into regions where they had no or weak services. This included Africa, the Middle East, and Europe. Pan American had pushed aggressively into Asia in the few years prior to the war: Hawaii, Manila, Singapore, and Hong Kong were connected by 1941, and both the airline and the government expected this network to expand greatly at the end of the war. The government was committed to supporting the growth of US international civil aviation, seeing in this expansion both commercial and military benefit as well as an extension of the country's political and cultural influence.[68]

This opportunity was presented in liberal, internationalist, and integrative terms. In his vision of 'The Next Ten Years in Air Transportation' presented at a meeting of the Academy of Political Science in April 1944, L. Welch Pogue (Chairman of the Civil Aeronautics Board) announced that:

a network of world airways will have a tendency to produce throughout the world unifying political and economic influences similar to those which the railroads produced in the development of the United States .... Air transportation can become the political, economic and social instrument of an internationalism with new power and promise. Used by a people freed from provincial concepts of the world, the free growth of air transportation holds forth new promises in the development of world civilization.[69]

This view of aviation resonated across the United States. Leading Republican Wendell Willkie's 1942 round-the-world trip appeared to portend a future in which Americans could not only travel seamlessly from airport to airport around the world, but also look down and comprehend the world in its global interconnected entirety like never before.[70] On his return his speech at a *New York Herald Tribune* forum on 'Economic Freedom for the World' presented modern civil aviation as part of the broader 'vibrant forces of modern science and industry' that were 'awaiting the chance to break forth into ever-widening streams of well-being for all mankind'.[71] By the end of the war the United States was awash with a new 'air-age' geography which emphasized the integrity and interconnectedness of the world, especially when approached through

[67] Diary entry, 27 April 1945, in *The Price of Vision: The Diary of Henry A. Wallace, 1942–1946*, ed. John Morton Blum (Boston: Houghton Mifflin, 1973), 437.
[68] For more on this see: Van Vleck, *Empire of the Air*, chapters 3 and 4.
[69] L. Welch Pogue, 'The Next Ten Years in Air Transportation', *Proceedings of the Academy of Political Science* 21, no. 2 (January 1945): 17–27.
[70] Wendell Willkie, *One World* (New York: Limited Editions Club, 1944). Also: Van Vleck, *Empire of the Air*, 103–118.
[71] Wendell Willkie, 'Economic Freedom for the World', in *Representative American Speeches: 1942–1943*, ed. A. Craig Baird (New York: HW Wilson), 105–115.

aviation. The Museum of Modern Art put on an exhibition called *Airways to Peace: An Exhibition of Geography for the Future* that displayed various different globes and maps in order to show that 'over our heads the airways have woven a web of intimacy, a new scene of mutual advantages, a world-brotherhood'. Publisher Henry Luce's well-known announcement to seize the opportunities and responsibilities of the 'American Century' included a call to 'make a truly American internationalism something as natural to us in our time as the airplane or the radio'. This 'logic of the air' propelled US aerial expansion.[72]

The aircraft industry demanded that this logic be pursued. Eugene Wilson, President of the Aeronautical Chamber of Commerce of America and President of United Aircraft Corporation, emphasized the need for unimpeded aerial commercial development, and the economic and social benefits this would bring to the United States and the world.[73] This was the theme too of a Public Affairs pamphlet authored by Waldemar Kaempffert, science editor for the *New York Times*, titled *The Airplane and Tomorrow's World*.[74] The country's 'Empire of the Air' was to be a liberal internationalist empire. Although this propaganda depicted the aeroplane not as an instrument of war but of peace, the aviation industry nevertheless felt the wartime public needed to be reminded of this. In *Flying* in 1943 the President of Consolidated Vultee Company called for increased government 'education of the people' so that they 'realize that the airplane is an instrument of peace, commerce and international understanding ... We need promotional and advertising campaigns to point out that the airplane must be used in pursuing the interests of peace'.[75]

Vice-President Wallace stood out as one of the few to publicly call for the international control of civil aviation, which he married to his calls for an international air force. His most detailed proposal reached the public through a February 1943 article in *American Magazine* which called for

---

[72] Samuel Zipp, 'Dilemmas of World-Wide Thinking: Popular Geographies and the Problem of Empire in Wendell Willkie's Search for One World', *Modern American History* 1, no. 3 (2018): 295–319; Henry R. Luce, 'The American Century', *Life*, 17 February 1941, 61–65; Van Vleck, 'The Logic of the Air'. On Luce's notions of modernization and technology see: Donald W. White, 'The "American Century" in World History', *Journal of World History* 3, no. 1 (Spring 1992): 105–127.

[73] See for example: Eugene P. Wilson, 'Fundamentals of Freedom: Initiative and Enterprise Protect Real Security', Ibiblio, 28 May 1942, accessed 2 August 2020, www.ibiblio.org/pha/policy/1942/1942-05-28b.html. See also: Eugene P. Wilson, *Air Power for Peace* (New York: McGraw-Hill, 1945).

[74] Waldemar Kaempffert, *The Airplane and Tomorrow's World* (New York: Public Affairs Committee, 1943).

[75] Harry Woodhead, 'A Fighting Plan for Peacetime Production ... .', *Flying*, November 1943, 24–25, 85–96.

a 'network of globe-girdling airways' controlled by the United Nations which would serve as an international air police in times of war. He also called for the internationalization of airports, the establishment of a United Nations air authority, and the carving up of the air into spheres of influence.[76] Within a broader 'new deal' of international prosperity and justice 'airways' were to be interconnected together, allowing 'improved transportation' to 'unlock the resources of the vast undeveloped regions of the world'.[77] He even suggested a job-creating construction programme for a series of international airports around the world for the use of an international air force, 'in somewhat the same way as the New York port authority or the TVA'.[78] Aviation, modern communications and transport, industry, technical skill, and planned organization were aspects of 'modern science' and 'technology' which had to be turned to the welfare of the 'common man'.[79]

Wallace's proposals had some wider impact in 1943, when they won a smattering of support within the Democratic Party, and led some internationalists to call for the UN to regulate key international airports. These ideas, however, largely disappeared from public view once Wallace was dropped as a potential running mate for Roosevelt in 1944.[80] Yet they were prominent for a while, during which time they were attacked by opponents. For one journalist writing in the *Reader's Digest* internationalist New Dealism was 'boondoggling on a global basis', an 'exaggerated internationalism' that was 'as dangerous, foolhardy, and destructive as narrow isolationism'.[81] Most famously Republican Congresswoman Clare Boothe Luce (and husband of conservative magazine magnate Henry Luce) coined the word 'globaloney' to describe Wallace's 'global thinking' in her first speech before the House of Representatives in February 1943.[82] Wallace in the *New York Times* responded that those

---

[76] Henry Wallace, 'What We Will Get out of the War', *American Magazine* 85, no. 3 (February 1943): 98.

[77] Henry Wallace, 'Business Measures', in *Democracy Reborn* (New York: Reynal & Hitchcock, 1944), 208–214.

[78] 'Wallace Tells New Ideas for Dream World', *Chicago Tribune*, 1 January 1943; Wallace, 'Business Measures'.

[79] Henry Wallace, *The Century of the Common Man* (New York: Reynal & Hitchcock, 1943); Henry Wallace, 'Industrialization of the World', *Vital Speeches of the Day* 11, no. 19 (15 July 1945): 603–605.

[80] Internationalist Democrats tried (and failed) to insert the formation of an international air force into the election manifesto at the July 1944 Democratic Convention; see: Divine, *Second Chance*, 212. Also: Commission to Study the Organization of Peace, *For This We Fight: Program No. 5 Making the World Secure*, 6.

[81] Henry J. Taylor, 'Boondoggling on a Global Basis', *Reader's Digest*, August 1943, 1–6.

[82] 78 Cong. Rec. H759-764 (9 February 1943). See: (New York: Harcourt, Brace, 1944), 3–19; Van Vleck, *Empire of the Air*, 103–118.

who wanted to spread the tentacles of the American air force around the world, the self-styled 'realists', were in fact 'new American imperialists' prepared to 'seize and hold every island in the Pacific' for 'domination in the air'.[83]

Freedom of the Air

There was no appetite for internationalization amongst mainstream US internationalist organizations, though there were calls for a new international regulatory regime.[84] The Carnegie Endowment's call for the amalgamation and 'strict government supervision' of international US airlines was as far as the foundation was willing to go. Eichelberger reassured readers in his August 1944 *Time Has Come for Action* that international civil aviation policy would remain firmly in the hands of the government. His suggested regulatory 'international air authority' would not 'compel the nations to do anything' and would not be incorporated into the General International Organization.[85] The respondents to the Universities Committee's 1944 poll voted unanimously against internationalization, and instead supported weaker international regulation. 'If only it is given reasonable freedom, air transportation will develop so rapidly as to revolutionize living in the coming era much as rail transportation has done in the past.'[86] The CSOP omitted any discussion of civil aviation in its November 1943 report, and its section on 'Economic and Social Collaboration' instead focused largely on existing and recently formed international organizations such as the International Labour Organization, the United Nations Organization for Food and Agriculture, and the United Nations Relief and Rehabilitation Administration.[87]

Members of the War and Peace Studies group of the CFR also showed little enthusiasm. For most discussants the only possible argument in

[83] Henry Wallace, 'Freedom of the Air – A Momentous Issue', *New York Times*, 27 June 1943.

[84] Some internationalists with strong connections to Europe continued to support internationalization through the war: Randall, *Improper Bostonian*, 373–374; Henri Bonnet, *The United Nations: What They Are and What They May Become* (Chicago: World Citizens Association, 1942), 74; Henri Bonnet, *Outlines of the Future: World Organizations Emerging from the War* (Chicago: World Citizens Association, 1943), 39–55.

[85] Eichelberger, *Time Has Come for Action*, 12, 22; Arthur E. Traxler, 'International Air Transport Policy of the United States', *International Conciliation* 21 (December 1943): 616–636.

[86] Universities Committee on Post-War International Problems, 'Summaries of Reports of Cooperating Groups: Problem XIII-International Air Traffic After the War'.

[87] Commission to Study the Organization of Peace, 'Fourth Report of the Commission to Study the Organization of Peace', particularly 51 to 67.

support of internationalization was that it could prevent the misuse of civilian aircraft for military purposes. This argument, however, carried little weight due to recent technical progress. Widespread support in Europe in 1932 was explained away as a response to the dangers of aviation due to convertibility.[88] Now, CFR internationalists argued, the world needed open access to international routes and competition amongst international airlines. The Armaments Group of the Council recommended an international organization with basic aerial regulatory powers. The most radical proposals made in the Council, and certainly far ahead of what Assistant Secretary of State Adolf A. Berle, Jr. and others in the State Department were at this time considering, was that the proposed international organization could allocate international air routes and set international fares.[89]

There were nevertheless a few internationalists willing to contemplate internationalization in Europe. In their 1940 textbook *Contemporary International Politics* political scientists Grayson Kirk and Walter Sharp noted in passing that it would be required as part of the post-war political consolidation of Europe.[90] In 1942 Quincy Wright, in a research paper circulated to the CSOP and published soon thereafter, imagined an interim period after the war, before the formation of a full-blown global United Nations Organization, in which a 'European union' would be required for post-war reconstruction. It would establish a pan-European air force, eliminate all military air forces, and control 'railroads and waterways, air communication and telegraph systems'. A limited internationalization would operate at the global level through 'some sort of World Assembly, Council and Secretariat' which would regulate commerce, radio, and 'intercontinental air communication'. They could also control major military sea vessels and large seaports.[91]

Although internationalists generally believed that civilian aircraft could not be converted to fighting aircraft, commercial facilities were still deemed useful for military purposes. In a January 1942 *Foreign Affairs*

---

[88] 'Armaments Series: Memorandum of Discussions. Thirty-Seventh Meeting', Studies of American Interests in the War and the Peace: Council on Foreign Relations, 12 July 1943, W-37-A-A37, fiche 368, CFR Records; G. Kirk, 'International Policing (A Survey of Recent Proposals)', 3 October 1941, W-83-A-B30, fiche 376, CFR Records; William M. Franklin, *Areas of Agreement in the Preparatory Disarmament Commission and the General Disarmament Conference* (New York: Council on Foreign Relations, 1940), 4. Also: Philip E. Mosley, *Alternatives to Absolute National Sovereignty of the Airspace* (New York: Council on Foreign Relations, June 1940).

[89] Armaments Group of the Council on Foreign Relations, 'Postwar American Policy in Relation to Civil Air Transport', July 1943, W-146-A-B93, fiche 385, CFR Records, page 1.

[90] Sharp and Kirk, *Contemporary International Politics*, 169, 761.

[91] Quincy Wright, 'Political Conditions of the Period of Transition', *International Conciliation* 21 (April 1942): 264–278.

article Kirk listed a number of ways in which 'civilian air lines perform an important military function', including gathering data on weather conditions and developing pilot experience.[92] *The Nation*'s financial editor, Keith Hutchison, in his 1944 Public Affairs pamphlet *Freedom of the Air*, expounded the commonly held view that the war had accelerated the 'differences between military and civil types.' Although 'bombers can be converted into transport planes … The reverse process, converting transport planes into bombers, is less satisfactory'.[93] The consensus amongst the respondents to the 1944 Universities Committee on Post-War International Problems polls was also that 'because of increasing functional specialization of aircraft design, planes for civil air traffic cannot be directly converted to military use', but 'aircraft production facilities, aircraft mechanics and air crews can be'.[94] A June 1940 memorandum by E.P. Warner for the CFR noted that commercial aircraft could be converted to military transports, but argued that internationalization was not the solution. Instead he suggested a limit on the number of commercial aircraft allowed for each country: 'it is hardly likely that any country will need more than 1,500 transports of all types … within the next ten years'. He also suggested a restriction on the speed and range of commercial aircraft, though noted that this may be harder to police.[95]

The 3 July 1943 CSOP-organized talk on NBC network radio, *Making the World Secure*, centred on an international police force and included a debate on the extent to which civil aviation would need to be internationally controlled. Eichelberger intimated that there existed some type of significant military potential within civil aviation which would require far greater international civil aviation regulation than under the League. Eagleton suggested that an international air force would only work if the international police also controlled civil aviation (to prevent it being used for military means). Senator Joseph H. Ball responded that a transport plane could not be converted into a 'modern bomber', which was a 'very specialized weapon'. International regulation of commercial aviation was needed but could be worked out through reciprocal arrangements between individual nations.[96]

These internationalist discussions indicate that the convertibility of airliners to bombers continued to exist in the popular imagination.

---

[92] Grayson Kirk, 'Wings over the Pacific', *Foreign Affairs* 20, no. 2 (January 1942): 293–302.

[93] Keith Hutchison, *Freedom of the Air* (New York: Public Affairs Committee, 1944), 13.

[94] Universities Committee on Post-War International Problems, 'Summaries of Reports of Cooperating Groups: Problem XIII-International Air Traffic After the War'.

[95] Edward P. Warner, *Possibilities of Controlling or Limiting Aircraft Suitable for Offense against Ground Objectives* (New York: Council on Foreign Relations, 1940).

[96] Commission to Study the Organization of Peace, *For This We Fight: Program No. 5 Making the World Secure*, 9–10.

Certainly De Seversky's *Victory Through Air Power* noted that convertibility existed as a 'popular notion'. This was, however, a 'constant source of confusion in the public mind'. Civil aircraft, he pointed out, were useful only for transport, though he emphasized the importance of 'commercial-aviation facilities' for military aviation, and argued that they 'must be scientifically meshed into the military-aeronautical structure'. Some experts may have believed in convertibility in the early years of the war: a 1940 *Foreign Affairs* article by legal scholar Oliver Lissitzyn on 'The Diplomacy of Air Transport', whilst emphasizing the importance of commercial aviation as a 'reservoir of equipment and personnel for military aviation', also claimed that 'modern airliners' could be converted into 'fairly efficient bombers'. E.P. Warner's June 1940 CFR memo claimed that some commercial transports could be converted into bombers by replacing as little as 30 percent of their structures.[97]

Within the administration Berle, Pogue, and indeed most others were committed, albeit in varying degrees, to 'Freedom of the Air', a nebulous concept which emerged as a rallying call for those opposing strong international regulation. The most common understanding was perhaps that expressed by Hutchison in *Freedom of the Air*: that airlines, private or nationalized, be allowed to travel and carry loads freely between international destinations. Proponents of freedom of the air, such as Hutchison, argued that the current aerial reality was that of a 'closed sky' in which national sovereign boundaries extended upwards into the air. These artificial and harmful barriers could, Hutchison conceded, be dismantled through complete internationalization along the lines suggested by David Davies and Henry Wallace. This idealistic solution was, however, unrealistic. It would not be accepted by states keen to uphold their sovereignty, nor by US and British 'military and aviation circles' who 'seem decidedly hostile to such ideas'. He instead called for a 'wider exchange of air rights' – a system of bilateral or multilateral agreements.[98] Most in government agreed with this suggested policy. Berle would note in a memo to Welles and Wallace in March 1943:

I am pretty clear that neither Britain nor ourselves would agree to pooling air interests so that air transport should be run by a single internationalized corporation. In many ways such an arrangement is logical, but I think it is a good way

---

[97] De Seversky, *Victory Through Air Power*, 295–6; Oliver J. Lissitzyn, 'The Diplomacy of Air Transport', *Foreign Affairs* 19, no. 1 (October 1940): 156–170; Warner, *Possibilities of Controlling or Limiting Aircraft*.

[98] Hutchison, *Freedom of the Air*. On Freedom of the Air see: Van Vleck, *Empire of the Air*, 169–174.

off... I myself doubt whether any of the countries most concerned are yet ready to internationalize their civilian air establishment.[99]

The CFR's most detailed study on international civil aviation, Lissitzyn's *International Air Transport and National Policy* (published 1942), also supported 'freedom of the air'. It was, the legal scholar argued, the natural and ideal state of international commercial transport, though now distorted by state meddling. Once instituted it would increase competition between airlines, leading to the demise of inefficient government-subsidized airlines and allowing the more efficient ones to flourish. Although internationalization could lead to this freedom, it was not achievable because of entrenched national interests, and anyhow required the 'internationalization of military and political power' to work. Lissitzyn attacked the 1932 French proposals for internationalization, and, citing Spaight, asserted that an international police force would not function unless 'a real world government' was first established.[100] Notwithstanding the ultimate goal of freedom of the air, Lissitzyn's book reflected the general sense in the CFR, and amongst internationalists, that commercial aviation was to play a central role in 'the future of world order'. Civil aviation would bring the nations of the world together leading to increasing political integration and, in response, the possibility of increased conflict. The United States needed to shape this process by consolidating its aerial hold on the 'western hemisphere' (North and South America) by developing its aerial network there and keeping other aerial interests out.[101]

The airline industry was unsurprisingly dead set against internationalization. In July 1943 heads of fifteen domestic airlines signed a declaration calling for them to be allowed to participate in open competition on foreign routes. Pan American and its supporters, meanwhile, called for the airline to be designated America's sole international flyer.[102] The aircraft industry was also keen that the United States acquire foreign bases for exclusive use of its military and civil aviation – a four-point programme issued by the Aeronautical Chamber of Commerce in mid-1944, for example, placed the acquisition of foreign air bases as the second point.[103] Aviation entrepreneurs and executives such as the First World War ace Eddie Rickenbacker, owner of Eastern Air Lines, argued against internationalization of

---

[99] Memorandum, A. A. Berle, Jr. to Sumner Welles, 2 March 1943, 800.796/258, Political Affairs – World Aviation 1940–1944, RG 59, NARA.

[100] Oliver J. Lissitzyn, *International Air Transport and National Policy* (New York: Council on Foreign Relations, 1942), 420–422, 419. Also: Lissitzyn, 'The Diplomacy of Air Transport'.

[101] Lissitzyn, *International Air Transport and National Policy*, 418, 420–422.

[102] Blair Bolles, 'The Future of International Airways', *Harper's Magazine*, January 1944, 97–106.

[103] E. E. Wilson, 'A Four-Point Program', *Flying*, August 1944, 46–47, 92, 95, 98, 101.

international US military air bases or their return to their previous owners. He called for existing base lease agreements to be renegotiated so that the bases could be utilized for the expansion of American commercial aviation overseas.[104]

Nevertheless some in the aviation industry saw benefit in calling for the creation of internationalized airports which would be open to all airlines. One aviation executive suggested 'free ports to be established throughout the world' which 'would be perhaps five miles square, containing hotels, restaurants, and various comforts for passengers changing from one airplane to another in the course of their journeys'.[105] The political scientist and Senator Elbert D. Thomas went further. Opposing internationalization, he called instead for the United Nations to establish 'free-air cities' or 'free-air islands' where states would continue to exercise full sovereignty on land but give it up in the air. Every country 'will want to have a free city or two, according to the size of the nation, for entrance of planes of all powers'. He extended this vision to waterways such as the Danube and the Rhine, which were 'already internationalized' and over which the air could be internationalized too. To this he added imperial possessions: 'the air over all the mandated territories ... should be free for United Nations planes, if not the planes of the world'.[106]

Although internationalization had little support, internationalists nevertheless feared that it might hold some popular appeal, especially following growing British interest and Wallace's interventions in support of it in 1943.[107] Warner launched an attack on internationalization in a 1943 article in *Foreign Affairs*. 'Airways for Peace' pointed out that British supporters of internationalization feared the detrimental effects of 'cut-throat competition' and a 'subsidy race'. These fears could be allayed by the creation of an 'international air commission of the judicial type' which would have 'powers over international service analogous to those exercised by the Civil Aeronautics Board in the United States'. Concerns that civil aviation could be misused for military purposes was misplaced, he argued. Although pilots are 'readily interchangeable', 'both types are too specialised for it to be

[104] 'Save Post-War Air Rights for America: Ricks', *Chicago Tribune*, 23 March 1944.
[105] Harold Evans Hartney, 'The Case for a Separate Air Force', *Flying and Popular Aviation*, September 1942, 38–42.
[106] Elbert D. Thomas, 'The Air Must be Free', *Flying* (May 1944, 51, 153). He also called for an international police force: Elbert D. Thomas, *The Four Fears* (Chicago: Ziff Davis, 1944), 153, 166.
[107] Prominently, the Governor of Minnesota, Harold E. Stassen, put the regulation of international aviation second in his seven-point plan for a United Nations government in 1943, with one 'United Nations Councilman' to run each of the seven areas. The seventh, as noted in Chapter 4, was an international police force. Harold E. Stassen, *A Proposal of a Definite United Nations Government* (St. Paul, MN: Foreign Policy Association, 1943).

possible to transform a transport fleet overnight into an efficient bombing unit'.[108] In an address to the American Bar Association in 1944 Warner suggested that the 'international commission' would only regulate 'technical' aspects such as runways; 'economic' matters such as carriage and landing rights should be left to bilateral agreement. International regulation of both was 'beyond the margins of present feasibility'.[109] Warner was, however, content to suggest internationalization where American interests were not directly involved. A 1943 article in *Foreign Affairs* on post-war German aerial disarmament suggested that internationalization would have 'obvious practical advantages' in Europe. This would allow for the international control of German civil aviation to prevent its military misuse (though he acknowledged that this was not a major concern because Germany would have no post-war aviation) and avoid 'wasteful international duplications of service and the heightening of commercial rivalries' in a continent with many sovereign nations 'in a total area barely half that of the United States'.[110]

The most protracted liberal critique of internationalization came from Parker Van Zandt, aviation entrepreneur, Civil Aeronautics Board consultant, and director of aviation research at the Brookings Institute. His 1944 *Civil Aviation and Peace* was a sustained and structured attack on strong international aerial regulation, especially internationalization proposals such as Mance's and the Labour Party's. His chief argument was that internationalization, a system of 'quotas, restraints, limitations, and prevention', would restrict the natural growth of aviation just when the world would need the economic stimulus that air travel could provide. Moreover, the military usefulness of civil aviation was 'grossly exaggerated'. Controlling world air routes had little bearing on international peace, and, of course, civilian aircraft were no longer convertible: 'as military science has progressed, the technical distinction between tactical and civil types has grown more marked'. The failed 1932 disarmament conference had demonstrated that using the internationalization of civil aviation to control military aviation would only work if internationalization was applied widely and deeply, which most states would not agree to. Delegates at the conference had realized, Van Zandt claimed, that internationalization would not produce the 'most efficient operating method' for international aviation – that 'such a huge monopoly would be inefficient and uneconomic'.[111]

---

[108] Edward P. Warner, 'Airways for Peace', *Foreign Affairs* 22, no. 1 (October 1943): 11–27.

[109] Edward P. Warner, 'Postwar Aviation', in *Proceedings of the Section of International and Comparative Law* (Chicago: American Bar Association, September 1944), 51–57.

[110] Edward P. Warner, 'Future Controls Over German Aviation', *Foreign Affairs* 21, no. 3 (April 1943): 427–439.

[111] J. Parker Van Zandt, *Civil Aviation and Peace* (Washington, DC: Brookings Institution, 1944), 26–28, 52, 62, 48, 93.

For Van Zandt these obstacles to the creation of a realistic scheme were as real now as they had been in the 1930s. Current schemes for international-ization, having descended from these earlier proposals, also fell 'into the old familiar pitfalls and repeat the same errors' leading 'straight back down the same blind alley travelled by the Disarmament Conference'. He pointed out that those countries calling for international control today were clearly attempting to restrict competition in order to assure themselves of a larger share of international aviation. Internationalization would anyhow be 'cat-egorically' rejected in the United States as it was inherently against American values: it involved 'an intolerable infringement of the personal liberty of its citizens. The dictation and overseeing inevitably involved are completely foreign to the American way of life'. Instead Van Zandt called for a multilateral agreement in which all airlines would have the right to transit across others' airspace, but not the right to pick up and drop cargo and passengers, which would then be negotiated on a case-by-case basis. This involved the least regulation and would, he claimed, allow for the fullest possible unhindered spread of international civil aviation, and so the spread of global peace and prosperity.[112]

Debates over post-war aviation were prominent enough to catch the attention of the Left. Socialist Irving Howe saw the battle for post-war civil aviation as a struggle amongst imperialist members of the United Nations. In April 1943 he wrote in the New York-based Trotskyite magazine *The New International* that the current thinking about post-war aviation was driven by the realization that 'success in obtaining world air bases, routes and trade, with the resultant need for continued transport production and the maintenance of a giant air industry, is a pleasing prospect for the imperialist planners'. And so 'The only signifi-cance which all the talk about "freedom of the skies" and "closed and open ports" and all the other aviation jargon can possibly have, therefore, is in light of the mounting rivalry between American and British imperi-alism'. Both Clare Luce's calls for 'national sovereignty' in the air and Wallace's for 'freedom of the skies' would help the country's aerial domination over the world. Aviation's potential could only be wasted under capitalism which can 'plan for it only the role of a mechanism for the furthering of post-war imperialist struggles'. It was only socialism which would be able to fully utilize aviation for the benefit of mankind.[113]

There was almost unanimous agreement within government that American military interests would be served by eschewing

---

[112] Ibid., 49, 52, 94–97.
[113] R. Fahan [Irving Howe], 'Struggle for Air Supremacy', *The New International* 9, no. 4 (April 1943): 103–106. On Irving Howe writing as Fahan see: Edward Alexander, *Irving Howe: Socialist, Critic, Jew* (Bloomington: Indiana University Press, 1998), 9.

internationalization. In January 1943 Berle, a strong proponent of American international aerial expansion, formed an Interdepartmental Committee on International Aviation, under his own chairmanship, to advise on aerial policy.[114] Berle and the committee's working assumption was that the effective global projection of US military power required a substantive US-controlled international civil aviation network. In a letter to Welles in March 1943 Berle explained that 'military air routes cannot be continuously effective unless they are continuously in use'. Civil aviation needed to 'keep "live" the air fields, supply stations, weather and wireless installations' which were 'essential to effective military use'.[115] That month Berle's committee's working subcommittee, chaired by Pogue, produced its first report on the nature of post-war international air transport. Rather than internationalization, it recommended that the United States bilaterally obtain as many landing and passage rights as possible from other nations in order to expedite the expansion of post-war American commercial aviation.[116] The JSSC considered this report on behalf of the JCS, and were positively disposed towards it. The JSSC/JCS agreed with the State Department in its contention that foreign air bases had dual military and commercial aviation value for the United States; though they emphasized the priority of military claims. The nation, they informed Berle, should secure international bases for its own military use under any agreement relating to international commercial aviation.[117]

The only aspect of internationalization in the subcommittee's proposals was the suggestion that a United Nations Airport Authority manage internationalized airports around the world. The Authority was initially to be under the command of the Combined Chiefs of Staff with a later transfer to international civilian control. Such internationalized airports, noted the Subcommittee, would also be useful for any future international air force, though the 'airports will continue to be of the utmost military importance whether or not an international air force is created'.[118] The JCS shot this proposal down immediately: they felt that

---

[114] Adolph A. Berle, Jr., diary entry, 2 January 1943 in, *Navigating the Rapids, 1918–1971*, eds. Beatrice Bishop Berle and Travis Beal Jacobs (New York: Harcourt Brace Jovanovich, 1973), 481.

[115] Memorandum, A. A. Berle, Jr. to S. Welles, 2 March 1943, 800.796/258, Political Affairs – World Aviation 1940–1944, RG 59, NARA.

[116] Interdepartmental Subcommittee on International Aviation, 'Preliminary Report as Adapted 1 March 1943', 1 March 1943, box 269, Central Decimal Files 1942–1945, RG 218, NARA.

[117] W. D. Leahy (on behalf of the JCS) to A. A. Berle, Jr., 16 March 1943, box 269, Central Decimal Files 1942–1945, RG 218, NARA.

[118] Interdepartmental Subcommittee on International Aviation, 'Preliminary Report on a United Nations Airport Authority', 11 June 1943, box 525, Central Decimal Files 1942–1945, RG 218, NARA.

they could gather all required airfields under the military's direct control without recourse to an international authority.[119]

The 'Splendid Dream' at Chicago

Internationalization remained a powerful enough idea to make an appearance at the November 1944 International Civil Aviation Conference in Chicago. Although there had been various meetings on post-war aviation in 1944, particularly during Berle's visit to London in April and Commonwealth discussions in Montreal in October, it was in Chicago that Britain, the United States and fifty other countries finalized the shape of post-war civil aviation.[120] A range of arrangements for a multilateral agreement were put forward; the most striking were those for complete internationalization advanced by the Australian and the New Zealand delegations. An extension of a January 1944 agreement in which the two countries had agreed to service the major air routes connecting them by an 'international air transport authority',[121] their proposal called for an international 'air transport authority' to operate air services on major trunk routes using its own aircraft and equipment.[122] The British delegation arrived with a proposal for limited internationalization consisting of an international organization which would allocate routes and set frequencies and tariffs. They quickly abandoned internationalization in favour of the Canadian proposal which agreed to four of the five freedoms demanded by the United States, but added a strong international regulatory authority. The US delegation insisted on an open skies policy from the very start of the conference. This led to deadlock and an eventual water-downed convention, though a permanent aviation organization was also formed. The lack of agreement on transit rights allowed the US government, over the coming years, to agree separate

---

[119] Letter, J. R. Deane for the JCS to A. A. Berle, Jr., 30 July 1943, box 525, Central Decimal Files 1942–1945, RG 218, NARA.

[120] Alan P. Dobson, 'The Other Air Battle: The American Pursuit of Post-War Civil Aviation Rights', *The Historical Journal* 28, no. 2 (June 1985): 429–439; David MacKenzie, 'Australia and Canada in the World of International Commercial Aviation', in *Parties Long Estranged: Canada and Australia in the Twentieth Century*, eds. Margaret MacMillan and Francine McKenzie (Vancouver: University of British Columbia Press, 2003), 99–123.

[121] *Agreement between His Majesty's Government in the Commonwealth of Australia and His Majesty's Government in New Zealand*, Cmnd. 6513 (1944), 4.

[122] 'Committee I: Multilateral Aviation Convention and International Aeronautical Body. Verbatim Minutes of Plenary Session, November 8', in Department of State, *Proceedings of the International Civil Aviation Conference*, vol. 1, part 2 (Washington, DC: Government Printing Office, 1948), document 117, page 540, accessed 2 August 2020, www.icao.int/ChicagoConference/Pages/proceed.aspx.

bilateral transit agreements with other states, setting the pattern for future international civil aviation.[123]

Internationalist rhetoric abounded at the conference. The staunchest statements were made by the New Zealand and Australian delegations, but even the US delegates referred to the integrative effects of the aeroplane.[124] Berle's opening speech, which laid out the American position, invoked the familiar trope of the transformative effects, both positive and negative, of modern science-based aviation on international relations. Although an outdated internationalist 'doctrine' from an earlier era of H.G. Wells, Kipling, and the Victorian poet Alfred Tennyson, internationalization nevertheless had some merit, a 'splendid dream'. It was, however, unachievable in the near future. International organization needed to be 'primarily consultative, fact-gathering, and fact-finding'.[125] The lack of an outright dismissal of internationalization (which would be repeated in a *Harper's Magazine* article the following March) did not endear Berle to nationalist opinion during the conference.[126] The *Chicago Tribune* joined a chorus of Republican and military criticism against him, fearing that he might 'compromise' or 'sell out US air rights' by allowing the creation of a powerful international regulatory regime. Critics took particular exception to the granting of reciprocal landing and other commercial rights to foreign airlines, a 'new deal position' which would jeopardize 'America's air future' and possibly even 'give Great Britain control of the skies after the war'.[127]

Generally, however, US policymakers and commentators welcomed the outcome of the conference. Aviation was now free to fulfil its destiny as a bringer of borderless trade and travel – a step 'in the direction of Wendell Willkie's *One World*'.[128] Warner, who would go on to become the first President of the newly formed International Civil Aviation Organization in 1947, highlighted the conference's accomplishments in the April 1945 issue

---

[123] David MacKenzie, 'An "Ambitious Dream": The Chicago conference and the quest for multilateralism in international air transport', *Diplomacy and Statecraft* 2, no. 2 (1991): 270–293; David MacKenzie, *ICAO: A History of the International Civil Aviation Organization* (Toronto: University of Toronto Press, 2010), 24–59.

[124] 'Verbatim Minutes of Opening Plenary Session, November 1', in Department of State, *Proceedings of the International Civil Aviation Conference*, vol. 1, part 1 (Washington, DC: Government Printing Office, 1948), document 33, pages 77–78, 80, accessed 2 August 2020, www.icao.int/ChicagoConference/Pages/proceed.aspx.

[125] Adolf A. Berle, Jr., 'International Civil Aviation Conference', *Vital Speeches of the Day* 11, no. 4 (1 December 1944): 124–128.

[126] Adolf A. Berle, Jr., 'Freedoms of the Air', *Harper's Magazine*, March 1945, 327–334.

[127] 'Sellout of U.S. Air Rights at Parley Feared', *Chicago Tribune*, 4 November 1944; 'Charge Berle is Giving up U.S. Air Rights', *Chicago Tribune*, 27 November 1944.

[128] Howard Osterhout, 'A Review of the Recent Chicago International Air Conference', *Virginia Law Review* 31, no. 2 (March 1945): 376–386.

of *Foreign Affairs*. The American vision of aviation, triumphant at the conference, offered a liberal internationalist future of 'more and easier intercontinental travel, better acquainted and friendlier peoples, and higher standards of living'. Proposals for internationalization, on the other hand, were impractical because 'the strength of national feeling makes it hard to conceive of employing personnel and selecting equipment without reference to nationality'.[129] The *New York Herald Tribune* correspondent recognized the Australia/New Zealand proposals as 'true internationalism at its highest degree', but nevertheless condemned the British proposal as 'reactionary . . . taking away more than it gave in return'.[130] Van Zandt's only regret was that the conference was not open enough, that 'unanimous agreement was not reached on the general right to fly and to trade along the world's airways'.[131] Policy advisor and one-time Vice-President of Pan American John C. Cooper similarly saw Chicago as a step towards 'the right to fly', but would have preferred that the five freedoms proposals be fully accepted. Individual states could still 'control world air trade routes' or 'affect world economy' by excluding others from their airspace.[132] One propaganda pamphlet issued by the UN Information Office presented the conference's results as the logical outcome of gradually developing international regulation, a culmination of the work begun at the Paris Convention (1919) and the Havana Conference of 1928.[133]

The few US internationalists who preferred internationalization declared the conference and the subsequent flurry of bilateral agreements a failure. For them Chicago was a missed opportunity not only for international aviation but for the rational organization of international affairs more broadly. Sumner Welles's 1946 *Where Are We Heading?* noted that 'aviation is one field where no success in international co-operation has yet been encountered'.[134] In May 1945 Quincy Wright regretted that the conference had pushed the world towards an aerial regime in which 'the three or four

---

[129] Edward P. Warner, 'Chicago Air Conference: Accomplishments and Unfinished Business' *Foreign Affairs* 23, no. 3 (April 1945): 406–421. On Warner see: Roger E. Bilstein, 'Edward Pearson Warner and the New Air Age', in *Aviation's Golden Age: Portraits from the 1920s and 1930s*, ed. William M. Leary (Iowa City, IA: University of Iowa Press, 1989), 113–126.

[130] Don Cook, *The Chicago Aviation Agreements: An Approach to World Policy* (New York: American Enterprise Association, 1945), 16, 32.

[131] Parker Van Zandt, 'International Air Conference', *Flying*, February 1945, 21–22, 150–152.

[132] John C. Cooper, 'Air Transport and World Organization', *The Yale Law Journal* 55, no. 5 (August 1946): 1191–1213; John C. Cooper, *The Right to Fly* (New York: Henry Holt, 1947).

[133] United Nations Information Office, *Towards Freedom in the Air* (New York: United Nations Information Office, 1944), 2–3.

[134] Sumner Welles, *Where Are We Heading?* (London: Hamish Hamilton, 1947), 28. First published in 1946.

great powers would regulate commercial aviation primarily in the interest of national and imperial power ... National aviation rivalries in both the military and commercial sphere would provide an index of national power rivalries and a prelude to new world wars'. He rebutted the analogy commonly made between freedom of the sea and freedom of the air by supporters of the new aerial regime, and pointed out that just as the seas had been regulated by a powerful hegemon (Britain), so too freedom of the air required first 'a world organization providing genuine security to nations and adequate regulation of international air navigation and commerce'. However, all was not lost: Dumbarton Oaks had set the foundation for a powerful UN organization which may use 'military air power' to 'promote international stability not national aggrandizement', and so, he hoped, also somehow take civil aviation out of the realm of national rivalries.[135]

In Britain liberal internationalists and the leftist press painted the Chicago Conference as a failure and condemned British negotiations during the conference. Journalist (and soon to be Labour MP) Michael Foot castigated the British government for not supporting the Australian/New Zealand proposals, which he noted were 'almost identical' to those of Labour MP Frank Bowles.[136] The *Daily Herald* also lauded them and dismissed Britain's 'pathetic appeasing role'.[137] De Madariaga clung to the idea that there could be no world government without internationalization of commercial aviation: 'without a world air authority', he informed the readers of *The Times*, 'the Dumbarton Oaks plan would be, as the Spanish saying goes, hare-pie without hare'.[138] Mance attempted to reconcile the Chicago outcome with his own pre-existing proposals by publishing in 1944 a detailed scheme for internationalization which incorporated the Chicago machinery. Outdated in several ways, the proposal even used the potential threat from convertibility of commercial to military aircraft as its primary rationale.[139]

Reactions from the liberal press were mixed. The liberal *News Chronicle*, supportive of British proposals, concluded that the work of the conference was 'useful and encouraging' and not a complete loss: 'International Aviation has been moulded a lot closer to the shape we want to see it take ... We have been champions of a progressive view which, sooner or later, will prevail'.[140] *The Times* too supported British proposals and

[135] Quincy Wright, 'The International Regulation of the Air', *The American Economic Review* 35, no. 2 (May, 1945): 243–248.

[136] Michael Foot, 'The Lion has Red, White and Blue Wings', *Daily Herald*, 14 November 1944.

[137] 'Aviation', *Daily Herald*, 27 January 1945.

[138] S. de Madariaga, 'Civil Aviation', *The Times*, 7 November 1944.

[139] H. O. Mance, *Frontiers, Peace Treaties, and International Organization* (London: Oxford University Press, 1946).

[140] Robert Waithman, 'What Happened at Chicago', *News Chronicle*, 8 December 1944.

sounded a positive note on the conference. A 'substantial advance' had been made and 'the hope of a world order in the air has moved some distance towards realization'.[141] The *Economist* declared the conference a success only on 'technical matters'. It called for Britain to form a rival 'British system' incorporating the Commonwealth and some Western European countries to counter that of the Americans. The British zone of aviation was to be based on the 'four freedoms' and the British were to rely upon converted bombers for transport.[142] The *Manchester Guardian* also declared the conference a success on the 'the technical side of aviation' but a failure with regard to 'relations between the United Nations'.[143] The Right was also unhappy with the British negotiations. The *Daily Mail*'s 'verdict' was that the conference was a 'draw'. It criticized Britain and the Commonwealth for not having agreed a singular plan to protect imperial and commonwealth aviation.[144] Some were pleased to see the British proposals defeated. C.G. Grey noted that a global 'internationally owned, controlled and operated Air Transport organisation' was 'utterly impractical' and would never have been agreed to. The Americans, Britain's 'gallant allies', saved international aviation from the aerial 'international socialism' proposed by the Australians.[145]

Conclusions

Internationalization emerged during the war as a realistic and popular solution to the problems and opportunities of post-war civil aviation – realistic and popular enough for the British state to consider it as official policy, and for opponents, particularly in the United States, to devote considerable effort to debunking it. Proposals for internationalization were not blindly utopian or disconnected from reality – rather their proponents were aware of opposition and used different strategies to overcome them. These proposals were peculiarly liberal internationalist in that they were particularly attractive to liberals and those in the middle of the political spectrum, and in that they incorporated liberal concerns about freedom to travel and trade as well as common liberal tropes relating to the internationalizing effects of aviation. Supporting arguments drew heavily on internationalist concepts of aviation and international relations dating back to the interwar years.

---

[141] 'Order in the Air', *The Times*, 4 December 1944; 'Aftermath of Chicago', *The Times*, 27 December 1944.

[142] 'The Air After Chicago', *Economist*, 30 December 1944, 860–862.

[143] 'Setback at Chicago – Air Conference not wholly fruitless', *Manchester Guardian*, 8 December 1944.

[144] 'Draw Ended Air "Battle" of Chicago', *Daily Mail*, 8 December 1944; 'Lesson of Chicago', *Daily Mail*, 27 November 1944.

[145] Grey, *The Civil Air War*, 135, 138–142.

These visions of internationalized aviation were, moreover, not disconnected from the wider internationalist thinking and rhetoric of the war years – rather they followed their ebbs and flows. In the United States there was limited support for internationalization as it was felt, amongst both state officials and internationalists, that US civil aviation stood to benefit from a less regulated international air space. Nevertheless, broader New Dealist visions of international relations drove some, such as Henry Wallace, to call for internationalization. His claims addressed other wider concerns: in what form and under what authority would the nation retain use of foreign military bases, and how could the aviation industry continue at its expanded levels into the post-war period? Nationalist solutions to these issues centred on US commercial aerial expansion on a bilateral basis, and the military's retention of its air bases. For the few internationalists such as Wallace who supported internationalization UN aviation could provide demand for US aeroplane production and make use of foreign bases.

Greater support in Britain built not only on the richer history of calls for internationalization but also on the belief that it would protect British civil aviation from the ravages of unrestrained US commercial competition after the war. Another important driver was the increasing sense that integration in technical matters, particularly transport, would bring war-torn European countries closer together, and thus act as a force for peace. Post-war European reconstruction opened the possibility of reconstruction of, and reconstruction based around, aviation.

The divergence between proposals for the international control of military aviation and of civil aviation was perhaps one of the most marked shifts in internationalist technological rhetoric from pre-war years. At one level this reflected the realization that aerial war machines were now differentiated enough from civilian machines for the convertibility argument to lose its persuasiveness. Internationalization of civil aviation was consequently no longer needed for international security or aerial disarmament. Newly emergent commercial concerns and opportunities drove internationalization in a different direction too. Internationalization came to be seen by supporters in Britain and opponents in the United States as a way of shielding the British aviation industry from US competition. Although the idea of internationalization died away as the international civil aviation regime developed towards a system of bilateral agreements from 1944 onwards, the aeroplane continued, and indeed continues today, to be seen as the great connector and integrator in world affairs.[146]

---

[146] Marc Dierikx, 'Shaping World Aviation: Anglo-American Civil Aviation Relations, 1944–1946', *Journal of Air Law and Commerce* 57, no. 4 (1992): 795–840.

6 A Battle for Atomic Internationalism: United States and the International Control of Atomic Energy

'But as a vast threat, and a new one, to all the peoples of the earth, by its novelty, its terror, its strangely promethean quality, it has become, in the eyes of many of us, an opportunity unique and challenging.'
– J. R. Oppenheimer, speaking November 1945 at the American Philosophical Society's Symposium on Atomic Energy and Its Implications.[1]

As a potentially new and powerful force in international affairs, atomic weapons (and 'atomic energy' more broadly) demanded to be absorbed into rhetoric, thinking, and activism on international relations. Intellectuals, activists, and policymakers scrambled to make sense of the atomic bomb when its existence was revealed to the world on 6 August 1945. What would its effects on international relations be? What implications did it have for the United Nations, and international organization? For internationalists the answer was clear: atomic energy needed to be internationally controlled, its power used for the further organization of international affairs. Internationalists mobilized their institutions and resources to push state policy towards international governance.

This chapter is not a comprehensive study of atomic internationalism; many of its aspects have already been explored through a number of detailed works on internationalist activity and proposals for international control in the early post-war years.[2] Instead in this chapter I add to this literature by making three arguments about the nature of this internationalism. First, I argue that it built upon existing internationalist thinking about science, technology, and international relations, and ported across many of the existing assumptions and arguments about aviation onto the newer technological marvel. Although the atomic bomb was undoubtedly crucial in sparking this internationalism, it also had origins in these earlier

---

[1] J. R. Oppenheimer, 'Atomic Weapons', *Proceedings of the American Philosophical Society* 90, no. 1 (January 1946): 7–10.
[2] For example: Smith, *A Peril and a Hope*; Herken, *The Winning Weapon*.

worldviews. The atomic bomb, rather than creating an entirely new internationalism, was a new factor in US liberal internationalism's continuing struggle for legitimacy and impact. Second, aerial internationalism did not immediately disappear: it persisted for a while in the form of proposals for an international air force (which for some was to be armed with atomic bombs). Third, I argue that the bomb did not elicit a unified response from internationalists. They and their differing visions clashed as they scrambled to incorporate it into their thinking on science, war, and international relations.

The chapter focuses on three broad movements: the supporters of the UN, the world government and federalist movement, and the atomic scientists' movement. These groups attempted to co-opt the atomic for their own projects on international relations: to nuclearize their visions.[3] Although there were some overlaps and intersections between them, they nevertheless vied with each other to produce solutions to the problem of atomic energy. Atomic scientists advocated international ownership and control of mining and manufacturing facilities through the UN or a more powerful international organization. The supporters of the United Nations worked to strengthen the organization and proposed a regime of UN licensing and inspection as the key to international control. The world government movement called for international control as part of wider visions of world federation.

Some important themes emerge from these arguments. First, there was significant continuity between the internationalism surrounding the atomic bomb and earlier internationalist activity. This connection has been obscured in the historical literature which has tended to see the internationalism surrounding the atomic bomb as a novel and innovative reaction driven largely by scientists and their inherent scientific internationalism.[4] This chapter shows that scientists shared much

---

[3] On nuclearity as a powerful but historical and socially contingent attribute see: Gabrielle Hecht, *Being Nuclear: Africans and the Global Uranium Trade* (Cambridge, MA: MIT Press, 2012), 1–46.

[4] Most crucially: Smith, *A Peril and a Hope*. See also: Robert Gilpin, *American Scientists and Nuclear Weapons Policy* (Princeton, NJ: Princeton University Press, 1962); Lawrence S. Wittner, *The Struggle Against the Bomb*, Vol. 1, *One World or None: A History of the World Nuclear Disarmament Movement Through 1953* (Stanford: Stanford University Press, 1993), 69; Joseph Manzione, '"Amusing and Amazing and Practical and Military": The Legacy of Scientific Internationalism in American Foreign Policy, 1945–1963', *Diplomatic History* 24, no. 1 (Winter 2000): 21–55; Baratta, *The Politics of World Federation*, vol. 1, 180. Histories of atomic culture have followed this literature, for example: Boyer, *By the Bomb's Early Light*; and Spencer Weart, *Nuclear Fear: A History of Images* (Cambridge, MA: Harvard University Press, 1988). One recent exception is: Fritz Bartel, 'Surviving the Years of Grace: The Atomic Bomb and the Specter of World Government, 1945–1950', *Diplomatic History* 38, no. 20 (2015): 275–302. A continuity between the interwar and post-war political engagement of scientists and their understandings of the relationship

more, ideationally, with their fellow internationalists than is generally supposed. Second, like aviation before it the internationalist reaction to the bomb was not simply based on fear but included a powerful and positive programme to reshape world order. Internationalists saw hope in the bomb: it allowed them to envisage the empowerment of international organization, the nurturing of promising new scientific research and development, and for some even the creation of a world state. Third. the politics of atomic energy were deeply intertwined with the domestic politics of the United Nations. The debate about international control was also a debate about the legitimacy of the United Nations as the carrier of internationalists' aspirations. Atomic energy gave some the opportunity to strengthen the nascent organization and bolster its standing in the United States, whilst others saw the opportunity to critique it or push for it to be replaced altogether by something more powerful.

Revolutionary Facts and Unrevolutionary Ways

The significant internationalist reaction to that atomic bomb by long-standing supporters of the United Nations began early and quickly after the atomic bombings. On 9 August, three days after the Hiroshima bombing, and on the day of the Nagasaki bombing itself, James T. Shotwell telegrammed the President calling for the international control of atomic energy through the United Nations. 'The United Nations Charter', he assured Truman, 'provides the framework through which adequate controls can be inaugurated'.[5] On 12 August Shotwell wrote an article for the *New York Herald Tribune* calling for international control, and mobilized the CSOP to prepare detailed proposals. A memo circulated to the CSOP suggested that the organization could provide the state with the expertise required to deal with the bomb: he warned that because the government might 'blunder' and 'rush forward with ill-considered plans' the Commission was needed to provide the 'competent body of men' to solve the 'immensely complicated' problem of atomic energy 'in the interest of universal peace'.[6] Shotwell was also quick to associate the CSOP with the prominent atomic scientists' movement. At the end of

---

between science and society has been noted in: Jessica Wang, *American Science in the Age of Anxiety: Scientists, Anticommunism, and the Cold War* (Chapel Hill: University of North Carolina, 1999), 4.

[5] Commission to Study the Organization of Peace, *A Ten Year Record 1939–1949* (New York: Commission to Study the Organization of Peace, 1949), 34. Another early call for internationalisation by him was: James T. Shotwell, 'The Atom and the Charter', *New York Herald Tribune*, 12 August 1945.

[6] James T. Shotwell, Memorandum, 'Preliminary Suggestions for the Study of the Control of Atomic Energy', 8 September 1945, folder 3, box 285, Carnegie Endowment for

October physical chemist Harold Urey and members of the Association of
Oak Ridge Engineers and Scientists (AORES) and the Association of
Cambridge Scientists explained atomic energy to the newly constituted
CSOP committee on atomic energy at the University of Columbia, result-
ing in a statement calling for UN international control.[7]

By October scientists had already mobilized to tackle what by now was
widely called the 'problem of atomic energy' and had created dedicated
associations to do so. Shotwell realized that without its own dedicated
effort the CSOP and the Carnegie Endowment would be left behind in
the emerging race to influence policy. One clear rationale for the formation of a non-scientists' internationalist organization on atomic energy
was that scientists lacked the requisite understanding of the political and
social world. An October 1945 memo reviewed various statements made
by the atomic scientists and concluded that they themselves 'demand
cooperation from the political and social sciences and from experienced
statesmen as well as the world of business management and military
science. They call for the mobilization of intelligence on this great challenge to intelligence and insist, with compelling logic, that there is no time
to be lost'.[8] Shotwell held a conference (attended by Urey and physicist
Leo Szilard amongst others) in November to plan further, and the formation of a Committee on Atomic Energy was officially agreed by the
trustees of the Carnegie Endowment in December 1945.[9]

Shotwell's December 1945 report to the trustees on the formation of
the Committee presented the bomb's effect on international relations in
dramatic terms which echoed his own earlier wartime rhetoric on the
impact of science and aviation. 'The bomb dropped on Hiroshima has
exploded within the structure of international relations as well', and so the
nation-state, in the face of this 'new era' of 'science', had to relinquish its
claim to sovereignty.[10] He emphasized the need to apply expertise in
shaping society and international relations to cope with this new science
and technology; the 'reassertion of the control of intelligence so that the
new products of science will no longer threaten civilization but contribute

International Peace Records, 1910–1954, Rare Books and Manuscript Library,
Columbia University (hereafter cited as Carnegie Records).

[7] 'Atomic Control by United Nations Sought by Educators, Scientists', *New York Times*,
29 October 1945.

[8] 'Memorandum on the Political and Economic Problems Arising from the Control of
Atomic Energy', 6 October 1945, folder 6, box 276, Carnegie Records.

[9] Smith, *A Peril and a Hope*, 241; Memorandum, 'Advisory Committee on Atomic Energy,
Meeting of November 7, 1945', folder 7, box 276, Carnegie Records; 'Atom Study Set
Up by Carnegie Group', *New York Times*, 11 December 1945.

[10] See: the 'General Statement' in 'Advisory Committee on Atomic Energy',
6 December 1945, folder 7, box 279, Carnegie Records.

to it'. Shotwell once again relied on the status and authority of scientists to make his argument. Physical scientists, usually unconcerned (he claimed) with the social and political implications of their science, were now clamouring for 'direction' from the political sciences. The physical and political sciences were for the first time working together for the good of the world, this 'linking together of the great techniques is now progressing strongly throughout the country and is bound to have a great influence not merely within the sciences but upon public opinion everywhere'. Shotwell's message was clear: the Carnegie Endowment had to continue to participate in this great meeting of minds.[11]

The trustees allocated an initial amount of $41,400 to a proposed programme of work which consisted of approaches to other American and British internationalist organizations, at least one conference, and three preliminary studies: one on international control of atomic energy, another on 'the economic and social adjustments due to the discovery of atomic energy', and a third on the 'transformation of international law into the law of Nations'.[12] The Committee on Atomic Energy went on to formally meet six times up to May 1946, and informally a further two times: in November and December 1946.[13] Although a rather long list of participants had initially been announced, the regulars on the main committee consisted of Shotwell as chairman, Clark Eichelberger, the eminent international lawyer and judge Manley O. Hudson, prominent electrical engineer Gano Dunn, and international lawyers George A. Finch and Herbert L. May. No scientists attended on a regular basis, though several did in an ad hoc manner.

The most significant action of the Committee was one of its earliest: the organization of the conference on international control in early January 1946. Although organized by Shotwell's Committee and funded through its budget, the official reports and publicity noted two better-known organizations as co-organizers: the Federation of Atomic Scientists (soon reconstituted as the Federation of American Scientists) and the CSOP. Forty-six scientists and twenty-five non-scientists attended this meeting in New York. All delegates agreed on the need for some form of international control, though opinions differed as to the type of international organization which would institute this regime. Some called for lesser forms of international control such as international inspections, whereas others suggested outright ownership of atomic

---

[11] Ibid.

[12] Memorandum, 'Carnegie Endowment for International Peace. Division of Economics and History', 28 November 1945, folder 7, box 276, Carnegie Records.

[13] Committee on Atomic Energy, 'Report of Progress', 5 April 1946, folder 7, box 276, Carnegie Records.

facilities by the UN or some other international organization. Scientists I.I. Rabi and Irving Langmuir even announced that they would readily support a complete ban on the development of atomic energy 'until such a time as the world state which we project got well under way and had its complete legislative machinery and precedents'; the benefits forgone being outweighed by the security created.[14]

The conference was an important part of Shotwell's continuing attempt to convince the atomic scientists and other internationalists that international control could most realistically be instituted through the current UN rather than through a drastically altered one or through world government. Unlike the atomic scientists, for whom attempts at international control had to be a break from the (as they saw it) failed diplomatic peace initiatives of the past, Shotwell emphasized that these efforts were a renewal of interwar attempts at disarmament and the outlawry of war. Just as previous peace initiatives should have worked through the League, so these efforts were to be carried out through the United Nations. He chided those internationalists and atomic scientists wanting international control through organizations other than the UN, and argued against what he saw as unrealistic versions of international control, such as, for example, proposals for international ownership of mines and atomic plants. For Shotwell radical new international organizations or internationalized ownership were tantamount to a world government, which, he reasoned, would not be acceptable to US policymakers and the public. Writing privately in November 1945 following a lecture and discussion at the American Academy of Science, he noted that he came away 'with an added sense of the seriousness of the situation. The scientists are so scared that they are jumping at world government as though it could be achieved overnight'.[15] His December 1945 memo to the trustees of the Carnegie Endowment charged atomic scientists with a certain naiveté: 'The irresponsible suggestions put forth immediately after Hiroshima, that a world government must be improvised overnight, have received wide publicity and a certain

---

[14] Notes, 'January 5th, 1946 – Saturday morning', n.d., folder 9, box 32, Series IV, Atomic Scientists of Chicago Records 1943–1955, Special Collections Research Center, University of Chicago Library (hereafter cited as Atomic Scientists Records). 'Report and Discussion on Problems of War and Peace in the Atomic Age based on Proceedings of a Joint Conference arranged by The Committee on Atomic Energy of the Carnegie Endowment for International Peace with the Federation of Atomic Scientists and with The Commission to Study the Organization of Peace', 1946, pages 183 and 188, folder 1, box 10, Eugene I. Rabinowitch Papers, Special Collections Research Center, University of Chicago Library. Rabinowitch made a similar suggestion: 'Technical Feasibility of Atomic Energy Controls', *Bulletin of the Atomic Scientists* 1, no. 2 (24 December 1945): 1.

[15] Shotwell to Frederick McKee, 21 November 1945, folder 1, box 280, Carnegie Records.

degree of acceptance among physical scientists; but they have never been taken seriously in authoritative circles'. His proposed committee was to correct this lack of serious internationalist proposals in the atomic policy sphere – to 'deal with a revolutionary fact in an unrevolutionary way'.[16]

From November 1945 onwards Shotwell began to express these views publicly, and in January 1946 he presented them directly to scientists in a widely publicized speech given at the January 1946 annual meeting of the American Physical Society and the American Association of Physics Teachers.[17] His speech was published in the *American Journal of Physics* and the *Bulletin of the Atomic Scientists* and noted in the *New York Times* with the provocative headline: 'Shotwell Assays "One World" Set-Up. Tells Scientists That Such a Government May Threaten Democratic Structure'.[18] The headline was no exaggeration. Shotwell did indeed warn that 'world government' would 'threaten the whole structure of democratic life throughout the world'. Instead the world needed an 'international system for peace and security' to guard against 'the danger of scientific warfare' and 'enhance … the growth of domestic freedom' which was the 'modern political problem created by the scientific revolution' of the atomic bomb. He harked back to the work of the Geneva Conference on disarmament and called for the current UN Charter to be retained but buttressed by additional treaties and UN organizations. Most importantly a 'Tripartite Advisory Body' was now required to advise the Security Council on atomic matters. As well as the pre-existing UN Military Staff Committee this was to include a 'technical body' of scientists. But their expertise was not enough to deal with the impact of atomic energy; political scientific expertise was required too. 'It would not be wise', he opined, 'to leave the whole problem entirely in the hands of scientists, for many of their technics do not apply without modification to the problems peculiar to politics'. International control of atomic energy also called for 'men who are experienced in the conduct of public affairs and especially in the working of international organization' – that is political scientists such as himself. Their expertise was

[16] See 'General Statement' in 'Advisory Committee on Atomic Energy', 6 December 1945, folder 7, box 279, Carnegie Records.

[17] For example: James T. Shotwell, 'Implementing and Amending the Charter: An Address by James T. Shotwell, November 11, 1945', *International Conciliation* 23 (December 1945): 811–823; and his lecture at the November 1945 Symposium on Atomic Energy and Its Implications: James T. Shotwell, 'The Control of Atomic Energy under the Charter', *Proceedings of the American Philosophical Society* 90, no. 1 (January 1946): 59–64.

[18] James T. Shotwell, 'The International Implications of Nuclear Energy', *American Journal of Physics* 14, no. 3 (May-June 1946): 179–185; James T. Shotwell, 'The Atomic Bomb and International Organization', *Bulletin of the Atomic Scientists* 1, no. 7 (15 March 1946): 8–9; 'Shotwell Assays "One World" Set-Up. Tells Scientists That Such a Government May Threaten Democratic Structure', *New York Times*, 25 January 1946.

required as international control was 'far more complicated than that of pre-atomic armaments, both because of the far reach of atomic power and because of the pioneering nature of the experiments which we have made so far'.[19]

Shotwell's attempts to position political scientists and the Carnegie group in particular as experts on the impact of atomic energy left him struggling to build institutional links with atomic scientists. Following the January 1946 conference his Committee resolved to continue cooperation with scientists in the form of a '(Recurring) Science Conference on War and Peace' with an executive committee which included himself, Eichelberger, the sociologist Edward Shils, I. I. Rabi, the chemist Ray Crist, and the atomic scientists John A. Simpson and Irving Kaplan. This organization, however, did little other than prepare and distribute the proceedings of the January 1946 conference.[20] Instead, the bulk of the work of the Committee on Atomic Energy was carried out through several subcommittees, though physical scientists did have some input. The subcommittees produced various reports and memos, the most significant of which was the Draft Convention of the Carnegie Endowment Committee on Atomic Energy, prepared by the Legal Subcommittee and published in final form in June 1946. It was mostly the work of, and reflected the views of, Shotwell and George A. Finch, director of the Division of International Law of the Carnegie Endowment. It proposed international control through the United Nations Atomic Energy Commission, which reported to the Security Council. Most crucially, the production and ownership of atomic bombs, as well as atomic plants and mines, was to remain under national ownership – though licensed, inspected, and policed by the UN.[21]

The report's limits on international control made it controversial for many internationalists. Under-Secretary of State Dean Acheson and David Lilienthal, who played significant roles in the production of the State Department-sponsored Acheson-Lilienthal Report on international control in March 1946, reacted negatively to drafts of the report circulated to them by the Committee on Atomic Energy.[22] Their report had

[19] Shotwell, 'The International Implications of Nuclear Energy'.
[20] See: M. Olson, 'Notes on Meeting of Executive Committee of Science Conference on War and Peace January 11, 1946', n.d., folder 13, box 278, Carnegie Records; I. Kaplan, 'Report of Meeting of Monday, January 7, 1945', n.d., folder 13, box 278, Carnegie Records.
[21] Committee on Atomic Energy of the Carnegie Endowment for International Peace, *Utilization and Control of Atomic Energy: A Draft Convention Prepared by the Legal Subcommittee in consultation with other Legal, Political and Scientific Experts* (New York: Carnegie Endowment for International Peace, June 1946).
[22] Josephson, *James T. Shotwell and the Rise of Internationalism in America*, 273.

a more radical vision, based on direct ownership of all raw material and plant by an international organization, and the 'denaturing' of raw materials. An Atomic Development Authority (ADA) answering to the Security Council was to control all fissionable raw materials and have a monopoly on all 'dangerous' activities (that is those with possible military applications). States would shut down such activities, and atomic bombs would be transferred to the UN. Peaceful development could continue in nation-states. The Authority would own and operate all mining, refining, and production of fissionable raw materials. Existing mines, plants, and factories (e.g. at Hanford and Oak Ridge) were to be transferred to its control. It would also set up its own research and development centres and conduct research on peaceful and warlike uses of atomic energy. The Authority would dispense 'denatured' fissionable raw materials (that which could not be used for 'dangerous' purposes) to individual nations for their nuclear power plants, and license and inspect operating civilian nuclear facilities in nation-states. The United States would begin a phased transition of its bombs, material, and facilities to the Authority once it had been set up, but not cease atomic operations prior to the setting up of the ADA. The Acheson-Lilienthal Report placed significant importance on the internationalist, cooperative, and apolitical attitudes of the international group of scientists working at the Authority to ensure that international control worked.[23]

Quincy Wright, who was then working on his own convention on international control, also thought the Carnegie Plan was not radical enough to be effective. Writing to Shotwell in May 1946, he took issue with two fundamental aspects of the plan. International inspection of national atomic facilities would not work, he maintained, and reliance on 'quotas of permissible atomic weapons and plants' would be problematic as they would be harder to negotiate than outright UN ownership. He expressed disappointment that the Carnegie plan 'should lag far behind the boldness' of the Acheson-Lilienthal Plan which included 'thoroughgoing' international control and the complete elimination of atomic weapons and production facilities. Once the official US government proposal for international control, the so-called Baruch Plan, was unveiled, he expressed his preference for that too, instead of the Carnegie convention.[24]

---

[23] Chester I. Barnard, J. R. Oppenheimer, Charles A. Thomas et al., *A Report on the International Control of Atomic Energy* (Washington, DC: Department of State, 1946).

[24] Q. Wright to J. T. Shotwell, 1 May 1946, folder 17, box 280, Carnegie Records; Memorandum, Q. Wright, 'The Carnegie Endowment's Draft of April 23, 1946 on Atomic Weapons and the Control of Atomic Energy', n.d., ca. 1 May 1946, folder 17, box 280, Carnegie Records; Q. Wright to J. T. Shotwell, 5 July 1946, folder 17, box 280,

For some critics the Committee on Atomic Energy's proposal that UN Security Council members be allowed to veto international control and related atomic matters was also problematic. In June 1946 Bernard Baruch, the US representative on the United Nations Atomic Energy Commission (UNAEC), revealed the official US proposal for international control (the Baruch Plan). Although based largely on the earlier report endorsed by the State Department (the Acheson-Lilienthal Report), it differed in several crucial ways. Most significantly it demanded that the veto not be applicable to the UN's control of atomic energy. Although some welcomed the removal of the veto as a necessary protection against Soviet perfidy, others (especially supporters of the Acheson-Lilienthal Report) saw it as an unnecessary stumbling block in negotiations.[25] Shotwell also came under pressure to follow suit and remove the veto from the Carnegie Plan. To one critic (Everett Case, President of Colgate University and a member of the Committee) his riposte was that its removal would lead to both a 'head-on conflict' with the Soviets as well as opposition from Congress.[26]

The atomic scientists' movement, broadly supportive of the Acheson-Lilienthal Plan and already largely suspicious of the Carnegie initiative on atomic energy, was disappointed with the cautious approach of the Carnegie convention. W. A. Hinginbotham, Chairman of the Federation of Atomic Scientists, wrote to Henry DeWolf Smyth (a Palmer Physical Laboratory atomic scientist on the Committee on Atomic Energy and author of the Smyth Report) in May 1946 urging Smyth to push the Committee to support the Acheson-Lilienthal proposals instead of publishing their own.[27] Smyth revealed that both he and Rabi had opposed the Carnegie Plan but were outvoted. 'I do not like the Carnegie Report, and I will continue to try to get it reconsidered, or at least not pushed, but I doubt whether I can do anything'.[28] Hinginbotham also prepared a letter in May 1946 urging those atomic scientists who attended

Carnegie Records. Also: Quincy Wright, 'Draft for a Convention on Atomic Energy', *Bulletin of the Atomic Scientists* 1, no. 8 (1 April 1946): 11–13. On the Baruch Plan see: Herken, *The Winning Weapon*, 159–181.

[25] Herken, *The Winning Weapon*, 160–174. On reaction see also: Smith, *A Peril and a Hope*, 464–474; Shane J. Maddock, *Nuclear Apartheid: The Quest for American Atomic Supremacy from World War II to the Present* (Chapel Hill: University of North Carolina Press, 2010), 56–60.

[26] J. T. Shotwell to E. Case, 20 June 1946, folder 6, box 279, Carnegie Records; E. Case to J. T. Shotwell, 18 June 1946, folder 6, box 279, Carnegie Records; E. Case to J. T. Shotwell, 25 June 1946, folder 6, box 279, Carnegie Records.

[27] W. A. Hinginbotham to H. Smyth, 27 May 1946, folder 2, box 13, Federation of American Scientists Records, Special Collections Research Center, University of Chicago Library (hereafter cited as FAS Records).

[28] H. Smyth to W. A. Hinginbotham, 30 May 1946, folder 2, box 13, FAS Records.

the January 1946 Carnegie conference to write in protest to the Committee on Atomic Energy.[29]

Shotwell responded to scientists' criticism in a letter to California-based economist Eugene Staley in August 1946. The Carnegie Convention, he claimed, presented a pluralistic approach which recognized that international control could be achieved in different ways in different countries, whereas the atomic scientists were envisaging one method of international control, international ownership, for all countries. Their approach was 'dogmatic and unrealistic'. Shotwell pointed to the perhaps unintended but nevertheless plausible outcomes of the Acheson-Lilienthal Plan, such as the United States building plants in the USSR and 'the elimination from American sovereignty of about one-third of the State of Colorado'.[30] Staley responded with two concerns that were increasingly being voiced by opponents of international control through inspection. Inspections, he feared, may not be robust enough to prevent a 'totalitarian government' surreptitiously arming and getting the 'drop' on the rest of the world. Anyhow, proposals for international control should aim high, and not be formulated with a built-in compromise.[31]

Privately, Shotwell did not shy away from describing his battle against other proposals for international control as a battle against the Left. In a letter to the editor-publisher W.S. Sherman in June 1946 he described the Acheson-Lilienthal Report as well as the 'world government' and scientist critics of his Carnegie Convention as demanding a 'world system of socialized government ownership'.[32] In July 1946 the Columbia University physicist Henry A. Boorse attended various conferences in England on atomic energy, including the International Conference of the Association of Scientific Workers. His report to the Carnegie Endowment noted 'large Communist influence', and Shotwell did not hesitate in circulating this 'disturbing' report to the trustees in order to drum up support for his Committee's proposals. Shotwell warned that it was the scientific background and approach of the World Federation of Scientific Workers, their lack of 'political experience', and the technical nature of the problem of international control itself, which made such internationalists particularly susceptible to Communist influence. This problem was worsened by the need to engage with 'technical fields' which confused the public and led it to 'lose interest and leave a problem which is political in essence in the hands of scientific experts'. He warned that

---

[29] Draft Letter, FAS to FAS members, 27 May 1946, folder 2, box 13, FAS Records.
[30] J. T. Shotwell to E. Staley, 9 August 1946, folder 9, box 280, Carnegie Records.
[31] E. Staley to J. T. Shotwell, 22 August 1946, folder 9, box 280, Carnegie Records.
[32] J. T. Shotwell to W. S. Sherman, 13 June 1946, folder 9, box 280, Carnegie Records.

'This is a situation which opens the door to demagogues and to their well-intentioned but confused followers' and called on the Carnegie Endowment to 'perform a major service in the analysis of policy and its clarification'.[33]

Shotwell and the Carnegie group's fear of increased state ownership and industrial regulation through international control was shared by corporate leaders. Bernard Baruch and his advisers (some of whom were businessmen) professed similar concerns during their meetings, with one participant fearing that the ADA envisaged by the Acheson-Lilienthal Plan would be 'the first start to an international socialized State'. This concern shaped the final Baruch Plan, which limited UN involvement to inspection, licensing, and punishment.[34] Although strong enough to shape Baruch's thinking and his Plan, this concern was nevertheless not strong enough to detract from the popularity of the atomic scientist-backed Acheson-Lilienthal Plan, or indeed build strong support for Shotwell's Convention. This is because US private industry was in fact not particularly enthusiastic about developing nuclear technology for civilian use or participating in continued military development. For companies such as General Electric and Du Pont greater profits waited in their core civilian businesses, and they were busy investing to boost production to meet greater post-war demand in these areas. Moreover, there was widespread scepticism at the near-term possibilities of atomic power generation. This outlook led to less industrial opposition to proposals such as the Acheson-Lilienthal Plan, which, if enacted, may have restricted private industrial profits and participation in atomic development.[35]

Modern Man and His Cosmic Gadgets

Ranged against Shotwell and supporters of the United Nations were advocates of world federation or world government. Having grown in prominence towards the end of the war, and dissatisfied with the United

---

[33] J. T. Shotwell to Trustees of the Carnegie Endowment for International Peace, 2 October 1946, folder 5, box 279, Carnegie Records; H. A. Boorse, 'Report on International Scientific Conferences held in England, July 1946', 1 October 1946, folder 5, box 279, Carnegie Records. A summary was published as: Henry A. Boorse, 'Two International Scientific Meetings in England', Bulletin of the Atomic Scientists 2, nos. 5–6 (1 September 1946): 5–6; and Henry A. Boorse, 'A World Federation of Scientific Workers', Bulletin of the Atomic Scientists 2, nos. 5–6 (1 September 1946): 7.

[34] Herken, The Winning Weapon, 161, 164.

[35] Brian Balogh, Chain Reaction: Expert Debate and Public Participation in American Commercial Nuclear Power, 1945–1975 (Cambridge: Cambridge University Press, 1991), 48, 49, 64–66; Pap A. Ndiaye, Nylon and Bombs: DuPont and the March of Modern America (Baltimore, MD: Johns Hopkins University Press, 2007), 179–187.

Nations, they incorporated atomic energy into their rationale and proposals for world government. International control was one significant rubric through which they did this. Robert M. Hutchins, the internationalist Chancellor of the University of Chicago, was one of the earliest public advocates for international control – on 12 August he demanded that a world government be formed to control the atom.[36] He organized a conference on the political implications of atomic energy at the University of Chicago in September 1945, and later formed an Office of Inquiry into the Social Aspects of Atomic Energy, headed by the sociologist Edward Shils and Dean Robert Redfield of Chicago's Division of Social Sciences, and gave $10,000 to support the atomic scientists' movement.[37]

An early world federalist conference in Dublin, New Hampshire, in October 1945 helped build broad support for international control amongst this internationalist community. Organized by the lawyer and world federalist Grenville Clark (and wartime personal assistant to Secretary of War Henry L. Stimson), the conference concluded that international control of the bomb and 'other major weapons' was the only way to institute world peace whilst also presenting the opportunity for world federation. Assembled delegates had their disagreements of course, the major one being between a majority who called for control to occur through a global federation, and a minority led by Clarence Streit who maintained that as a realistic and interim step the United States needed to form a 'nuclear union with nations where individual liberty exists', that is Britain and her white ex-colonies, and perhaps Western Europe too.[38]

Cultural and social lag arguments emerged as important ways in which supporters of world federation understood atomic energy and made cases for their internationalist solutions. Norman Cousins, an executive editor at the *Saturday Review of Literature* (and soon to be vice-president of the United World Federalists), penned the most prominent exposition of the lag thesis in relation to atomic energy in the form of his widely cited essay 'Modern Man is Obsolete' (published in the *Saturday Review of Literature*

---

[36] 'Hutchins Urges World State Now', *New York Times*, 13 August 1945.
[37] David E. Lilienthal, *Journals of David E. Lilienthal* Vol. II: The Atomic Energy Years, 1945–1950 (New York: Harper & Row, 1964), Appendix A, 637–645; folder: 'Atomic Energy Conference at Chicago: Sept. 19, 20, 1945', box 28, Series III, Atomic Scientists Records; Smith, *A Peril and a Hope*, 94–96.
[38] 'Declaration of the Dublin, N.H., Conference', *New York Times*, 17 October 1945. See also: Baratta, *The Politics of World Federation*, vol. 1, 150–155; Gerald T. Dunne, *Grenville Clark: Public Citizen* (New York: Farrar, Strauss, Giroux, 1986), 140–143. There were subsequent world federalist conferences; another early significant conference was held at Rollins College in March 1946: Baratta, *The Politics of World Federation*, vol. 1, 156–157.

in August 1945 and in October expanded as a short book). 'Modern man' himself was now lost in an atomic world of 'gaps', 'gaps between revolutionary science and evolutionary anthropology, between cosmic gadgets and human wisdom, between intellect and conscience'. In the 'Atomic Age' the world was 'united geographically ... strict national sovereignty was an anomalous holdover from the tribal instinct in nations'. 'National man' needed to transform himself into 'world man'. This required economic and intellectual changes, but most urgently the creation of world government to 'control' and 'police' the atom. The alternative to this, and the only other way to save mankind from atomic destruction, was to destroy 'everything relating to science and civilization ... destroy all machines and the knowledge which can build or operate those machines'.[39]

Many proponents of world government did not envisage its immediate foundation and were content to advocate a gradual approach to international governance. Cousins for example was satisfied to see the Acheson-Lilienthal Report 'as the basis for immediate discussion and action' even though it was clearly far from world government.[40] In a letter to AORES in October 1946 he explained that the reduction of the likelihood of war should be the immediate goal rather than a world government, which would be more concerned with welfare.[41] Cord Meyer, Jr., the founder and first President of the United World Federalists, similarly noted in a letter that 'we must ask for no more than is the absolute minimum necessary to prevent the immediate danger of war. If we attempt more, there is great danger that we shall achieve nothing'.[42] Meyer did, however, disagree with Cousins on the Baruch Plan. Cousins, like most Americans, supported it, whereas Meyer thought it impractical and unlikely to be accepted by the Soviet Union. Eliminating the veto, he argued, would effectively revive the discredited

---

[39] Norman Cousins, 'Modern Man is Obsolete', *Saturday Review of Literature*, 18 August 1945, 5–9; Norman Cousins, *Modern Man Is Obsolete* (New York: Viking, 1945). On Cousins see: Allen Pietrobon, 'Peacemaker in the Cold War: Norman Cousins and the Making of a Citizen Diplomat in the Atomic Age' (PhD diss., American University, Washington, DC, 2016), 14–81. Another prominent call for world government using the lag argument was: Walter Lippmann, 'International Control of Atomic Energy', in *One World or None*, eds. Dexter Masters and Katharine Way (New York: McGraw Hill, 1946), 66–75.

[40] Norman Cousins and Thomas K. Finletter, 'A Beginning for Sanity', *Saturday Review of Literature*, 15 June 1946, 5–9, 38–40.

[41] N. Cousins to J. L. Balderston, Jr., 4 October 1946, folder 10, box 8, Series II, Association of Oak Ridge Engineers and Scientists Records, Special Collections Research Center, University of Chicago Library (hereafter cited as AORES Records).

[42] C. Meyer, Jr. to J. L. Balderston, 12 October 1946, folder 1, box 9, Series II, AORES Records.

League system. The world federation leadership was in fact split on the Baruch Plan: Grenville Clark supported it, but Hutchins was opposed. The United World Federalists' official comment was that the Plan did not go far enough, but nevertheless 'almost any world control of atomic energy is better than none. An international control plan would at least provide a period of truce'.[43]

Ultimately for Cord Meyer, Jr., as indeed for most if not all world federalists, the problem of atomic energy could only be permanently solved as part of a broader internationalist solution which dealt with international security and international organization. In a speech to the United World Federalists and the Federation of American Scientists in April 1947 Cord Meyer, Jr. explained that effective international control could not occur until the UN was strengthened with an international police and additional powers, allowing it to administer laws binding on individuals of each country, giving it the power to arrest and try individuals in world courts, and allowing it to institute a wide-ranging system of international inspection of armaments.[44] He explained his reasoning to Hinginbotham in a correspondence exchange which followed the speech: 'Of course, I agree with you that the control of atomic energy is the most immediate problem ... But ... there is no hope in attempting to gain security by "the one weapon at a time" approach'. Rather, 'Without world law, courts and preponderously power police, enforcement is impossible without world war ... In other words, I think that the atomic control problem is not separable from the security problem as a whole'. He hoped that 'the atomic scientists will come to see this fact and abandon their belief that atomic energy is a unique problem that can be solved in some airtight compartment of its own. Einstein sees this clearly as do some of the others'.[45]

Emery Reves, author of the best-selling world federalist tract *Anatomy of Peace*, also held that international control would not work without first instituting world federation: he called for this agglomeration to begin first amongst interested countries with a focus on control of world security and trade, including international control of atomic energy.[46] Edith Wynner, of the New York-based International Campaign for World Government

[43] Wittner, *The Struggle Against the Bomb*, vol. 1, 69–70.

[44] Cord Meyer, Jr., *The Search For Security* (New York: United World Federalists, 1947).

[45] C. Meyer, Jr. to W. A. Hinginbotham, 30 April 1947, folder 3, box 25, FAS Records.

[46] Reves put this point directly to the atomic scientists in January 1946: 'Emery Reves on World Government', *Bulletin of the Atomic Scientists* 1, no. 4 (1 February 1946): 6. See also Emery Reves, 'National Sovereignty – the Road to the Next War', *World Affairs* 109, no. 2 (June 1946): 109–116; and the postscript to his later editions of *Anatomy of Peace*, for example: Emery Reves, *The Anatomy of Peace*, 8th ed. (New York: Harper & Brothers, 1946), 277–293.

and supporter of the Dublin statement, also called for world government first, though she preferred to see one based on some form of people's convention rather than one signed between national governments.[47] Writing to Oak Ridge scientist John L. Balderston of the World Government Committee of AORES in June 1946, she noted that, although she had been 'one of the earliest to hail your entry into world affairs and also to comment enthusiastically on the State Department plan', the atomic scientists' proposals for international control were 'a glamorous but still futile blind-alley'. The Baruch Plan was devised to 'put the onus for stopping international control on Russia ... Then, our Government will turn helplessly to the people and to you, scientists, saying we have no choice but to go all out in preparation for future atomic, bacteriological and what-not war'.[48]

Atomic Scientists and the Opportunity Unique and Challenging

Given that they were not the only ones advocating international control, atomic scientists worked hard to differentiate their message from that of others. Only atomic scientists, they claimed, could bring to bear the skills and knowledge required to deal with the implications of atomic energy – including time already spent thinking about the bomb's implications, unique technical knowledge, rational and scientific methods of problem solving, and a perspective untainted by politics.[49] They also emphasized that they did not exclusively support world government or federation, or the UN, as other internationalists did. Rather their proposals for peace through international control were compatible with both.[50]

Atomic scientists' pronouncements on atomic energy, in fact, shared crucial assumptions about the nature of international relations and the effects of modern science and machines with interwar and post-war internationalism. Scientists saw the atomic bomb, and thus the problem of the atomic, as stemming directly from the modern science of 'atomic

[47] Baratta, *The Politics of World Federation*, vol. 1, 169, fn 26.
[48] E. Wynner to J. L. Balderston, 18 June 1946, folder 2, box 9, Series II, AORES Records.
[49] See for example: 'Appendix B: The Franck Report June 11, 1945', in Smith, *A Peril and a Hope*, 560–575; James Franck, 'The Social Task of the Scientist', *Bulletin of the Atomic Scientists* 3, no. 3 (March 1947): 70; Robert R. Wilson and Philip Morrison, 'Half a World ... and None: Partial World Government Criticized', *Bulletin of the Atomic Scientists* 3, no. 7 (July 1947): 181–182; Eugene Rabinowitch, Editorial, 'Scientists and World Government', *Bulletin of the Atomic Scientists* 3, no. 12 (December 1947): 345–346.
[50] See for example the mission statements of the Association of Los Alamos Scientists: 'The Policy of the A.L.A.S.', *Bulletin of the Atomic Scientists* 1, no. 1 (10 December 1945): 2.

energy'. These problems (including primarily the destructive effects of the atomic bomb itself) were unique and unlike any that had gone before, and so required radical solutions which otherwise would have been unthinkable. This monolithization of the atomic around the science of atomic energy was built into the DNA of the atomic scientists' movement. The newly formed Atomic Scientists of Chicago, for example, expressed its aims in December 1945 as: '(1) To explore, clarify and formulate the opinion and responsibilities of scientists in regard to the problems brought about by the release of nuclear energy; and (2) To educate the public to a full understanding of the scientific; technological and social problems arising from the release of nuclear energy'.[51] The preamble to the 1946 constitution of the Federation of American Scientists noted: 'The value of science to civilization has never been more clear, nor have the dangers of its misuse been greater ...'[52] A December 1946 policy statement by the Federation noted that its aim was to 'establish a system for the international control of atomic energy, which will eliminate atomic armaments, will remove the competitive pressure toward developing such armaments, and will do these things by building a framework which will, at the same time, provide for the growth of the new technology in a form that is beneficial to the world'.[53]

Atomic scientists, like other internationalists, also thought of war in terms of the destructiveness of the latest technical weapons, and of the development of warfare through the decades in terms of the development of these weapons. Scientific research, they believed, drove this development. The Franck Report on the impact of the atomic bomb (a 'foundational text for the scientists' movement' produced by the Manhattan Project physicist James Franck and other Met Lab scientists in June 1945 for the Secretary of War) noted: 'Scientists have often before been accused of providing new weapons for the mutual destruction of nations, instead of improving their well-being. It is undoubtedly true that the discovery of flying, for example, has so far brought much more misery than enjoyment or profit to humanity'. Yet atomic energy was a different category of science altogether. It was:

---

[51] Atomic Scientists of Chicago, 'The Atomic Scientists of Chicago', *Bulletin of the Atomic Scientists* 1, no. 1 (10 December 1945): 1.

[52] Pamphlet, Federation of American Scientists, *The Federation of American Scientists* (Washington, DC: Federation of American Scientists, 1946), folder 1, box 1, FAS Records.

[53] The Federation of American Scientists, 'Policy Statement by the Administrative Committee', 9 December 1946, folder 5, box 21, FAS Records. On the Federation's activism see: Megan Barnhart Sethi, 'Information, Education, and Indoctrination: The Federation of American Scientists and Public Communication Strategies in the Atomic Age', *Historical Studies in the Natural Sciences* 42, no. 1 (February 2012): 1–29.

fraught with infinitely greater dangers than were all the inventions of the past ... In the past, science has often been able to provide adequate protection against new weapons it has given into the hands of an aggressor, but it cannot promise such efficient protection against the destructive use of nuclear power. This protection can only come from the political organization of the world.[54]

Urey's public writings and speeches presented the most detailed exposition of such narratives. He focused on war as the central problem of international relations, and narrated histories of inventions from prehistoric times through the nineteenth century to the First World War and beyond. In these histories aviation was the forerunner to atomic energy: able to bring great peacetime dividends yet used to make war to great destructive effect.[55]

Abstract forces such as anarchy and nationalism were not particularly prominent in the rhetoric of the atomic scientists' movement. Scientists instead tended to focus on the atomic bomb itself as the issue or a structured arms race between the US and the Soviet Union. Nevertheless, they were there, and linked strongly to outside influences. Albert Einstein, an important figurehead for the movement (he was Chairman of the Emergency Committee of Atomic Scientists), applied much of his interwar internationalist sentiments to atomic energy.[56] He aptly summarized his thoughts in a conference address published in the *Bulletin* in 1948:

In smaller entities of community life man has made some progress toward breaking down anti-social sovereignties ... But in relations among separate states complete anarchy still prevails. I do not believe that we have made any genuine advance in this area during the last few thousand years. All too frequently conflicts among nations are still being decided by brutal power, by war.[57]

Einstein's statements such as 'As long as we have states clinging to unrestricted sovereignty we will undoubtedly have to expect bigger and better wars fought with bigger and better technical weapons' could easily

---

[54] 'Appendix B: The Franck Report June 11, 1945', in Smith, *A Peril and a Hope*, 560–575. On the Report as a foundational text see: Matt Price, 'Roots of Dissent: The Chicago Met Lab and the Origins of the Franck Report', *Isis* 86, no. 2 (June 1995): 222–244.

[55] Harold C. Urey, 'Atomic Energy, Master or Servant?', *World Affairs* 109, no. 2 (June 1946): 99–108; Harold C. Urey, 'Atomic Energy, Aviation, and Society', *Air Affairs* 1, no. 1 (September 1946): 21–29.

[56] On Einstein's interwar internationalism see: David E. Rowe and Robert Schulmann, eds., *Einstein on Politics: His Private Thoughts and Public Stands on Nationalism, Zionism, War, Peace, and the Bomb* (Princeton, NJ: Princeton University Press, 2007), chapter 4; Richard Crockatt, *Einstein and Twentieth-Century Politics: 'A Salutary Moral Influence'* (Oxford: Oxford University Press, 2016), 62–78.

[57] Albert Einstein, 'A Message to the World Congress of Intellectuals', *Bulletin of the Atomic Scientists* 4, no. 10 (October 1948): 295, 299.

have come directly from David Davies or some other interwar liberal internationalist.[58] In February 1948 Einstein responded to some pointed Soviet critiques of liberal internationalism by accusing his critics of being 'passionate opponents of anarchy in the economic sphere, and yet equally passionate advocates of anarchy, e.g., unlimited sovereignty, in the sphere of international politics'. Whilst conceding that capitalism posed problems for the world, the ultimate danger arose from the new 'state of international affairs' which 'tends to make every invention of our minds and every material good into a weapon and, consequently, into a danger for mankind'. Perhaps the most prominent scientist in the United States to support world government, he argued that world government was the only way of correcting this state of affairs and 'avoiding total destruction'.[59] Manhattan Project chemist Harrison Brown, in his 1946 *Must Destruction be our Destiny?*, argued that the atomic bomb was the latest in a number of recent technologies which had now made 'political nationalism' a 'senseless concept'. He quoted Emery Reves's *Anatomy of Peace* in his argument that increasing clashes between sovereign states underlay warfare.[60]

Many of these internationalist ideas were present in the Acheson-Lilienthal Report, which was substantially drafted by Oppenheimer and strongly supported by the scientists' movement. More than just a scheme for international control, the report was a liberal internationalist manifesto on atomic energy which rallied internationalist opinion and helped popularize internationalist approaches to atomic energy. The report reproduced and publicized not only the ideas propounded in atomic scientists' rhetoric but longer-standing internationalist thinking on technology, war, and international relations. It characterized the atomic bomb as a 'revolutionary' weapon against which there was no defence – a 'surprise weapon', which could be 'secretly developed and used without warning' to destroy cities and eradicate populations. It emphasized the essential sameness of the technology underlying both military and 'peaceful' applications ('interchangeable and interdependent') and identified 'national rivalries' as the 'heart' of the atomic problem. The Report

---

[58] A. Einstein to Chancellor Robert Hutchins, n.d., folder: 'Atomic Energy Conference at Chicago: Sept. 19, 20, 1945', box 28, Series III, Atomic Scientists Records.

[59] Albert Einstein, 'A Reply to the Soviet Scientists', *Bulletin of the Atomic Scientists* 4, no. 2 (February 1948): 35–37; Sergei Vavilov, A. N. Frumkin, A. F. Ioffe, and N. N. Semyonov, 'Open Letter to Dr. Einstein – From Four Soviet Scientists', *Bulletin of the Atomic Scientists* 4, no. 2 (February 1948): 34, 37–38.

[60] Harrison Brown, *Must Destruction be our Destiny? A Scientist Speaks as a Citizen* (New York: Simon and Schuster, 1946), 91. Urey also used anarchy to make the case for international control: Harold C. Urey, 'Atomic Energy and World Peace', *Bulletin of the Atomic Scientists* 2, no. 10 (1 November 1946): 2–4.

described its proposal for the creation of a power international technical governance organization (the ADA) as something which was advanced, perhaps 'too advanced' for the world, just as the atomic bomb itself was 'too advanced' for mankind.[61]

The report was not simply driven by terror or alarm. It also articulated an enthusiasm and willingness to use the atomic to build an internationalist world order. It called for the grasping of this 'opportunity' to prevent future war, strengthen the UN, and establish 'new patterns of cooperative effort ... capable of extension to other fields, and which might make a contribution toward the gradual achievement of a greater degree of community among the peoples of the world'.[62] Oppenheimer recalling his work on the report a few months later noted that the international control of atomic energy was actually a 'code', because the real objective, the 'real problem', was the 'prevention of war'. The report's proposed ADA was also not simply to be a technical organization, but a miniature 'world government' in the field of atomic energy.[63] With State Department and scientists' support the report became the most widely supported and most definitive expression of the urge for international control in the United States. It helped spread scientists' internationalism but was also popular (in part) because it reflected and fed off internationalist ideas about technology, war, and international relations that were already widely present in the public domain.[64]

Atomic scientists' rhetoric, like that of Shotwell and the Carnegie internationalists, was technocratic in nature. Scientists too sought a central role for technical experts in not only advising but policymaking, and their schemes positioned themselves (rather than political scientists or industrialists) as central to any solution to the problems of atomic energy. The 'need for a more active political role of the scientists' was central to the 1946 constitution of the Federation of American Scientists; a December 1945 analysis emphasized the importance of supranational

---

[61] Barnard et al., *A Report on the International Control of Atomic Energy*, 1, 4, 5, 31.

[62] Ibid., 3.

[63] J. Robert Oppenheimer, 'A Future for Atomic Weapons', in *Science and Civilization*, by A. V. Hill et al., vol. 1, *The Future of Atomic Energy* (New York: McGraw-Hill, 1946), 77–90.

[64] *Life* magazine, for example, was an enthusiastic supporter: as late as December 1946 it would describe the Acheson-Lilienthal Plan as a natural and necessary part of the peaceful use of atomic energy: 'Peacetime Uses of Atomic Energy', *Life*, 2 December 1946, 97–103. The magazine was a strong supporter of international control: Editorial, 'Atomic Spring', *Life*, 15 April 1946, 36. On the atomic in *Life* see: Scott C. Zeman, '"To See ... Things Dangerous to Come to": Life Magazine and the Atomic Age in the United States, 1945–1965', in *The Nuclear Age in Popular Media: A Transnational History, 1945–1965*, ed. Dick Van Lente (New York: Palgrave Macmillan, 2012), 53–77.

scientific research and development as a necessary part of any system for international control. This included 'an international laboratory staffed with top-flight men working on the problem of atomic energy, to keep ahead of any single nation in developing new methods which may make inspection procedures obsolete'.[65] The Acheson-Lilienthal Report also gave atomic scientists and engineers, working through the ADA, a prominent role in international control. They would run the ADA, which would have policymaking and enforcement power (subject to a charter), and be recruited from the world's top scientists, ensuring that the ADA would stay ahead of national research programmes.[66] For Szilard the ADA was to point the way to the formation of other technical international organizations, each bound by a charter and dedicated to particular technical areas such as commodities, consumer industries, agriculture, and education.[67]

Liberal internationalist sentiments were evident in the earliest scientists' thinking on the social and political impact of atomic energy. During the war Manhattan Project scientists at the Metallurgical Laboratory branch of the project debated atomic energy's impact on international relations and produced the Jeffries and Franck Reports on this issue. These reports reflected their sense of the deep importance of scientific research for modern society, as well as their concern that scientific research should continue, well funded and with minimal government interference, after the war. But they also reproduced liberal internationalist views. Section VI of the Jeffries Report, written by Rabinowitch and physical chemist Robert Mulliken in October/November 1944, dealt with 'The Impact of Nucleonics on International Relations and the Social Order' and used the lag to explain the effects of atomic energy. The gap between 'continued technological progress and our relatively static political institutions', it warned, would be widened by atomic energy unless 'an international administration with police powers which can effectively control at least the means of nucleonic warfare' was formed.[68] The Franck Report, finalized June 1945 (with Rabinowitch once again taking a leading role in drafting), not only included the negative effects of atomic energy as a rationale for international control but also emphasized the opportunities inherent in this new science. The bomb offered the

---

[65] The Federation of American Scientists, *The Federation of American Scientists*; 'Atomic Power Control Problems. 2. General Analysis', *Bulletin of the Atomic Scientists* 1, no. 3 (10 January 1946): 2.

[66] Barnard et al., *A Report on the International Control of Atomic Energy*, 42, 44, 45.

[67] Leo Szilard, 'The Physicist Invades Politics', *Saturday Review of Literature*, 3 May 1947, 7–8, 31–34.

[68] Zay Jeffries et al., 'Appendix A: Prospectus on Nucleonics (the Jeffries Report)', in Smith, *A Peril and a Hope*, 539–559.

opportunity for international cooperation and organization: 'Among all the arguments calling for an efficient international organization for peace, the existence of nuclear weapons is the most compelling one'.[69]

There is some evidence that Danish physicist Niels Bohr's conversations with senior US scientists in 1944 may have played a role in directing them towards international control. His calls for international control were encased in liberal internationalist language with an added emphasis on the inherent internationalism of the scientist. Bohr's July 1944 letter to Roosevelt urged international control as part of a 'new approach to the problems of international relationship' through the institution of collective security and the international spreading of the boon of atomic energy. Scientists' international communication and cooperation offered a model for this internationalization. It was fortunate, he noted, that these 'unique opportunities' for 'international co-operation' created by 'the advancement of science' coincided with a general internationalism amongst the 'united nations'. Bohr later reworked these thoughts for an article published in *The Times* on 11 August (and subsequently republished in the *One World or None* volume). Although this article reduced the earlier emphasis on the uniting of nations, it nevertheless opened with a sentence on a lag between the new 'formidable power of destruction' and 'human society', and moved on to call for humanity to unite in 'averting common dangers and jointly to reap the benefit from the immense opportunities which the progress of science offers'. Scientists had a special role to play in this new internationalism as they were already 'a brotherhood working in the service of common human ideals'.[70]

Connections to wider internationalist currents are also apparent in scientists' support for world government and international federation. Urey, Smyth, and physicist Samuel King Allison attended the March 1946 Rollins conference on world government and added their names to its final declaration.[71] Urey gave a speech at the conference calling for world

---

[69] James Franck et al., 'Appendix B: The Franck Report June 11, 1945', in Smith, *A Peril and a Hope*, 560–572. Also: Price, 'Roots of Dissent'.

[70] Niels Bohr, 'Niels Bohr's Memorandum to President Roosevelt', Atomic Archive, July 1944, accessed 2 August 2020, www.atomicarchive.com/Docs/ManhattanProject/Boh rmemo.shtml; Niels Bohr, 'Energy from the Atom. An Opportunity and a Challenge, The Scientist's View', *The Times*, 11 August 1945; Niels Bohr, 'Science and Civilization', in Masters and Way, *One World or None*, ix–x. Also: Niels Bohr, 'Challenge to Civilization', *Science* 102, no. 2650 (12 October 1945): 363. For more on Bohr's activities see Smith, *A Peril and a Hope*, 4–13; Martin J. Sherwin, 'Niels Bohr and the First Principles of Arms Control', in *Niels Bohr: Physics and the World*, eds. Herman Feshbach, Tetsuo Matsui, and Alexandra Oleson (Abingdon: Routledge, 1998), 319–330; Finn Aaserud, 'The Scientist and the Statesmen: Niels Bohr's Political Crusade during World War II', *Historical Studies in the Physical and Biological Sciences* 30, no. 1 (1999): 1–47.

[71] 'The Atomic Bomb and World Government Conference', *Rollins Alumni Record* 14, no. 1 (March 1946): 8–9; Rollins College Conference on World Government, 'An Appeal to the Peoples of the World: Issued by the Rollins College Conference on World

government and welcomed the Acheson-Lilienthal Report in July 1946 as not only an effective plan for international control but also the first step towards such a polity. In June 1947 the *Bulletin of the Atomic Scientists* published an article by him advocating international federal union amongst Western democracies, and a few months later physicist Edward Teller asserted that scientists should support more than the Baruch Plan – they should support world government.[72] Einstein subscribed to Emery Reves's internationalist vision and reasoned in November 1945 that the Soviet Union, the United States, and Britain could form a world government, with other countries joining later. He continued to call for a world government into 1947, without the Soviets if necessary.[73]

The World Government Committee of AORES represented the most organized support for world government within the scientists' movement. Formed in March 1946, it lasted around 14 months, and was particularly active in the latter half of 1946 under the chairmanship of Oak Ridge scientist John L. Balderston. Although particularly vocal in its activism, it failed in its attempts to insert world government as a policy objective of AORES.[74] The Committee's proposal for international control envisaged the creation of an Atomic Development Authority of the type described by the Acheson-Lilienthal Report, the crucial difference being that this Authority was to be embedded within a world government and not the United Nations. The UN was too weak to enforce international control on powerful countries such as the Soviet Union; the new world government would be composed of all the peoples of the world and would be directly answerable to them, and would have the power to punish individuals for breach of atomic (and other) rules.[75] Although drawing on

Government (March 11–16, 1946, Winter Park, Fla.)', *World Affairs* 109, no. 2 (1946): 83–86.

[72] Urey, 'Atomic Energy, Master or Servant?'; 'Acheson Report Praised By Urey', *New York Times*, 12 July 1946; Harold C. Urey, 'An Alternative Course for the Control of Atomic Energy', *Bulletin of the Atomic Scientists* 3, no. 6 (June 1947): 139–142; Edward Teller, 'Atomic Scientists have Two Responsibilities', *Bulletin of the Atomic Scientists* 3, no. 12 (December 1947): 355–356.

[73] Albert Einstein, 'Einstein on the Atomic Bomb', *Atlantic Monthly*, November 1945, 43–45; see also Albert Einstein, 'Letter to Emery Reves, 29 October 1945', in Rowe and Schulmann, *Einstein on Politics*, 371; Albert Einstein, 'Atomic War and Peace', *Atlantic Monthly*, November 1947, 29–32.

[74] Memorandum, J. L. Balderston to T. Rockwell, 17 October 1946, folder 8, box 8, Series II, AORES Records. On the Committee see also: Baratta, *Politics of World Federation*, Vol. 1, 130–131.

[75] Report, World Government Committee of AORES, 'Effective International Control Requires World Government', 22 October 1946, folder 4, box 9, Series II, AORES Records; World Government Committee of the Association of Oak Ridge Engineers and Scientists, *Primer for Peace: World Government for the Atomic Age – An Analysis with Recommendations for Action* (Oak Ridge, TN: Association of Scientists for Atomic Education, November 1947), 4.

ideas from the broader world government movement, the Committee's vision of international control was more demanding than that of many world federalists. In response to a request from the Committee for comments on their draft proposals, Cord Meyer, Jr. responded that he wanted a world government with extensive powers as well, but there was 'little chance' of the United States accepting and 'no chance' of the USSR doing so. Instead AORES needed to demand and accept a more realistic and gradual approach to international organization.[76] Quincy Wright, also replying to the same query, thought that the Committee's proposals would lead to a 'strongly centralized world government' which would be unacceptable not only to the American public but worldwide.[77] Replying to Meyer, Balderston conceded that the '80–90%' that Meyer proposed may be enough, but insisted that his Committee wanted to go further and draft up a more comprehensive '100%' plan for world government.[78]

Atomic Police

The transition from aviation to atomic energy in the internationalist consciousness was consequently not a sharp break, rather a transition. This evolution is also apparent in the re-emergence of the international police force, this time embedded within proposals for the international control of atomic energy, and often equipped with atomic bombs. The notion of an atomic bomb armed police was popular enough, in the early months, to feature in an opinion poll on 22 August 1945 in which 14 percent supported the suggestion that 'the new United Nations Security Council use the atomic bomb to help keep peace by putting it under control of a special international air force'.[79] For those whose primary concern was bolstering the UN or empowering a new powerful world organization an international air force still seemed to be the only way to ensure global enforcement. Moreover, for many the atomic bomb was just another (albeit exceptionally powerful) addendum to the power of the aeroplane. Many in the US continued to think of the country as a global aerial power whose security and military might rested on its air force and global system of air bases.[80]

---

[76] C. Meyer, Jr. to J. L. Balderston, 12 October 1946, folder 1, box 9, Series II, AORES Records.

[77] Q. Wright to J. L. Balderston, 10 October 1946, folder 2, box 9, Series II, AORES Records.

[78] Letter, copy, J. L. Balderston to C. Meyer, Jr., 18 October 1946, folder 1, box 9, Series II, AORES Records.

[79] Hadley Cantril, *Public Opinion 1935–1946* (Princeton, NJ: Princeton University Press, 1951), 20–21.

[80] On perceptions of air power in the early post-war years see: Call, *Selling Air Power*, 33–37.

Prominent proponents included world federalist Cord Meyer, Jr., Ely Culbertson, Clyde Eagleton, the internationalist Senator Joseph H. Ball, Harold E. Stassen, Thomas Finletter, and Shotwell himself. Culbertson, Stassen, and Ball had called for an international police force in the early 1940s, and quickly incorporated the atomic bomb into their continuing calls in late 1945.[81] Stassen, a one-time Governor, delegate to the San Francisco Conference, and a leading Republican, attracted substantial media attention. One pronouncement, in particular, was widely reported: a November 1945 call at the Academy of Political Science for a 'United Nations Air Force of five bomber squadrons and ten fighter squadrons' based at five bases around the world and furnished with five atomic bombs for each bomber squadron 'to serve as the stabilization force for world order'.[82] In November 1945 Ball called for atomic bombs to be stored in an international police force base, and argued that this reformed United Nations force should be created immediately, irrespective of whether or not the Soviet Union participated.[83] Culbertson, meanwhile, quickly incorporated the atomic bomb into his pre-existing and widely known Quota Force Plan. Writing to Eichelberger in September 1945, he explained that his 'Quota Force method for the control of modern fighting machines and aggression' could easily be applied with 'certain adjustments and improvements' to atomic weapons. In this 'logical extension' atomic bombs were to be treated in a similar way to what he called 'heavy' armaments – predominantly bombers, submarines, and tanks. He called for the formation of a UN 'Atomic Arms Commission' to allocate quotas for atomic bombs to the only nations that would be allowed to own them: the United States, Britain, Canada, France, China, and the USSR. He also called for the formation of an 'Atomic Armament Trust' to produce atomic armaments for the smaller nations of the world, which would then

[81] H. E. Stassen, 'Blueprint for a World Government', *New York Times Magazine*, 23 May 1943, 3, 34; H. E. Stassen, 'We Need a World Government', *Saturday Evening Post*, 22 May 1943, 11. Stassen also supported proposals for a UN air force during the Dumbarton Oaks Conference; see 'Oral History Interview with Harold Stassen', Truman Library, accessed 2 August 2020, www.trumanlibrary.gov/library/oral-histories/stassen. He would continue to call for a 'U.N. legion' into the nineties; see H. E. Stassen, *United Nations: A Working Paper for Restructuring* (Minneapolis, MN: Lerner Publication, 1994).

[82] Harold Stassen, 'Atomic Control', *Proceedings of the Academy of Political Science* 21, no. 4 (January 1946): 103–112. First published as: Harold E. Stassen, 'From War to Peace: An Address by Harold E. Stassen, November 8, 1945', *International Conciliation* 23 (December 1945): 801–810. For publicity see: 'Outlawing of Atomic Bombs Advocated by Capt. Stassen', *New York Times*, 8 November 1945; 'World Control for Atomic Power Captain Stassen's Proposal', *The Times*, 10 November 1945.

[83] Joseph H. Ball, 'A Chance for Peace', in *A Chance for Peace and Responsibility for Peace*, by Joseph H. Ball and Harold C. Urey (Cincinnati, OH: Cincinnati Foreign Policy Association, November 1945).

be made available to a UN international police force, which would enforce this quota system on both atomic and 'heavy' arms.[84]

Wartime members of the CSOP also reiterated their calls for an international police force, now to be armed with atomic bombs. Shotwell called for a UN international air force with exclusive use of the atom bomb between August and October 1945 but dropped the call once he began work on the Carnegie Plan for international control. Notably, he defended Stassen's proposals for an atomic force at a November 1945 symposium on atomic energy.[85] Eagleton had called for an international police force in a series of lectures given at New York University in 1943,[86] and, although he only briefly mentioned it as a possibility in May 1946,[87] his revised and reissued 1948 edition of *International Government* reiterated this call in detail. His new sections on the United Nations called for the force to have a monopoly on 'scientific weapons', which included the 'atomic bomb and other modern weapons'.[88]

The October 1945 Dublin Declaration included a commitment to an international police force with exclusive use of the atomic bomb and 'other major weapons'.[89] Cord Meyer, Jr., a signatory, explained the logic behind his support to Balderston of the World Government Committee of AORES in October 1946. Although he agreed with the Committee that atomic bombs would ultimately have to be destroyed, in the short term 'it will be necessary at first to allow the world police to possess the bombs already in existence. This would create confidence among all members that no state would have a chance of successful revolt and aggression against the preponderant power of the world police. As time went on and mutual confidence began to grow, the world police could afford to dispose of the bombs, but in the first stages it should have them to discourage national revolt'.[90] Lawyer and wartime State Department official Thomas Finletter, Director of Americans United

---

[84] E. Culbertson to C. M. Eichelberger, 15 September 1945, folder: 'E 1945', box 59, Eichelberger Papers; Ely Culbertson, *How to Control the Atomic Threat* (New York: Total Peace, 1945). *Time Magazine* was a supporter of Culbertson, seeing his ideas behind various initiatives on atomic matters; see: 'Heads Up!', *Time Magazine*, 15 October 1945; and 'Perilous Fission', *Time Magazine*, 29 October 1945.

[85] A revised draft of his *Preliminary Suggestions*, dated 6 October 1945, made this suggestion (see Josephson, *James T. Shotwell and the Rise of Internationalism in America*, 265) as did: James T. Shotwell, 'The Atom and the Charter', *New York Herald Tribune*, 12 August 1945; James T. Shotwell, 'Control of Atomic Energy', *Survey Graphic* 34 (October 1945): 407–408, 417; Shotwell, 'The Control of Atomic Energy under the Charter'.

[86] Eagleton, *The Forces that Shape Our Future*.

[87] Eagleton, 'The Demand for World Government'.

[88] Eagleton, *International Government*, 2nd ed., 448.

[89] 'Declaration of the Dublin, N.H., Conference', *New York Times*, 17 October 1945.

[90] C. Meyer, Jr. to J. L. Balderston, 12 October 46, folder 1, box 9, Series II, AORES Records.

for World Organization and another prominent signatory to the Dublin Declaration, wrote in March 1946 that the UN should exclusively own and manufacture all 'weapons of mass destruction' which include 'atomic and bacteriological weapons, V-1s, V-2s, bombers and all bombs (especially the 50-tonners we have been told about), and all similar weapons already invented and to be invented'.[91] Proposals for an international police force were also expounded at international meetings of the various world federalist organizations. The manifesto of the United World Federalists (November 1947) and the Montreux Declaration from the World Movement for Federal Government (August 1947) both envisaged the formation of an international police force.[92]

Some scientists were also attracted to the idea of an atomic bomb armed police force. James B. Conant, an early exponent, laid out his ideas in a memo dated May 1944. He thought that a powerful supranational body, run by a fifteen-member commission with three representatives each from the United States, Britain, and the USSR could license national atomic plants and have full authority to inspect them. The commission would control an international nuclear arsenal located in Canada as well as an air force and 10,000 personnel. The commission would have the power to use its atomic weapons against violators of the international atomic control treaty. Conant responded positively to Culbertson's quota force scheme, including to the suggestion that the force be formed without the Soviet Union if required.[93] The physicist and administrator Arthur H. Compton was another prominent exponent. At a November 1945 Symposium on Atomic Energy he announced that 'World government has become inevitable' and called for the UN to 'enforce peace' through a 'virtual monopoly' of the three 'major means of warfare': 'airplanes, long range rockets, and atomic bombs'. Atomic disarmament would be a 'fatal mistake'; instead Compton called for all atomic weapons to be placed at the disposal of a Military Commission, based he suggested in Canada, under the control of the UN Security Council.[94] AORES called for atomic bombs to be surrendered to the UN for policing purposes in a statement read before a Senate hearing in

---

[91] Thomas Finletter, 'Timetable for World Government', *Atlantic Monthly*, March 1946, 53–60.
[92] Baratta, *The Politics of World Federation*, vol. 1, 545–550; First International Congress of the World Movement for World Federal Government, 'Montreux Declaration', 23 August 1947, folder 9, box 32, Series IV, Atomic Scientists Records.
[93] Hershberg, *James B. Conant*, 198, 205–207, 246–247.
[94] Arthur H. Compton, 'Franklin Medal Lecture: Atomic Energy as a Human Asset', *Proceedings of the American Philosophical Society* 90, no. 1 (January 1946): 70–79. See also: Arthur H. Compton, 'The Social Implications of Atomic Energy', *American Journal of Physics* 14, no. 3 (May–June 1946): 173–178.

October 1945, and the World Government Committee of AORES emphasized the importance of a world police force in its November 1947 *Primer for Peace.*[95]

Some prominent scientists called for an international police force but without explicitly equipping it with atomic weapons. Physicist Robert Andrews Millikan first made the call during the Second World War, and reiterated it in his autobiography, published in 1951. In a 1947 book on subatomic particles he called for international control to be enabled by an international air force for international reconnaissance.[96] Einstein, who had supported calls for an international police force in the early 1930s, also publicly supported an international force between 1945 and 1947; his essay in the 1946 *One Way or None* compilation was essentially a plea for the gradual denationalization of national military forces and their amalgamation into an international army.[97] The Rollins College world government conference declaration called explicitly for a world government armed with an international force. Alongside Urey and Smyth, signatories included Einstein, Joseph H. Ball, Cord Meyer, Jr., I.I. Rabi, Emery Reves, and Raymond Swing. Internationalist (and President of Rollins College) Hamilton Holt, the moving force behind the conference, explained that his experience with the League to Enforce Peace was central to his support for the proposal.[98] These calls were so prominent by early 1946 that even the Acheson-Lilienthal Report left open the possibility of UN use of atomic weapons. Although the Report did not explicitly address UN use, it did not call for the destruction of the current stock of US atomic bombs. They were instead to be transferred to the United Nations (perhaps in a dismantled form) which would decide what to do with them.[99] This lack of clarity was highlighted and criticized

[95] Smith, *A Peril and a Hope*, 155; World Government Committee of the Association of Oak Ridge Engineers and Scientists, *Primer for Peace*.

[96] Robert A. Millikan, *Science, War and Human Progress: Address by Dr. Robert A. Millikan* (New York: The National Association of Manufacturers of the United States of America, 1943); Robert A. Millikan, *The Autobiography of Robert A. Millikan* (London: MacDonald, 1951); Robert A. Millikan, *Electrons (+ and -), Protons, Photons, Neutrons, Mesotrons, and Cosmic Rays* (Chicago: University of Chicago Press, 1947), 440.

[97] On the thirties: Pemberton, *The Story of International Relations*, part two, 156–162; Albert Einstein, 'The Way Out', in Masters and Way, *One World or None*, 76–79. Also: Albert Einstein, 'Einstein on the Atomic Bomb', *Atlantic Monthly*, November 1945, 43–45; Albert Einstein, 'The Real Problem is in the Hearts of Men', *New York Times Magazine*, 23 June 1946, 7.

[98] George C. Holt, 'The Conference on World Government', *Journal of Higher Education* 17, no. 5 (May 1946): 227–235; Rollins College Conference on World Government, 'An Appeal to the Peoples of the World'; Hamilton Holt, 'How the Conference Came About', *World Affairs* 109, no. 2 (1946): 87–90.

[99] Barnard et al., *A Report on the International Control of Atomic Energy*, 48, 54, 56.

by some commentators who preferred complete abolition of atomic weapons.[100]

Calls for an atomic bomb armed international police force were prominent enough for some to speak out publicly against them. Urey pointed out the difficulties in deciding where the UN would store their atomic weapons, and anyhow would the US public really support a United Nations which might turn its atomic bombs against their country? Oppenheimer joined Urey in warning that atomic bombs were too destructive to be used for policing, and that one nation may seize the UN's atomic stockpile.[101] Even the Yale Institute of Advanced Studies' June 1946 *The Absolute Weapon*, remembered now for its realist stance towards the bomb (particularly by Bernard Brodie), expressed both support and criticism of atomic-armed police force proposals.[102]

Those arguing against a UN atomic force had some reason to be concerned. There was some support for the idea amongst hawkish policymakers. When Baruch's advisors turned to the military in April 1946 for advice on preparing the official international control proposal to be placed before the UNAEC, Army Air Force General Henry (Hap) Arnold suggested that the members of the UNAEC conclude a treaty amongst themselves that would outlaw the use of the bomb and agree to use it against any violator. Fred Searls, Jr., a mining engineer and one of Baruch's leading advisors, suggested that this idea be operationalized in the form of a UN air force. Armed with atomic bombs the force could be based at five strategic and far-flung air bases: the Azores, Cairo, Karachi, Burma, and the Philippines. Needless to say, this force was opposed by others in the Baruch delegation and many in the US military itself, which was horrified at the thought of losing control of atomic weapons in this way. It was not incorporated into the Baruch Plan.[103]

---

[100] Edward Shils, 'Some Political Implication of the State Dep't Report', *Bulletin of the Atomic Scientists* 1, no. 9 (15 April 1946): 7–9, 19.

[101] Harold C. Urey, 'The Atom and Humanity: An Address by Harold C. Urey, October 21, 1945', *International Conciliation* 23 (December 1945): 790–800; J. R. Oppenheimer, 'The Atom Bomb as a Great Force for Peace', *New York Times*, 9 June 1946; Smith, *A Peril and a Hope*, 155; Frederick Dunn, 'Controlling Atomic Power by Treaty', n.d., folder: 'Atomic Energy Conference at Chicago: Sept. 19, 20, 1945', box 28, Series II, Atomic Scientists Records.

[102] Arnold Wolfers, 'The Atomic Bomb in Soviet-American Relations', in *The Absolute Weapon: Atomic Power and World Order*, ed. Bernard Brodie (New York: Harcourt Brace, 1946), 111–147; Percy E. Corbett, 'Effect on International Organization', in Brodie, *The Absolute Weapon*, 148–165; William T. R. Fox, 'International Control of Atomic Weapons', in Brodie, *The Absolute Weapon*, 169–203. On the study see: Fred Kaplan, *The Wizards of Armageddon* (Stanford: Stanford University Press, 1991), chapter 2.

[103] Herken, *The Winning Weapon*, 165–167.

A Mess of Things

In August 1946 editor and intellectual Clifton Fadiman wrote a piece in *The Saturday Review of Literature* on the death of H.G. Wells earlier the same month. Fadiman grieved at the passing away of not only Wells but a certain type of 'Wellsian energy' or 'Wellsian hope' which looked to the power of mankind's rationalistic intelligence to make the world a better place and to save it from the horrors of mechanized warfare:

> The nationalism that Wells attacked with such fury thirty years ago is fiercer today than it has ever been. The leaders, the great captains of industry and men of action in whom he placed his occasional trust, have, for the most part, proved frail reeds. ... Perhaps, a few hundred years from now, a few men will painfully arise from the ruins of a cracked and seamed and fissured planet, and begin to rediscover, very slowly, the basic concepts of civilization. And perhaps one of them will unearth from the rubble of what was once a library a copy of 'The Outline of History', or 'The Shape of Things to Come', and read it, and wonder how his ancestors, with such books to warn them, could ever have made such a mess of things.[104]

The 'mess' that Fadiman referred to was the failure to institute international control. The Soviets rejected the Baruch Plan laid before the UNAEC in June 1946, and instead made a counterproposal that envisaged the complete abolition of atomic bombs. Baruch refused to budge from his proposal, and the Soviets from theirs, and by August hopes for international control were slipping. Negotiations between the United States and the Soviet Union continued through to the end of 1946, but no agreement was reached. The Baruch Plan was effectively vetoed by the USSR in the Security Council in December 1946. By 1947, although some internationalists continued to call for international control, and to hope that negotiations may be revived, the US state's interest in international control had passed, and the US public also lost interest. Instead of cooperation, concerns over a possible war with the Soviet Union now occupied the public mind.[105] Calls for an international police armed with atomic bombs died away too. By late 1946 Stassen no longer spoke of an atomic armed force, and Culbertson reissued his quota force plan but without reference to atomic bombs. His scheme now introduced the category of 'scientific weapons' which included his earlier 'heavy' weapons, as well as atomic, chemical, and 'bacteriological' weapons and any

---

[104] Clifton Fadiman, 'H.G. Wells Sees It Through: 1. The Passing of a Prophet', *Saturday Review of Literature*, 31 August 1946, 3–6.

[105] Herken, *The Winning Weapon*, 169–170–175, 186–191; Melvyn P. Leffler, 'Strategy, Diplomacy, and the Cold War: The United States, Turkey, and NATO, 1945–1952', *Journal of American History* 71, no. 4 (March 1985): 807–825.

new 'physical or chemical devices of highly destructive power' that may be developed in the future. His international force this time was only to be armed with heavy weapons, not scientific weapons such as the atomic bomb.[106] Cord Meyer, Jr.'s October 1947 *Atlantic Monthly* article on international relations restated his call for a powerful UN force to be armed with all 'modern weapons', but not, explicitly, the atomic bomb.[107] Although still the 'absolute weapon', the atomic bomb was no longer seen as an internationalist tool of peace, but as a nationalist tool for deterrence or even war.[108]

[106] Culbertson, *Must We Fight Russia?*, 54, 6, 58.
[107] Cord Meyer, Jr., 'Peace Is Still Possible', *Atlantic Monthly*, October 1947, 27–33.
[108] Brodie, *The Absolute Weapon*.

7   A Blessing in Disguise: Britain and the International Control of Atomic Energy

> 'We are thus brought to realize, only a few weeks after the signing of the United Nations Charter, that the world order established at San Francisco is nothing like sufficiently strong or sufficiently watertight. There must be evolved from it, as speedily as possible, a world sovereignty recognized everywhere as supreme. If this new invention, which opens up such appalling vistas, forces the nations to accept this basic truth, it will have proved a blessing in disguise'.
> – *News Chronicle* (8 August 1945).[1]

In Britain too there was a widespread liberal internationalist response to the unveiling of the atomic bomb. This response came not only or primarily from scientists, but rather from a broad range of individuals, organizations, and press located at the centre and centre-left of the political spectrum. Between late 1945 and 1948 liberal internationalists (many also vocal supporters of the UN) emerged as the leading source of pronouncements on the effects of atomic energy on international relations. Many worked through internationalist organizations such as the New Commonwealth Society, the United Nations Association, and Chatham House. Although their analysis and rhetoric emphasized the uniqueness of the threats and opportunities latent in atomic energy, and they argued that their proposals were similarly new and novel, these proposals, like in the US case, built upon long-standing internationalist ideas, including an international police force and science's ability to transform international relations. British atomic internationalism was not the same as its US cousin, however. Some constituencies were more prominent than others, and an international atomic police force was more strongly supported in Britain.

Recognizing this atomic internationalism allows us to reconstruct important elements of post-war British atomic culture now lost or forgotten. In Britain liberal internationalism is generally associated with

---

[1] 'An End or a Beginning?', *News Chronicle*, 8 August 1945.

interwar and wartime intellectual and political culture, and is generally missing from our histories of the late 1940s.[2] Internationalist activity surrounding the atomic bomb is usually seen as being driven not by a widespread liberal internationalism but by scientists and their scientific internationalism, or by pacifists and peace movements.[3] This chapter argues that there was in fact a broader liberal internationalist reaction to the bomb, much of it from pre-existing internationalists and their organizations. Although in the historical literature an international police force is generally thought to be an interwar proposal, post-war cultural responses to the bomb and nuclear imaginaries incorporated and further embellished the police force concept.[4] Recognizing this response's appeal to science, its militant nature, and its willingness to envisage a world with atomic weapons may help to explain why it has gone unrecognized by historians.[5]

A Widespread Response

Calls for the international control of atomic energy were widespread and reached their peak in mid-1946. They were always couched in terms of strengthening the United Nations, and often included the notion of a UN

---

[2] See for example: Long and Wilson, *Thinkers of the Twenty Years' Crisis*; Sylvest, *British Liberal Internationalism*; Rosenboim, *The Emergence of Globalism*; Pugh, *Liberal Internationalism*. A partial exception is: R. M. Douglas, *The Labour Party, Nationalism and Internationalism, 1939–1951* (London: Routledge, 2004), chapter 5.

[3] On Science: Jones, 'The Mushroom-Shaped Cloud'; Greta Jones, *Science, Politics, and the Cold War* (London: Routledge, 1988), chapter 5; Christoph Laucht, 'Transnational Professional Activism and the Prevention of Nuclear War in Britain', *Journal of Social History* 52, no. 2 (2018): 439–467; Christoph Laucht, *Elemental Germans: Klaus Fuchs, Rudolf Peierls and the Making of British Nuclear Culture 1939–59* (Houndmills: Palgrave Macmillan, 2012), chapter 6. Peace movements: Holger Nehring, *Politics of Security: British and West German Protest Movements and the Early Cold War, 1945–1970* (Oxford: Oxford University Press, 2013), 16–62.

[4] On early post-war British atomic culture: Matthew Grant, 'The Imaginative Landscape of Nuclear War in Britain, 1945–65', in *Understanding the Imaginary War: Culture, Thought and Nuclear Conflict, 1945–90*, eds. Matthew Grant and Benjamin Ziemann (Manchester: Manchester University Press, 2016), 92–115; A. Bingham, '"The monster"? The British popular press and nuclear culture, 1945–early 1960s', *British Journal of the History of Science* 45, no. 2 (2012): 609–625; Christoph Laucht, '"Dawn – Or Dusk?" Britain's Picture Post Confronts Nuclear Energy', in *The Nuclear Age in Popular Media: A Transnational History, 1945–1965*, ed. Dick van Lente (New York: Palgrave Macmillan, 2012), 117–148; Nehring, *Politics of Security*, 16–62; Jonathan Hogg, *British Nuclear Culture: Official and Unofficial Narratives in the Long Twentieth Century* (London: Bloomsbury, 2016).

[5] That the atomic bomb came to be seen as a weapon for internationalism may also help to explain why there was so little opposition (noted by Gowing) to the bomb in Britain: Margaret Gowing, *Independence and Deterrence: Britain and Atomic Energy, 1945–1952*, Vol. 1, *Policy Making* (London: Macmillan, 1974), 50.

international police force. For liberal Britain international control meant a strong system of UN collective security. One January 1946 poll gave a 74 percent approval rating for 'control of atomic energy under the United Nations Security Council'.[6] Two Mass Observation surveys on 'World Organisation and the Future' found that 52 percent of people polled favoured a UN international police force in February 1946, and 65 percent in June 1946.[7] Mass Observation observed with some surprise that 'Opinion breaks down along very definite political lines', with Labour supporters, rather than Conservatives, more likely to support an international police force.[8] This cleavage was also evident in press reaction to the atomic bombings. The left-leaning press called for international control as soon as the atomic bomb was used in August 1945. The centre-right press, by contrast, did not. Indeed, an internationalist magazine, publishing in October 1945 a 'cross-section of opinion' on atomic energy, could only present editorials supporting international control from the *Manchester Guardian*, the *Observer*, and the *New Statesman and Nation*.[9]

On the Centre-Left the *Manchester Guardian* emerged as the strongest supporter of international control. The newspaper had strong liberal internationalist leanings in its foreign affairs coverage and was (with some reservation) an early supporter of the United Nations. It is perhaps not surprising then that soon after the atomic bombings the newspaper called for the UN to take ownership of atomic raw material and plant, and for atomic bombs to be handed over to a UN police force. Indeed, this needed to be the new organization's 'first task'.[10] Moreover, as atomic bombs needed to be delivered by bombers, the newspaper was able to reiterate its support for the international air force provisions of the UN Charter. 'The argument for an International Air Force', it concluded, 'is now overwhelming'. To be able to effectively utilize the bomb the UN 'must have a separate territory for its manufacture and a separate Air Force for its delivery'.[11] Ownership of existing atomic bombs was not enough of course; after all they could be produced by any nation with research capabilities. Consequently, atomic science itself had to be internationally controlled with 'atomic plants and large-scale laboratories ... centralized somewhere outside the borders of the larger nations'. This would require, in addition, a 'world-wide intelligence system to ensure

[6] 'The Quarter's Polls', *Public Opinion Quarterly* 10, no. 1 (Spring 1946): 104.
[7] Mass Observation, 'World Organisation and the Future', March 1946, and 'World Organisation and the Future – Changing Attitudes – February to June', June 1946, Mass Observation File Reports 2370 and 2397, Mass Observation Archives.
[8] Mass Observation, 'World Organisation and the Future', 40.
[9] 'The Bomb', *The New Commonwealth* (October 1945), 324–327.
[10] 'The Charter', *Manchester Guardian*, 23 August 1945.   [11] Ibid.

that its monopoly was not infringed'.[12] For the *Manchester Guardian* supporting international control was not simply about providing the UN with teeth, it was also an opportunity to further develop international organization towards world government. The newspaper used the term 'world government' regularly from August 1945 onwards; when the Baruch Plan was announced it hailed the proposed ADA as a 'world government' in the 'atomic sphere' which could eventually become so powerful that it could 'supersede the Security Council and become the world government'.[13]

Although not all left-leaning newspapers called for international control (the popular daily the *Daily Mirror*, for example, did not mention it all), there was significant support in the *News Chronicle*, the *Observer*, and the *Daily Herald*. Lord Astor's *The Observer* explained in some detail why atomic energy now made UN military forces practical. Here at last was the modern efficient 'weapon against which no criminal could stand':

It is so powerful that it could be used effectively by quite a small, specially instructed Arm. It cuts out the need for elaborate interweaving of complex national and varied forces with all their separate traditions, jealousies, and suspicions. Thus it enormously simplifies the recruitment and organisation of a World Police.[14]

The *News Chronicle* and the *Daily Herald* took more radical stances by calling for the UN to be further strengthened through the reduction of national sovereignty. The *News Chronicle* reasoned that the UN was too weak to effectively control atomic energy, and called for it to be 'evolved' into a stronger 'world sovereignty': 'If this new invention, which opens up such appalling vistas, forces the nations to accept this basic truth, it will have proved a blessing in disguise'.[15] The *Daily Herald* similarly called for the UN to reduce national sovereignty with the power to inspect and regulate military and industrial activities.[16] Like the *Manchester Guardian* these newspapers also later greet the Acheson-Lilienthal and Baruch plans with enthusiasm.[17] Many left-of-centre magazines also supported

---

[12] 'The Atomic Committee', *Manchester Guardian*, 22 August 1945.
[13] 'The Control Plan', *Manchester Guardian*, 17 June 1946. Also: 'The Challenge', *Manchester Guardian*, 8 August 1945; 'The Atomic Committee', *Manchester Guardian*, 22 August 1945; 'The Charter', *Manchester Guardian*, 23 August 1945; 'Atomic Realism', *Manchester Guardian*, 9 February 1946.
[14] 'The Bomb', *The Observer*, 12 August 1945.
[15] 'An End or a Beginning?', *News Chronicle*, 8 August 1945.
[16] 'This New Power', *Daily Herald*, 7 August 1945; 'The Threat', *Daily Herald*, 10 August 1945.
[17] 'The American Plan', *Manchester Guardian*, 9 April 1946; 'Atomic Proposals', *Manchester Guardian*, 3 June 1946, 'The Control Plan', *Manchester Guardian*, 17 June 1946; 'Atomic Progress Report', *Economist*, 4 May 1946, 697–698; 'The Atomic Plan', *Economist*,

international control. The country's leading illustrated magazine, *Picture Post*, for example, devoted significant space to atomic energy in 1945 and 1946 (including the whole of its 25 August 1945 issue) and called for UN control of the atomic bomb.[18] The *Spectator* similarly devoted a notable amount of space to atomic energy, and called for UN control as early as its 10 August 1945 issue.[19]

Newspapers of the centre and centre-right, by contrast, strongly supported the development of a British atomic bomb and took an ambivalent stance towards atomic energy. The *Daily Mail*'s immediate reaction was a call for Britain to develop an atomic arsenal, and for cooperation between Canada, Britain, and the United States. Rather than international control, countries needed to 'establish their own controls of its use'.[20] The newspaper's air correspondent, meanwhile, argued that it was indeed possible for scientists to develop new technologies to protect cities from atomic destruction.[21] The Beaverbrook-owned *Daily Express* eschewed any comment on international control, though it did call for international cooperation, and even the sharing of the atomic bomb with the Soviet Union, in its earliest opinion pieces.[22] The *Economist* was sceptical about the possibility of international control in late 1945 but supported the Acheson-Lilienthal and Baruch plans once they were announced in May and June 1946.[23] Although *The Times* did sometimes signal its support for international control, it more often than not either abstained from mentioning international control in its editorials on atomic energy, or weakly called for greater international cooperation, supervision, or a 'spirit of common understanding and action'. Its response to both the Baruch and Acheson-Lilienthal plans was guarded: it expressed support but warned that the proposed ADA would be a step towards 'world government', with all its tyrannical

22 June 1946, 995–996; 'Atom Plan', *News Chronicle*, 15 June 1946; 'All-Nations Atom Power Corporation. US Plan for UNO', *Daily Herald*, 29 March 1946.

18 'Man Enters the Atom Age', *Picture Post*, 25 August 1945, 7; 'Atom Bomb: UNO's Problem and its Inspiration', *Picture Post*, 26 January 1946, 7. On the *Picture Post* and the atomic see: Laucht, '"Dawn – Or Dusk?" Britain's Picture Post Confronts Nuclear Energy'.

19 'A Crisis of Civilisation', *Spectator*, 10 August 1945, 117; Francis Williams, 'Control of the Bomb', *Spectator*, 17 August 1945, 146–147; 'The World's Last Hope', *Spectator*, 7 September 1945, 211–212.

20 'Control or Perish', *Daily Mail*, 8 August 1945; 'We Must Develop It', *Daily Mail*, 9 August 1945; 'Future of the Atom', *Daily Mail*, 22 August 1945.

21 Courtnay Edwards, 'There is an Answer to the Atomic Bomb', *Daily Mail*, 13 August 1945.

22 'Opinion', *Daily Express*, 7 August 1945; 'Opinion', *Daily Express*, 31 October 1945.

23 'The Control of Destruction', *Economist*, 18 August 1945, 225; 'Controlling the Atom', *Economist*, 25 August 1945, 256–257; 'Atomic Progress Report', *Economist*, 4 May 1946, 697–698; 'The Atomic Plan', *Economist*, 22 June 1946, 995–996.

possibilities.[24] In July 1946 it called for international cooperation through agencies such as the ILO, UNESCO, and FAO as it was 'easier to find common ground in the daily affairs of human beings than in strategic concepts, easier to agree about safety measures in the mines or standards of housing than about atomic bombs or the future of Italian colonies'.[25]

The Same Old Problem Dressed Up in a New and More Devilish Guise

Internationalists mobilizing in support of international control often saw atomic energy as the new aviation. Philip Noel-Baker informed his audience at the inaugural meeting of the Preparatory Commission of the United Nations Organization in November 1945 that nations had ignored 'science' for long enough: 'For 40 years we have had aviation, and hitherto it has been nothing but an unmitigated curse to all mankind. Atomic energy may be controlled, or it will end us'.[26] Stafford Cripps similarly pointed out that the world had failed to control the previous 'Frankenstein monster', the aeroplane, and that the atomic bomb would be 'ten thousand times worse'.[27] Internationalist Arthur Salter (more on him later) drew parallels between the League of Nations and aviation on the one hand and the atomic bomb and the United Nations on the other during the 22 August 1945 Commons debate on UN Charter ratification. The security provisions of the League of Nations had been devised, he noted, before the military effect of aviation became evident. Just as these provisions, and eventually the League itself, were undone by the arrival of the bomber in the 1930s, so too were the security provisions of the United Nations, suited to deal with military aviation, now threatened with obsolescence unless they were adapted to deal with the atomic bomb.[28]

This turn from aviation to atomic energy was evident too in the activism of the New Commonwealth Society. The Society, subdued following the death of David Davies in 1944, was quick to use the bomb to resuscitate

---

[24] 'Fulfilment', *The Times*, 16 August 1945; 'Controlling the Atom', *The Times*, 16 August 1945; 'Peace and the Atom', *The Times*, 9 October 1945; 'The Explosive Atom', *The Times*, 31 October 1945; 'Survival and the Price', *The Times*, 28 November 1945; 'A Nuclear Plan', *The Times*, 20 April 1946; 'Atomic Prospects', *The Times*, 17 June 1946.

[25] Editorial, 'I.L.O. and U.N.', *The Times*, 27 July 1946.

[26] 'Organization to End War. Mr. Noel Baker on British Faith', *The Times*, 26 November 1945.

[27] Stafford Cripps, 'Either Achieve "Utopia" – or Dies Horribly', *Daily Herald*, 24 September 1945.

[28] Hansard, HC, vol. 413, cols. 695–697, 22 August 1945.

its activities and its vision of a heavily armed international organization enforcing global order. Many in the Executive viewed atomic destructiveness as an extension of aerial destructiveness: in the words of New Commonwealth supporter Vivian Carter, the bomb 'only represents an extension of aerial warfare to the nth degree (total annihilation) ... In short, it's the same old problem dressed up in a new and more devilish guise, not a new problem'.[29] In December 1945 or early 1946 the Society's Executive passed a resolution adopting as policy the proposal that the atomic bomb be preserved for the exclusive use of an international air force.[30] The New Commonwealth was quick to organize meetings and lectures on atomic energy, and it sponsored the Mass Observation reports mentioned earlier. It also brought in scientists to explain atomic energy and its impact. In October 1945 the New Commonwealth's journal published an article by physicist N.F. Mott (who in 1946 would become the first President of the Atomic Scientists' Association) describing the impact of the atomic bomb on warfare, and in November 1945 scientists M.L. Oliphant and William Chadwick explained the science behind atomic energy to a meeting of the New Commonwealth Parliamentary Group.[31]

Individuals associated with the New Commonwealth Society also put the atomic bomb into their pre-existing arguments for the formation of an international air police. A prominent example is R.E.G. Fulljames, onetime RAF officer and active member of the Society from the 1930s to the late 1940s. He was a public proponent of an international air force in the 1930s and particularly during the war, when he lobbied the Foreign Office and was a leading mover for a resolution calling for an international air force at the Liberal Party Assembly in February 1945.[32] After retiring from the air force in 1945 in order to contest a seat for the Liberal Party, he continued his internationalist activities, working through the New Commonwealth, the Federal Union, and the United Nations Association. In 1946 he published

---

[29] V. Carter to N. B. Foot, 27 June 1946, B23/3- NC correspondence 1945 – 'C', Davies Papers.

[30] N. B. Foot to G. Cockerill, 18 January 1946, B23/3 – NC correspondence 1945 – 'C', Davies Papers.

[31] N. F. Mott, 'The Atomic Bomb and World Affairs', *London Quarterly of World Affairs* 11 (October 1945): 187–190; 'Minutes of the New Commonwealth Parliamentary Group', folder: 'NC Parliamentary Group Meeting', box 101, Davies Papers.

[32] Fulljames, 'An International Air Police Force'. For wartime letters to government urging internationalization, and response, see: R. Law MP to R. E. G. Fulljames, 25 May 1944, FO 371/40795, TNA; R. E. G. Fulljames to R. Law MP, 30 May 1944, FO 371/40795, TNA. He also lobbied Noel-Baker: R. E. G. Fulljames to P. Noel-Baker, 13 October 1942, 3/15, Noel-Baker Papers. On his sponsorship of the Liberal Party international air force resolution: R. E. G. Fulljames, *An International Police Force* (Southampton: Shirley Press, 1946).

an update of his long-standing proposals for an international air force, which this time was to be armed with atomic bombs and rockets. He presented his new proposals as an urgent solution to a novel problem – atomic energy had 'revolutionised warfare', made 'mighty fleets of giant bombers unnecessary' and 'opened out new possibilities of terror weapons'.[33] Another addition to his pre-existing proposals was international control of all scientific research and development in order to forestall further invention and innovation in scientific weaponry. An 'International Inspectorate and Intelligence Corps' was to 'enter and inspect all potential arms factories throughout the world' and an 'Armament Development Corps', staffed with both scientists and military men, was to develop these modern weapons. Important aspects of his earlier proposals that he retained included 'international control of air power' through an international air force and 'a measure' of internationalized civil aviation.[34]

The Society proved remarkably successful in attracting energetic and newly elected MPs into its ranks. The best known of these was Raymond Blackburn, an ardent anti-communist on the right of the Labour Party (he would leave it in 1950 to sit as an independent). Blackburn pushed for international control in Parliament, lobbied Ministers directly, toured the United States as a representative of the New Commonwealth (meeting, amongst others, the Federation of American Scientists), and participated in conferences and BBC discussions on international control.[35] By late 1945 Blackburn had carved out a reputation as the leading proponent of international control within Parliament. He was the principal mover behind the first tabled motion calling for international control, put to the House of Commons on 22 August 1945. Backed by nine other MPs, his amendment to a motion for approval of the UN Charter called for all research, development, and production of atomic energy to be placed exclusively in the hands of a UN 'international centre' manned by 'international scientists and experts'.[36] The New Commonwealth Society's muscular internationalism was reflected in Blackburn's reactions to the atomic bomb. Although he himself did not explicitly call for an international atomic force, he greeted the Baruch Plan by emphasizing the need to back its proposed inspections by force, and to force hesitant

---

[33] Fulljames, *An International Police Force*. Also published as: R. E. G. Fulljames, 'The Problem of Security from the Fear of War in the Atomic Age', *Fighting Forces* 24, no. 2 (June 1947): 84–86. A similar proposal was published in the journal of the New Commonwealth-affiliated London Institute of World Affairs: R. E. G. Fulljames, 'Security Under the United Nations', *World Affairs* 1 (October 1947): 402–408.
[34] Fulljames, *An International Police Force*.
[35] Raymond Blackburn, *I Am an Alcoholic* (London: Allan Wingate, 1959), 86, 87.
[36] 'Use of Atomic Energy. Cabinet Acts. Advisory Panel of Experts', *Manchester Guardian*, 22 August 1945.

nations to sign up to the Plan.[37] He was also keen that the United Nations acquire as much of the world's supply of uranium and thorium as soon as possible – a call he made in Parliament in August 1946.[38]

The high point of the Society's advocacy occurred in March 1947 when it sent a statement calling for international control, signed by a number of MPs and scientists, to the Prime Minister and Foreign Secretary. Signatories included Blackburn, Lord Brabazon, Clement Davies, Michael Foot, Martin Lindsay, Harold Nicholson, Arthur M.F. Palmer, and Arthur Salter, as well as the scientists Harrie S.W. Massey, Marcus L. Oliphant, and Rudolf E. Peierls.[39] In response, the Prime Minister (in a letter subsequently made public by the Society) suggested that 'the statement is somewhat optimistic' and that the Soviets were then unlikely to accept internationalization.[40] The Committee in reply suggested that Attlee bypass the United Nations altogether and approach Stalin and Truman directly.[41] This exchange received significant attention from the press – the *Manchester Guardian*, in particular, issued an editorial welcoming the 'note of urgency' in the approach to Attlee.[42] In May 1948, following the news that the UNAEC was to suspend its discussions on international control, Blackburn reiterated in Parliament this call to bypass the UN.[43]

The United Nations Association (UNA, until late 1945 known as the League of Nations Association) was initially cautious in its response, calling for UN inspection rather than international ownership of atomic energy. It was soon pushed in a more radical direction by the Chairman of its Executive Committee, Victor Bulwer-Lytton (the second Earl of Lytton, a civil servant and colonial administrator who chaired the Commission that produced the 1932 Lytton Report on Japanese aggression in China).[44] Lytton's proposal for international control, which

---

[37] Raymond Blackburn, letter to the Editor, 'Atomic Energy', *The Times*, 20 June 1946.

[38] 'Parliament Atomic Energy Control, Merits of The Two Plans, House of Commons', *The Times*, 3 August 1946.

[39] 'Banning of Atomic Weapons Request for International Authority', *The Times*, 8 March 1947.

[40] C. Attlee to R. Blackburn, 10 April 1947, folder: 'NC Observer, World Army by Liddell Hart, NC Atomic Energy Committee 1947', box 101, Davies Papers.

[41] 'Control of Atomic Energy Mr. Attlee's Reply to Committee', *The Times*, 16 June 1947.

[42] 'A New Atomic Start', *Manchester Guardian*, 16 June 1947. See also: 'Control of Atomic Energy Mr. Attlee's Reply to Committee', *The Times*, 16 June 1947; and 'Atomic Energy Deadlock. New Commonwealth Committee Suggests Meeting', *Manchester Guardian*, 16 June 1947. It even reached the *Bulletin of the Atomic Scientists*: 'Prominent Britons Urge Attlee to call Big Three Meeting', *Bulletin of the Atomic Scientists* 3, no. 8 (August 1947): 218, 220.

[43] 'Parliament Atomic Energy Control, "Final" Approach Suggested, House of Commons', *The Times*, 15 May 1948.

[44] 'Resolutions Adopted by the General Council of the League of Nations Union at its Annual Meeting in London on September 6th and 7th, 1945', 14 September 1945, folder

included an atomic bomb armed international police, was circulated to UNA branches in November 1945, and accepted and adopted by the Executive in December.[45] The UNA's published proposal, based largely on Lytton's, called for an international Board to 'own and to control the production of atomic energy, and to establish and maintain the central and most highly equipped research organization for the development of atomic energy'. Existing bombs were to be handed over to the Board, which would also produce new atomic weapons for a newly constituted Authority armed with international forces distributed around the world.[46] The report was propagated by leading UNA members such as Herbert Samuel (leader of the Liberal Party in the House of Lords), published in *United Nations News*, formed the basis of an Executive Committee-sponsored resolution at the June 1947 General Council meeting, and received some press coverage.[47] Following Lytton's resignation in late 1946 (to be replaced, incidentally, by Air Vice-Marshal Donald C.T. Bennett, who had been a strong proponent of an aerial international police force), UNA interest in international control declined and the atomic subcommittee ceased to function by February 1947.[48]

Although both the UNA and the New Commonwealth proposed the formation of a powerful atomic bomb armed international police force, they were divided on how this force was to be used. The UNA suggested that atomic bombs only be used against those countries that violated

4X/124: 'UNA 1946 Onwards', Noel-Baker Papers; 'Controlling Use of Atomic Bomb L.N.U. Council's Hope of International Action', *The Times*, 7 September 1945.

[45] League of Nations Union, *Annual Report for 1944* (London: League of Nations Union, 1945), 11; Victor Bulwer-Lytton, *Can We Survive the Atomic Bomb?* (London: Hutchinson, 1945); 'Minutes of a Meeting of the Provisional Executive Committee ... on Thursday, November 15th, 1945 at 2.15 PM', 3/1/1, Papers of the United Nations Association of Great Britain and Northern Ireland, Archives and Special Collections, British Library of Political and Economic Science, London School of Economics and Political Science (hereafter cited as UNA Papers); 'Minutes of a Meeting of the Provisional Executive Committee ... on Thursday, December 6th, 1945 at 2.15 PM', 3/1/1, UNA Papers; Memorandum, Earl of Lytton, 'Appendix A: International Control of Atomic Energy. Stages by Which International Control of Atomic Energy can be Established', n.d., ca. 1 December 1945, 3/1/1, UNA Papers.

[46] United Nations Association, 'Atomic Energy Sub-Committee Report to the Executive Committee', 5 December 1946, 3/1/1, UNA Papers.

[47] 'Appeal For Outlawing Of Atomic Bomb. Lord Samuel's Proposal', *The Times*, 25 March 1947; United Nations Association, 'Agenda of the Second Annual General Meeting', June 1947, 2/1, UNA Papers; 'Atomic Energy Report on Control and Production, International Board Suggested', *The Times*, 7 January 1947.

[48] 'Minutes of a Meeting of the Executive Committee ... on Thursday, 20th February, 1947 at 11 AM', 3/1/2, UNA Papers; Earl of Lytton, 'Memorandum by the Earl of Lytton', n.d., ca. September 1947, 3/1/2, UNA Papers; D. C. T. Bennett, 'International Security by Police Control', 6 June 1944, 8/1030, Chatham House Papers; D. C. T. Bennett, *Freedom from War* (London: Pilot Press, 1945); D. C. T. Bennett, *Pathfinder: A War Autobiography* (London: Frederick Muller, 1958), 266.

international control provisions.[49] The New Commonwealth's force on the other hand was to implement global law and order on a comprehensive basis – particularly through the enforcement of the decisions of an international tribunal. General Secretary N.B. Foot explained what he understood to be the difference thus:

As you know, UNA have embraced an idea of Lytton's that an International Force should be formed simply and solely to serve as a sanction for the outlawry of the Atom bomb ... Our main thesis, as we say in the Resolution, is that the only satisfactory course is, in a sense, to make use of this 'Atomic Opportunity' to solve the problem once and for all and to bring into being a really satisfactory international machinery.[50]

Even within the UNA, though, there was a constituency which preferred a wider scope of activities for the international police force and its atomic bombs. Fulljames, a member of both the Commonwealth and the UNA, wrote to the *United Nations News* in May 1946 protesting that under the UNA's plan 'a crisis might arise in which the International Force would stand by with folded arms during a major war purely because none of the contestants had used atomic weapons. Surely this is not our intention?' Instead, he proposed that the force be ready to use 'all necessary weapons' against any aggressor.[51] This constituency tried, in November 1945, to bring the two organizations to cooperate together on international control, and in February 1946 to get its proposals adopted as official UNA policy. They failed on both counts.[52] In fact, organization members expended almost as much effort urging unity amongst internationalist organizations as they did urging unity amongst nations. 'We must unite or perish', pleaded one letter to *Federal News*, referring not to antagonistic nations but to the UNA and Federal Union.[53]

Other non-scientist internationalist organizations were much less interested in international control. The Federal Union, the National Peace Council, and the Crusade for World Government took little initiative with regard to it.[54] Of the three, Labour MP Henry Usborne's newly

[49] 'General Council Decisions', *International Outlook* 1 (March 1946): 11–12.
[50] N. B. Foot to G. Cockerill, 18 January 1946, B23/3 – NC correspondence 1945 – 'C', Davies Papers.
[51] R. E. G. Fulljames, letter to the editor, 'International Force', *United Nations News* 1 (May 1946): 26.
[52] 'UNA's First General Council', *International Outlook* 1 (March 1946): 7–12; 'Minutes of a Meeting of the Provisional Executive Committee ... on Thursday, November 29th, 1945 at 2.15 PM', 3/1/1, UNA Papers.
[53] Ronald E. Gundry, 'Letters: Two Roads – No Goal', *Federal News* 129 (December 1945): 12.
[54] For a sense of the Federal Union and the National Peace Council's thinking on atomic energy see: F. L. Josephy, 'Federate or Perish', *Federal News* 126 (September 1945): 3;

formed Crusade for World Government was the only one to call for international control, alongside an atomic international force, in its publications. It did little other advocacy on atomic energy.[55] Like Federal Union, the prime aim of the Crusade was international federal union, though with an emphasis on European federation first. Unlike the Union, Usborne wanted an international atomic bomb armed force formed immediately under the UN, rather than waiting for a world federation. The UN also required, he reasoned, the ability to further develop atomic weapons, and the right to control through inspection nationally owned atomic plants. His logic, as explained in a National Peace Council pamphlet in 1946, was that a European federal union would antagonize the Soviet Union because a pan-European army would then face off against the USSR. To circumvent this the union would need to be disarmed at the time of its creation, and for the European nations to agree that the UN would first need to be bolstered through control of atomic energy.[56] Although the Crusade was able to generate significant publicity and receive support from a number of well-known public figures, as well as a number of younger Labour MPs, its influence, even at its peak in 1947, was minimal.[57]

The widespread support for international contro in late 1945 and early 1946 cannot be solely or even largely attributed to the activities and pronouncements of British internationalist scientists. They only organized themselves in late 1945; individual scientists speaking on atomic control in 1945 tended to speak through internationalist platforms organized by non-scientist internationalist groups. The November 1945 World Unity Movement symposium, for example, was attended by physiologist Leonard Hill, crystallographer Kathleen Lonsdale, journalist Ritchie Calder, and medical geneticist Lionel Penrose.[58] Although scientists did advise government on international control, Ministers already had opinions against or in support of it.[59] The secondary literature has focused on the Association of Scientific Workers (AScW) and the British Atomic Scientists' Association as the most significant scientists'

F. L. Josephy, 'The Federal Implications of the Atom', *Federal News* 126 (September 1945): 1–2, 11; Mosa Anderson, 'The Approach to World Government', *One World* 1 (June 1946): 3–6.

[55] It is noteworthy that Usborne defected to the Liberals in 1962.

[56] Henry Usborne, *Towards World Government – The Role of Britain* (London: National Peace Council, 1946); Henry Usborne, 'What Holds us Back', *Federal News* 135 (June 1946): 2, 5.

[57] Douglas, *Labour Party Nationalism and Internationalism*, 159–162.

[58] Leonard Hill et al., *Atomic Energy: Science v. Sovereignty* (London: World Unity Movement, 1946).

[59] Gowing, *Independence and Deterrence*, Vol. 1, 70–72.

organizations in relation to international control.[60] Yet the Atomic Scientists' Committee of the AScW, which had been meeting since late 1945, garnered little publicity that year. The most public interventions were a letter to *The Times* and a press release.[61] The Atomic Scientists' Association was not launched until the spring of 1946, and its most significant public intervention, the travelling train-based exhibition (dubbed the Atom Train), launched in 1947 and only briefly mentioned international control in one text panel at the end.[62] The major scientists' conferences on the implications of atomic energy took place in July 1946, after the announcement of the Baruch Plan.[63] The liberal newspapers at the forefront of calls for international control in late 1945 barely referred to scientists at all in this regard. Instead, newspapers chose to print statements by politicians or leading liberal intellectuals such as the economist and newspaper editor Walter Layton on the international relations-transforming properties of atomic energy, or turn to H.G. Wells for an understanding of this scientific predicament.[64]

New Political Thinking

A wide swathe of liberal intellectual opinion engaged with atomic energy and understood its interaction with international relations through an internationalist lens. The interwar precedents to some of this reaction are most evident in the thinking of two prominent liberal intellectuals: philosopher Bertrand Russell and technocrat Arthur Salter. In both cases their interest in atomic energy was not simply a response to the

---

[60] Jones, 'The Mushroom-Shaped Cloud'; Laucht, 'Transnational Professional Activism and the Prevention of Nuclear War in Britain'.

[61] J. D. Bernal, letter to the editor, *The Times*, 5 September 1945; Association of Scientific Workers, Press Statement, 'Scientists and the Atomic Bomb', 1 November 1945, MSS.79/ASW/3/3/1, Records of the Association of Scientific Workers, Modern Records Centre, University of Warwick (hereafter cited as AScW Records).

[62] The Association worked closely with government and national NGOs (including local chapters of the UNA) in operationalizing the exhibition and related Atomic Energy Weeks wherever the Atom Train called. Christoph Laucht, 'Atoms for the People: the Atomic Scientists' Association, the British State and Nuclear Education in the Atom Train Exhibition, 1947–1948', *British Journal for the History of Science* 45, no. 4 (December 2012): 591–608. On the Association see: Laucht, 'Transnational Professional Activism'; Laucht, *Elemental Germans*, chapter 6.

[63] The AScW conference in London 20–21 July; the International Physical Conference at Cambridge 22–27 July; and the International Conference of the Atomic Scientists' Associations at Oxford 29–31 July. On these see: Boorse, 'Two International Scientific Meetings in England'; Boorse, 'A World Federation of Scientific Workers'.

[64] 'Atom: Boon or Doom? H.G. Wells Prophesied it in 1894', *Daily Herald*, 9 August 1945; Editorial, 'The Lesson of Hiroshima', *Daily Herald*, 5 September 1945; Walter Layton, 'Hiroshima', *News Chronicle*, 9 August 1945; 'A Spectator's Notebook', *Spectator*, 17 August 1945, 144.

destructiveness of the atomic bomb or the threat of the Soviet Union, but rather reflected their long-standing interests in modern science, international peace, and international economic planning. Russell's reaction to the bomb was immediate. In an article published on 18 August 1945 he turned to internationalization as a solution to the 'new and appalling problems with which the human race is confronted by its conquest of scientific power'. He called for humanity to 'bring to bear upon social, and especially international, organization, intelligence of the same high order that has enabled us to discover the structure of the atom'. Uranium ore, atomic plants, and atomic bombs were to be placed under the direct control of an 'irresistible' atomic bomb armed international authority which would then enforce peace. If the Soviet Union did not agree to these then it should be forcibly disarmed through a pre-emptive war.[65] He stuck remarkably close to this early reaction over the following four years, when he was a prominent supporter of international control and an atomic bomb armed international organization. An armed organization was part of his deeper support for a powerful international organization 'that really governs', not 'an amiable façade like the League of Nations, or a pretentious sham like the United Nations'. In order to be truly preponderant the force would need to be armed with atomic bombs, an air force (for bombing and for reconnaissance), battleships, and a standing army.[66] Other prominent calls for international control and an atomic bomb armed police force included speeches in the House of Lords in 1946 and 1947, where he was vocally supported by Viscount Samuel and A.D. Lindsay (who was linked to the New Commonwealth: he was President of the *London Quarterly of World Affairs*, previously the *New Commonwealth Quarterly*), and during a lecture tour to Holland and Belgium in late 1947 on behalf of the New Commonwealth Society (which made full use of Russell's celebrity: his lecture was published in their magazine and released as a pamphlet).[67]

---

[65] Bertrand Russell, 'The Bomb and Civilization', *Glasgow Forward* 39 (18 August 1945): 1, 3.

[66] Bertrand Russell, 'The Atomic Bomb and the Prevention of War', *Bulletin of the Atomic Scientists* 2, nos. 7–8 (1 October 1946): 19–21. See also: Bertrand Russell, 'Values in the Atomic Age', in *The Atomic Age*, by M. L. Oliphant et al. (London: George Allen and Unwin, 1949), 81–104.

[67] Hansard, HC, vol. 139, cols. 1259–1260, 7 March 1946; Hansard, HC, vol. 147, cols. 244–289, 30 April 1947; Bertrand Russell, 'International Government', in *Towards World Government* (London: New Commonwealth, 1948), 3–12; Bertrand Russell, 'International Government', *The New Commonwealth* 9 (January 1948): 77–80. See also: Ray Perkins, 'Bertrand Russell and Preventative War: A Reply to David Blitz', *Russell: The Journal of Bertrand Russell Studies* 22, no. 2 (Winter 2002–03): 161–72; and Ray Monk, *Bertrand Russell 1921–70. The Ghost of Madness* (London: Jonathan Cape, 2000), 298–305.

Although Russell termed his reaction to the bomb 'new political thinking', in reality it was a reconfiguration of his, and other liberal internationalists', earlier thinking on technology and international relations.[68] Russell himself had already called for world government and an international air police in the late 1930s. In *Which Way to Peace?* (1936) he had joined liberal internationalist opinion in asserting that aviation had transformed war, giving attackers an insurmountable advantage. Aviation, along with gas and bacteriological weapons, needed to be handed over to an international force for its exclusive use. Permanent peace required a 'single supreme world government, possessed of irresistible force, and able to impose its will upon any national State or combination of States'.[69] This belief had continued into the war. In January 1941, writing to Gilbert Murray, he had pointed out that 'There can be no permanent peace unless there is only one Air Force in the world with the degree of international government that that implies. Disarmament alone, though good, will not make peace secure'.[70] His *Scientific Outlook* (1931), citing David Davies's *Problem of the Twentieth Century*, had argued that the scientific outlook, the key to mankind's success, required a world state armed with 'a single highly efficient fighting machine, employing mainly aeroplanes and chemical methods of warfare, which will be quite obviously irresistible, and will therefore not be resisted'.[71] The notion that modern science and science-based technologies were transforming the world, and moving it towards a world state, were a recurring theme in Russell's writings from the 1920s to the early 1950s. The 1949 second edition of *Scientific Outlook*, for example, added atomic bombs alongside military aviation to the dangerous products of the scientific approach, and to the list of technologies (which included railways, telegraph, and aeroplanes) now pushing the world inexorably towards a singular world state.[72]

Arthur Salter had emerged as a leading expert on international trade and transport in the interwar years. During the two World Wars he had worked on the organization of shipping between Britain and the United

[68] Russell, 'The Bomb and Civilization'.
[69] Bertrand Russell, *Which Way to Peace?* (London: Michael Joseph, 1936), 80–81.
[70] Letter, Bertrand Russell to Gilbert Murray, 18 January 1941, in Bertrand Russell, *Autobiography* (London: Routledge, 2010), 471.
[71] Bertrand Russell, *The Scientific Outlook* (London: George Allen & Unwin, 1931), 34–35.
[72] Bertrand Russell, *The Scientific Outlook*, 2nd ed. (London: George Allen & Unwin, 1949), 35–36, 217–8. On his thinking on science see: Casper Sylvest, 'Technology and Global Politics: The Modern Experiences of Bertrand Russell and John H. Herz', *The International History Review* 35, no. 1 (2013): 121–142; Peter Denton, *Bertrand Russell on Science, Religion, and the Next War, 1919–1938* (New York: State University of New York Press, 2001).

States, and as Director of the Economic and Finance Section of the League between 1922 and 1931. A prominent internationalist, he had spoken in support of the League and international economic organization in the 1920s and 1930s, and in support of European unification and post-war international organization during the Second World War.[73] After the atomic bombings, he linked his calls for international control to wider proposals for the strengthening of the United Nations in the political and economic domains. During the August 1945 Commons debate on UN Charter ratification he called for the 'shock' created by the bomb to be used by statesmen to institute international control and strengthen the UN, and emphasized the importance of scientists in managing this control.[74] A speech to a joint Federal Union and UNA meeting in October 1945, widely reported in Britain and also published in the United States, described how the current 'international anarchy' required supranational authority and the dilution of national sovereignty. The atomic bomb, he announced, provided a 'new lever of power to compel action' that could be used to this end, and if done so 'its power to achieve a constructive purpose will thus be as great as its destructive potency'. Calling for an end to the 'fratricidal dispute' amongst supporters of world federation and the UN, he pointed out that this powerful supranational authority could be created through either a strengthening of the Charter or the formation of a whole new organization, and could begin first with the few 'like-minded countries with a preponderance of collective power'. He called for all atomic bombs and facilities to be handed over to the UN, and to be operated, owned, and policed (through an international force) by the organization.[75] In a November 1945 episode of the BBC radio discussion programme *The Brains Trust*, Salter informed listeners that 'atomic energy is after all a very powerful forcing hothouse to the development of political institutions', and called for the creation of a powerful international organization to control armaments and foreign policy. In this discussion he was opposed by Russell who argued against such a 'thorough-going' international organization; Russell suggested instead just the creation of a powerful international force armed with 'all the really important weapons of war'.[76]

---

[73] Leonie Holthaus and Jens Steffek, 'Experiments in international administration: The forgotten functionalism of James Arthur Salter', *Review of International Studies* 42, no. 1 (2016): 114–135.

[74] Hansard, HC, vol. 413, cols. 694–697, 22 August 1945.

[75] 'Safeguard against Atomic Bomb Sir Arthur Salter's Proposal', *The Times*, 9 October 1945; 'Sir Arthur Salter's Proposal', *Manchester Guardian*, 9 October 1945; Arthur Salter, 'The United Nations and the Atomic Bomb', *International Conciliation* 24 (January 1946): 40–48.

[76] 'Extract from "The Brains Trust"', *Federal News* 130 (January 1946): 4–5.

Salter had a history of pronouncements on modern industry and international relations stretching back to the 1930s. He believed, in keeping with many other economists and industrialists, that the 'whole tendency of modern industrial development is towards a large scale organization' which was now moving beyond national borders. The international economy that was now emerging required technical experts ('whether they are scientists, or schoolmasters, or financiers, or industrialists, or trade unionists') to cooperate at the international level and form international technical organizations. This technical cooperation was leading to a widening and deepening of international contacts, allowing the 'basis of international relations' to be 'broadened', and to a reduction in international political disputes.[77] These ideas mirrored his prescription for national economies, which was that more technical expertise was required to manage and co-ordinate them (for example through expert Economic Advisory Councils).[78]

Salter also subscribed to widely held ideas about the social impact of scientific and industrial change. In an April 1933 lecture delivered at McGill University (and published by Oxford University Press as *Modern Mechanization and its Effects on the Structure of Society*), Salter, like Shotwell, positioned the industrial revolution as the defining moment in human history.[79] The lag thesis was used to explain the impact of science and industry on society:

The pace set by progress in scientific invention and improved industrial technique is too hot for man's regulative control to overtake. And when it lags behind, every new progress in specialized activity is a new danger; every new access of power threatens destruction to what he has more than it promises increase. That is why mechanization is compelling, and will compel, profound changes in the whole structure of society.[80]

In a series of talks for the BBC in the early 1930s, Salter mobilized techno-globalist ideas and notions of the dual nature of science in support of a call for the League to be strengthened into a world government. He explained how 'Transport and Telegrams Transform the World', and how the need for 'world government' had been brought about by 'scientific inventions, and especially those which have improved transport and the transmission of news' (the train, steamship, motor-car, aeroplane, telegraph, telephone,

---

[77] Arthur Salter, 'The Future of Economic Nationalism', *Foreign Affairs* 11, no. 1 (October 1932): 8–20.

[78] Arthur Salter, *The Framework for an Ordered Society* (Cambridge: Cambridge University Press, 1933), 14, 41–60.

[79] Arthur Salter, *Modern Mechanization and its Effects on the Structure of Society* (London: Oxford University Press, 1933), 9.

[80] Ibid., 8–9.

and wireless). All was not rosy, however, because science had also produced weapons, which, if uncontrolled, threatened to destroy civilization:

Science has given us a monster of illimitable power, to serve us or to destroy us. If
we allow it, like a Frankenstein, to escape the control of our regulative wisdom, it
will destroy us. But if we can keep control it can be the instrument of our salvation.
The task of our age is to see that this Frankenstein remains our servant and does
not become our master.[81]

His BBC talks even imagined a techno-internationalist future in which
delegates to the League of Nations in 1957 would zoom around the
world in aeroplanes and remain in instantaneous contact through wireless
receiving-sets in their waistcoat pockets.[82] In the late 1930s Salter accepted
an appointment as a Vice-President of the New Commonwealth Society,
though was not a particularly active member.[83]

The emphasis on international technical cooperation in Salter's
approach to international control was also evident in discussions on
atomic energy at the international relations think tank Chatham House.
Long interested in aviation and international relations, Chatham House
now turned its focus on atomic energy, particularly through a formation
of an Atomic Energy Study Group to discuss and then produce a report
on the impact of atomic energy on international relations. The Group
held its first meeting in September 1946, and continued to meet
roughly every two months through to mid-1948. It was chaired by
physiologist Henry Dale and brought together scientists, civil servants,
academics, and military men. Regular attendants included Oliver
Franks, Air Marshal Roderic Hill, H.S.W. Massey, Arthur Salter, and
Charles K. Webster. Others attending included Lord Hankey, James
Chadwick, Marcus Oliphant, R.E. Peierls, Cecil Edgar Tilley, and J.D.
Cockcroft. Some, such as Arthur Salter, had been involved in earlier
discussions on atomic energy at the New Commonwealth Society, but
were attracted to the more agenda-free environment on offer at
Chatham House.[84] The Group published its report on atomic energy
and international relations in January 1948, though by that stage its
impact was limited due to declining interest in internationalist
approaches to atomic energy.[85]

---

[81] Arthur Salter, 'World government', in *The Modern State*, eds. Mary Adams (London:
George Allen & Unwin, 1933), 253–316.
[82] Ibid.   [83] Pemberton, *The Story of International Relations*, part two, 131.
[84] 'Dinner and Discussion on Atomic Energy', 6 May 1946, folder 9/42b: 'Atomic Energy
Subcommittee. Minutes Correspondence', Chatham House Papers.
[85] Royal Institute of International Affairs, *Atomic Energy. Its International Implications.
A Discussion by a Chatham House Study Group* (London: Royal Institute of International
Affairs, 1948).

The two most vocal supporters of international control in the Group were Arthur Salter and Charles Webster. Webster had held professorships in international relations at the University of Wales, Aberystwyth, and the London School of Economics in the 1920s and 1930s, and during the war worked with the Foreign Research and Press Service and eventually with the Foreign Office on UN planning. A strong supporter of the League and subsequently the UN, his last post prior to retirement in April 1946 was as Special Advisor to Philip Noel-Baker, the Minister of State responsible for United Nations Affairs. Webster, like many other supporters of the UN, was a strong proponent of international control, believing that it afforded the 'opportunity' to 'see the United Nations transformed by action, into something much more powerful than it is today'.[86] Like Salter and many others he also saw international control as the start of the evolution of the United Nations towards a powerful world government. He pushed this view within the Chatham House Group, and, perhaps mindful of the organization's connections to political scientist David Mitrany (by then well known for this 'functional' approach to international relations), characterized this evolution from international control to 'World State' as an example of 'functional development'.[87] Webster emphasized that an international controlling authority's power would derive not only from its ability to control the manufacture of atomic bombs but also through its ability to control the distribution of power generated by atomic energy – power which held 'great potentialities for the welfare of the world'.[88]

In the Group's final report a chapter by Arthur Salter, Charles Webster, and Oliver Franks propounded a plan for international control. They maintained that this control was important not only because it controlled the bomb and atomic energy and allowed for the production of abundant electricity, but also because it would lead to cooperation spreading into other domains. It would be the harbinger of a new international 'efficient system of direction and control' and would make possible the 'creation of a world community with practical institutions strong enough to control national States'.[89] The fact that someone like Oliver Franks would add his

---

[86] Charles Webster, 'The Making of the Charter of the United Nations', in *The Art and Practice of Diplomacy* (London: Chatto & Windus, 1961), 70–91. On Webster and the early UN see: Ian Hall, *Dilemmas of Decline: British Intellectuals and World Politics, 1945–1975* (Berkeley: University of California Press, 2012), 72–75.

[87] C. Webster, 'The International Aspect of Atomic Energy Control', 25 November 1946, folder 9/42e, Chatham House Papers. For more on Mitrany and his functional approach see Chapter 1.

[88] Ibid.

[89] Charles Webster, Arthur Salter and Oliver Franks, 'The Control of Nuclear Energy and the Development of International Institutions', in Royal Institute of International Affairs, *Atomic Energy*, 91–101.

name to this proposal for internationalization alongside Salter's and Webster's is of some significance. Franks, a career diplomat, was at that time chairing the European organization formulating requirements for the Marshall Plan, and serving on the government's Advisory Committee on Atomic Energy. He would later serve as ambassador in Washington, DC. He was not otherwise a vocal supporter of international control, and so his involvement signifies a certain attraction within mainstream British liberalism – that is outside the world of overt or activist liberal internationalism – to American-led proposals for international control.

The other internationalist chapter in the Chatham House volume was produced by the aeronautical engineer Harry Wimperis, formerly director of scientific research at the Air Ministry and President of the Royal Aeronautical Society. In early 1946 Wimperis had already published, under the auspices of Chatham House, a book calling for comprehensive UN-based international control. He had advocated UN control of all atomic bombs and their use for peace-keeping, as well as UN scientific development of the peaceful uses of atomic energy. The Security Council veto was to be abolished for atomic matters.[90] By late 1947, when Wimperis penned his chapter for the Study Group report, discussions on international control had been stalled for a year. He, much more so than Salter, Franks, and Webster, reflected this awkward reality in his chapter by turning away from international control and instead making less radical functional-type arguments. Although based on similar assumptions about the internationalizing and integrative nature of science and technical work, he added the presumption that international control would not be agreed, and took up a suggestion made by Air Chief Marshal Roderic Hill during Group discussions that control attempts instead focus on the formation of a large UN atomic research and development laboratory. Nation-states would be allowed to retain their own atomic research programmes, but these would be overshadowed by the UN effort, and so wither away over time.[91] Echoing arguments made in the Acheson-Lilienthal Report, Hill had claimed that the working methods of scientists would gradually lead to international peace as their cooperation spread to administrators and politicians.[92] Wimperis argued in his chapter that scientists' atomic cooperation could become a template for international technical cooperation in other areas.[93]

[90] H. E. Wimperis, *World Power and Atomic Energy: The Impact on International Relations* (London: Royal Institute of International Affairs, 1946), 1, 77.
[91] R. Hill, 'Memorandum', 15 January 1947, folder 9/42e, Chatham House Papers.
[92] 'Verbatim Report of the Second Meeting of the Atomic Energy Group', 10 December 1946, folder 9/42e, Chatham House Papers.
[93] H. E. Wimperis, 'International Co-operation in the Development of Atomic Energy', in Royal Institute of International Affairs, *Atomic Energy*, 102–111.

Opposing Atomic Douhetism

There was of course significant opposition to and scepticism about international control in those early post-war years. One important argument which emerged from both the right and the left was that the impact of the atomic bomb, like aviation before it, was exaggerated. Spaight, the long-standing advocate for national air power and opponent of the aerial alarmism of the 1930s, made this argument in his 1948 *The Atomic Problem*. The atomic bomb was not a radical break in warfare but simply a continuation of wartime mass bombing (which he condemned).[94] The most prominent leftist intellectual writing on war and atomic weapons, physicist P.M.S. Blackett, called this exaggeration of the bomb's military impact 'atomic Douhetism'. Although the bomb was a great new weapon, it did not, he reasoned, transform warfare. It could not win wars solely or largely by itself: other weapons would still be required, especially military aviation. Just as in the earlier case of aerial bombing it was not true that there was no defence against atomic bombs. It was possible to take precautions against the worst effects of atomic bombings, and huge numbers of atomic weapons would be required to make a decisive difference in warfare. The 'neo-Douhets' or 'atomic-Douhets' were as wrong as the prominent interwar air power theorist Giulio Douhet himself had been about the ability of new weapons to deliver quick victory.[95] There was no need to adjust the UN Charter, remove the veto, or give greater sanction power to the organization. The Security Council rightly reflected the power balance amongst the world powers.[96] Blackett also dismissed the Baruch Plan as a US diplomatic and propaganda manoeuvre.[97]

George Orwell drew a different, albeit also anti-internationalist, conclusion by comparing the atomic bomb to both civilian aircraft and the bomber. Although it had commonly been said that the civilian aeroplane and radio would reduce national boundaries, he noted, they had actually strengthened them. So too with the atomic bomb. Writing in the *Tribune* in October 1945, he claimed that rather than threatening civilization or

---

[94] J. M. Spaight, *The Atomic Problem* (London: Arthur Barron, 1948), vii, 6–7, 13, 79.

[95] P. M. S. Blackett, *The Military and Political Consequences of Atomic Energy* (London: Turnstile Publications, 1948), 1–8, 38–39, 45, 56–57, 80–82; P. M. S. Blackett, 'The Military and Political Consequences of Atomic Energy', in Oliphant et al, *The Atomic Age*, 32–51.

[96] P. M. S. Blackett, *The Atom and the Charter* (London: Fabian Publications, 1946), 4–5.

[97] Blackett, *The Military and Political Consequences*, 143–44. For more on Blackett and Orwell's response see: John Callaghan and Mark Phythian, 'Intellectuals of the Left and the Atomic Dilemma in the Age of the US Atomic Monopoly, 1945–1949', *Contemporary British History* 29, no. 4 (2015): 441–463. For more on Blackett see: Mary Jo Nye, 'A Physicist in the Corridors of Power: P. M. S. Blackett's Opposition to Atomic Weapons Following the War', *Physics in Perspective* 1 (1999): 136–156.

bringing anarchy the bomb, like the bomber before it, was creating a world of large powerful police-states ruled by small elites.[98]

A broader criticism, which included questioning international control's practicality, was put forward by Maurice Hankey, a member of the House of Lords and one-time prominent civil servant (especially as Secretary to the Cabinet). Hankey participated in the Chatham House Study Group and quickly emerged as the most vocal opponent of international control. A conservative in many ways, and a long-standing opponent of arms control and an international police force, Hankey had been dead-set against international control since the beginning, warning of the danger of 'premature disarmament' during a Chatham House lecture in October 1945, and informing the Group in November 1946 that international control was too intrusive to be acceptable, and even if successful would anyhow not prevent all warfare. For it to work international control would have to be extended to 'other scientific developments' (he highlighted biological warfare, rockets, and radar) which was unrealistic as it was 'too great a burden to place on the infant UNO'. Current fears about the atomic bomb were simply a replication of interwar fears of aerial bombardment, and civilization, he pointed out, would survive the coming of the atomic bomb just as it had survived the dangers of air bombing.[99] Hankey's (final) chapter for the published report, penned in May 1947, presented a short history of attempts at arms control to show that they had always failed. Britain and the United States needed to continue to develop atomic weapons, though perhaps, 'as a token of good intention', could sign a treaty banning their use in war.[100]

As the leading speaker against international control, Hankey laid out three sets of arguments, focused on power politics, during a February 1948 debate in the House of Lords on atomic energy. He pointed out that the UNAEC negotiations were always likely to fail given the growing rift between the Western and Eastern blocs and Soviet expansionism. Moreover, creating a workable system of control was unrealistic given states' unwillingness to throw open their territories and

[98] George Orwell, 'You and the Atom Bomb', in *The Collected Essays, Journalism and Letters of George Orwell*, eds. Sonia Orwell and Ian Angus, vol. 4, *In Front of Your Nose 1945–1950* (London: Secker & Warburg, 1968), 6–10.

[99] Maurice Hankey, 'The Control of External Affairs', *International Affairs* 22, no. 2 (March 1946): 161–173; Maurice Hankey, 'Control of Atomic Warfare: Political Aspects', 25 November 1946, folder 9/42e, Chatham House Papers. On Hankey's interwar opposition to disarmament see: Johnson, *Lord Robert Cecil*, 136; Kitching, *Britain and the Problem of International Disarmament*, 37; Maurice Hankey, 'The Study of Disarmament' and 'International Forces', in *Diplomacy by Conference: Studies in Public Affairs 1920–1946* (London: Ernest Benn, 1946), 105–119, 120–130.

[100] Maurice Hankey, 'The Human Element', in Royal Institute of International Affairs, *Atomic Energy*, 112–124.

facilities to foreign inspection, and the possibility of deceit, spying, or corruption. It would be just as easy, he suggested, to 'assume that war itself can be abolished'. Second, given Soviet participation in the UN and the continuing likelihood of conflict, the West could not entrust its security to the organization or abandon its atomic weapons. Third, he suggested that the failed disarmament process of the 1920s and 1930s demonstrated that to allow current talks on international control to continue would be not only fruitless but also counterproductive as it would increase friction between the two blocs.[101]

Nationalist International Control

Support for international control and the use of liberal internationalist language and perspectives also emerged in government in late 1945. Prime Minister Clement Atlee was probably the most prominent Labour politician to use such language. His Cabinet memorandum on international control, prepared before the November meeting on atomic cooperation with Truman and President Mackenzie King of Canada, laid out official British policy, and based it on a 'general thesis'. His key points were that: (1) the atomic bomb presented a 'new situation': a weapon against which there was no defence; (2) war would lead to the 'collapse of civilization'; (3) states needed to abandon 'all out-of-date ideas of power politics' and 'nationalistic ideas'; and (4) states needed to abandon war as an instrument of national policy.[102] Other prominent politicians using this language included the President of the Board of Trade Stafford Cripps and Secretary of State for Foreign Affairs Ernest Bevin. The 22–23 August 1945 House of Commons debate on ratification of the UN Charter saw several Labour MPs call for an internationalist approach to the atomic bomb centred on the UN. Many called for the Charter to be revised in a more internationalist direction to make the organization more powerful.[103]

The British state's initial approach to international atomic regulation was, however, much more nationalist and self-interested than the rhetoric of its Labour leaders suggested. Policymakers recognized the concentration of power in the hands of the Great Powers, especially the United States and the Soviet Union, whilst the United Nations was seen as politically weak and peripheral to international politics. Moreover, the perception of a looming

[101] Hansard, HC, vol. 153, cols. 1190–1198, 18 February 1948.
[102] C. Attlee, 'International Control of Atomic Energy', CP (45) 272 (5 November 1945), CAB/129/4, TNA.
[103] Hansard, HC, vol. 413, cols. 861–950, 23 August 1945; Hansard, HC, vol. 413, cols. 659–755, 22 August 1945; Douglas, *Labour Party Nationalism and Internationalism*, 146–149.

Soviet threat to Europe led policymakers to the very 'nationalist ideas' that they derided in public, including the strengthening of national power through atomic weapons. In late 1945 Attlee and most of the Cabinet concluded that, rather than being abolished, the atomic bomb should remain in national hands. Their perception of the United Nations as weak was borne out by the diplomatic manoeuvres within the UN Security Council itself once it was formed in January, when clashes between the Soviet Union and Britain seemed to indicate that the Council would be nothing more than a platform for the playing out of national rivalries.[104] The policy adopted by the Cabinet in October 1945, although called international control, contained only a weak form of international governance. Nation-states were to use their atomic weapons to create a system of collective security to support the fledgling organization and ensure international peace. The recourse to the language of atomic internationalism, then, reflected the desire to use the dominant discourse of the time, which contained ideas and terms such as international control that were ill-defined and flexible enough to be used in a number of different ways.[105]

Britain's national approach and geopolitical concerns were also evident at the November 1945 meeting on atomic cooperation between the United States, Britain, and Canada. Underlying Britain's stance at the meeting was the recognition that the country was second only to the United States in terms of technical expertise on atomic matters and was already on the path to atomic bomb development. Britain wanted to expand atomic cooperation with the United States, and so the British delegation suggested sharing of atomic information. The United States, meanwhile, suggested more restricted sharing, and that too only when UN control was firmly established. The British also argued against abolishing the bomb, and instead suggested that an international control treaty be signed which would sanction the use of the atomic bomb against any country which broke the treaty and attempted to develop atomic weapons. This, once again, reflected the belief that Western Europe could only be defended from a future Soviet attack through atomic weapons, either in actual use or through deterrence. Lastly, the British suggested that each country should monitor atomic programmes in its own territory without using foreign inspectors. This suggestion was included because it was believed that any treaty with foreign inspections would not be acceptable to the Soviet Union.[106]

---

[104] Ibid., 151–154.   [105] Attlee, *International Control of Atomic Energy*.
[106] James L. Gormly, 'The Washington Declaration and the "Poor Relation": Anglo-American Atomic Diplomacy, 1945–46', *Diplomatic History* 8, no. 2 (1984): 125–143; Septimus H. Paul, *Nuclear Rivals: Anglo-American Atomic Relations, 1941–1952* (Columbus: Ohio State University Press, 2000), 96–97; Matthew Jones, *The Official History of the UK Strategic Nuclear Deterrent*, vol. I, *From the V-Bomber Era to the Arrival of*

Although the November meeting produced an agreement closer to the US vision of international atomic cooperation rather than the British one, by December 1945 the United States was more willing to reach out to the Soviets for a way forward on international control. A December 1945 agreement between the United States and the Soviet Union agreed to the formation of a UNAEC under the Security Council, something the British had hoped to avoid as it implied that its deliberations could be vetoed by the Soviet Union.[107] Once the formation of the UNAEC was announced and the Acheson-Lilienthal Plan revealed, the British government's Advisory Committee on Atomic Energy produced its own proposal for international control. In comparison to British thinking in late 1945, this plan gave more power and agency to the United Nations. Although in some ways similar to the Acheson-Lilienthal Plan, one crucial difference was that it relied heavily on the international inspection of national facilities, whereas the US plan relied on the distinction between safe and dangerous activities.[108] British policymakers came to see the Acheson-Lilienthal and Baruch Plans as weak on inspections and so susceptible to surreptitious bomb development by the Soviet Union. Nevertheless, once the Baruch Plan had been presented at the UNAEC Britain supported it, however tentatively, and pondered over how to make it acceptable to the Soviet Union. Once it became clear that there was unlikely to be any agreement on international control, British thinking on international atomic cooperation once again turned to the possibility of US–British atomic cooperation, and perhaps even some sort of multilateral mutual security agreement between the United States, Britain, and Canada.[109] This did not emerge, however, and British policymakers made the formal decision to go ahead with an atomic bomb programme in January 1947, with the first atomic test bomb being carried out in 1952.[110]

Conclusion

The internationalist response to the atomic bomb in Britain was in many ways strikingly similar to the United States. Just as in the US there was

---

Polaris, 1945–1964 (Abingdon: Routledge, 2017), 1–12; John Baylis, *Ambiguity and Deterrence: British Nuclear Strategy, 1945–1964* (Oxford: Clarendon Press, 1995), 37–47.

[107] Gormly, 'The Washington Declaration'.

[108] Susanna Schrafstetter, '"Loquacious ... and Pointless as Ever"? Britain, the United States and the United Nations Negotiations on International Control of Nuclear Energy, 1945–48', *Contemporary British History* 16, no. 4 (Winter 2002): 87–108.

[109] Ibid.

[110] On British–US atomic relations see: Margaret Gowing, 'Britain, America and the Bomb', in *British Foreign Policy, 1945–56*, eds. Michael Dockrill and John W. Young (New York: Palgrave Macmillan, 1989), 31–46.

strong public support for international control in late 1945 and 1946. Although interwar and wartime understandings about science and international relations re-emerged reinvigorated after the war, there were nevertheless important differences to earlier thinking, with aviation being largely replaced by atomic energy as a new arena over and through which internationalists fought for their visions of international order. Powerful technologically superior military forces played an important role in many of these visions, as did the purported peace-spreading and integrative effects of international scientific and technical cooperation. British atomic culture, like its US cousin, held it to be self-evident that the atomic bomb heralded a new era of international political relations and warfare which required drastic transformations in societies, militaries, and international relations. Supranational organizations such as the UN assumed an even greater importance in the eyes of intellectuals and the public – as did the idea of supranational organization itself. The bomb was assumed to be a product of science, so scientists acquired a peculiarly high status as experts not only on scientific, technological, and industrial matters related to the bomb but also on its effects on society and international relations.

There were however several crucial differences between Britain and the United States. First, constituencies supporting international control in Britain did not coalesce into three powerful camps as prominently as they did in the United States. Scientists emerged as a more powerful force for international control in the US, where they were better organized, more cohesive in their actions and statements, and more prominent in shaping the public's understanding of atomic energy. In Britain, by contrast, scientists were politically and ideologically fractured (and so with different understandings of international relations), and less effective in shaping public discourse and government thinking. The atomic bomb was also then under development in Britain, and many atomic scientists were involved in the programme. This involvement and the general culture of secrecy surrounding atomic matters promoted by the state dampened open discussion.[111] The world government movement, in addition, was

---

[111] On divisions between the AScW and the British Atomic Scientists' Association, and within the Association itself, see: Laucht, *Elemental Germans*, chapters 6 and 7. On secrecy see: Gowing, *Independence and Deterrence*, 51; Simone Turchetti, 'Atomic Secrets and Governmental Lies: Nuclear Science, Politics and Security in the Pontecorvo Case', *British Journal for the History of Science* 36, no. 4 (December 2003), 389–415. Physicist Alan Nunn May was prosecuted in Britain for spying in 1946, further stifling scientists: Daniel W. B. Lomas, *Intelligence, Security and the Attlee Governments, 1945–51: An Uneasy Relationship?* (Manchester: Manchester University Press, 2017), 149–185.

generally weaker in Britain, and much less driven by the dangers of atomic energy.[112]

Second, the concept of an international police force was much more commonly heard in Britain. This was a legacy, no doubt, of earlier wartime and interwar interest in an international air force, and particularly the hope that such a force would represent and embody a special US–British relationship. Just as in the wartime years, internationalists continued to assume that such a force would be largely Anglo–American, representing the current preponderance of the two countries' aerial might and wartime military cooperation. There was greater support also for a UN system of collective security; and international control and an international atomic force fitted neatly into, and reinforced, such hopes for international security. Lesser support for an international force in the United States was probably also a reflection of the greater prominence in the country of scientists, who were on the whole less attracted to the idea. Third, discourse on international control was much more clearly demarcated along the political spectrum in the British press; and political and intellectual support tended to more clearly follow political and ideological divides, with the strongest support for international control coming from the middle of the spectrum, and opposition from the right and left. Liberal internationalist supporters of the United Nations, in particular, had more influence in Britain on atomic matters as compared to the United States, where the better-organized scientists dominated most discourses on atomic energy. Just as in the US the deepening Cold War destroyed hopes for international control and a UN police force.

---

[112] For a comparative study of world federalist movements see: Joseph Preston Baratta, 'The International Federalist Movement: Toward Global Governance', *Peace & Change* 24, no. 3 (July 1999): 340–372.

Conclusion: Science, Technology, and Internationalism into the Cold War and Beyond

The age of peace through technology never arrived. The facets of international relations that so horrified internationalists in the first half of the twentieth century remain with us today. International politics is as anarchic today as it was in the 1930s and 1940s: although there are socially constructed international norms and complex interdependence, strongly competing national interests continue to dominate.[1] Wars continue to bring suffering to millions; international organizations which once promised freedom from sovereign state interests remain largely beholden to these very same interests; and the ordered worlds of internationally planned resources or free trade remain ideals distant from the current jumble of bilateral and multilateral trade and economic treaties. The weak multilateral response to crises such as climate change and the recent Covid-19 pandemic point to little cooperation even in matters related to science.

Yet if international relations has not moved in the direction US and British internationalists hoped, their fears about aviation have in some ways come true. Although aerial bombing did not destroy civilization, or lay waste to internationalists' homelands, it nevertheless left a swathe of destruction in Germany, Japan, and eventually in colonial and post-colonial peripheries. A total of 635,000 tonnes of ordnance were dropped on Korea from 1950 to 1953, and more bombs were dropped on Vietnam and Cambodia than during all of the Second World War. Some 12 to 15 percent of North Korea's population was killed in the Korean War, most through bombing.[2] Bombing continues to be used to devastating effect today, for example in Yemen where it has brought much of the

---

[1] Robert O. Keohane and Joseph S. Nye, Jr., *Power and Interdependence*, 2nd ed. (New York: Harper Collins, 1989), chapter 2; Alexander Wendt, *Social Theory of International Politics* (Cambridge: Cambridge University Press, 1999). The state is central to the neorealist (or structural realist) theories of international relations, most influentially: Kenneth Waltz, *Theory of International Politics* (New York: Addison-Wesley, 1979).

[2] For an overview of twentieth-century bombing see: Thomas Hippler, *Governing from the Skies: A Global History of Aerial Bombing* (London: Verso, 2017).

country's population to the brink of starvation.[3] Nuclear weapons similarly have not brought (in Norman Cousins's words) 'the complete obliteration of the human species', but nevertheless the likelihood of their use, either in anger or by mistake, continues to grow as proliferation carries on and regional powers build up their nuclear arsenals.[4] Through the development and use of bombing and nuclear weapons (and in other ways), liberal militarism, a significant aspect of the technological internationalism explored in this book, has also continued to play a role in the development of national military forces and armaments, and their use.[5]

Civil aviation's promise of interconnectedness has also partially emerged. Middle classes around the world can now afford to move relatively effortlessly from one corner of the globe to the other, and up to the Covid-19 pandemic there was little let up in the annual increase in flying passengers. Air cargo has also increased in leaps and bounds, and pre-Covid-19 constituted an astonishing 35 percent of world trade by value.[6] Yet this interconnectedness has brought little of the harmony, cohesiveness, or one-worldness that internationalists hoped for, and the Covid-19 pandemic has demonstrated its fragility.[7] Atomic energy has provided even less of its promise. Although radiation plays an important role in medicine and scientific research, and nuclear reactors continue to be built (especially in Asia), cheap plentiful atomic power remains almost as distant now as it did in the late 1940s.[8] In both these cases the environmental effects of these machines, unaccounted for and unimagined by the internationalists encountered in this book, point to further limits on their civilian use.

It is no wonder then that aviation and atomic energy are no longer seen as peace-making technologies: they hardly invite the sort of radical calls for governance and order that surrounded them in the 1930s and 1940s. Aviation, as this book has shown, was quickly replaced in

[3] Martha Mundy, *The Strategies of the Coalition in the Yemen War: Aerial Bombardment and Food War* (London: World Peace Foundation, 2018).

[4] Cousins, 'Modern Man is Obsolete', 5–9.

[5] David Edgerton, 'Liberal Militarism and the British State'; Edgerton, *Warfare State*, 7–14, 285; Mabee, 'From "Liberal War" to "Liberal Militarism"'; Michael S. Sherry, *In the Shadow of War: The United States Since the 1930s* (New Haven, CT: Yale University Press, 1995), chapters 3–9.

[6] 'Air Cargo Matters', IATA, accessed 2 August 2020, www.iata.org/whatwedo/cargo/sustainability/Pages/benefits.aspx. Also: Peter S. Morrell, *Moving Boxes by Air: The Economics of International Air Cargo* (London: Routledge, 2011).

[7] On mass air travel: Marc Dierikx, *Clipping the Clouds: How Air Travel Changed the World* (Westport, CT: Praeger, 2008).

[8] For an overview of civilian nuclear power see: 'Nuclear Power in the World Today', World Nuclear Association, accessed 2 August 2020, www.world-nuclear.org/information-library/current-and-future-generation/nuclear-power-in-the-world-today.aspx.

internationalist visions by atomic energy in the late 1940s. Hopes for international control of atomic energy, in turn, dwindled away by 1948. Some calls for international control re-emerged in the 1950s as hopes for nuclear arms control increased, but disappeared once again by the 1960s. Perhaps the greatest impact proposals for international control of atomic energy may have had on atomic weapons was to strengthen the nuclear taboo surrounding their use.[9] There remain calls for aerial regulation, of course (particularly in relation to environmental impact), and nuclear arms limitations, but not radical governance through international organization.[10] Calls for a powerful international police force remain, but are no longer mainstream, and no longer hinge solely on aviation.[11]

A broader decline in liberal internationalism, especially of the more radical varities, itself also accounts for the disappearance of international control. The 1950s saw the disappearance of the types of centralized liberal internationalist world orders imagined in the first half of the twentieth century. The Cold War polarized the world, whilst decolonization disrupted the structure of metropolitan centre and colonial periphery which underlay internationalist thinking. Both made a tightly integrated world polity unimaginable. Even though the Cold War has passed, the multipolar world makes such a radical reordering of international affairs unimaginable and perhaps even undesirable. Liberal internationalism, to the extent that it exists, is understood today to mean greater international cooperation, either through multilateral military action or the type embodied in international organizations and institutions such as the United Nations or the World Trade Organization – an embedded or (for G. John Ikenberry) a diffused 'Liberal Internationalism 3.0'.[12]

[9] Nina Tannenwald, *The Nuclear Taboo: The United States and the Non-Use of Nuclear Weapons Since 1945* (Cambridge: Cambridge University Press, 2007), 99–100.

[10] For example: Stefan Gossling and Paul Upham, *Climate Change and Aviation: Issues, Challenges and Solutions* (Abingdon: Earthscan, 2009); Michael A. Levi and Michael E. O'Hanlon, *The Future of Arms Control* (Washington, DC: Brookings Institution Press, 2005).

[11] Augusto Lopez-Claros, Arthur L. Dahl, and Maja Groff, 'Completing the Collective Security Mechanism of the Charter: Establishing an International Peace Force', in *Global Governance and the Emergence of Global Institutions for the 21st Century* (Cambridge: Cambridge University Press, 2020), 145–180.

[12] G. John Ikenberry, 'Liberal Internationalism 3.0: America and the Dilemmas of Liberal World Order', *Perspectives on Politics* 7, no. 1 (2009): 71–87. Also: G. John Ikenberry, *Liberal Leviathan: The Origins, Crisis, and Transformation of the Liberal World Order* (Princeton, NJ: Princeton University Press, 2011); Sandrine Kott, 'Cold War Internationalism', in Sluga and Clavin, *Internationalisms*, 340–362; Hall, *Dilemmas of Decline*, 64–82.

Science, Technology, and Political Integration after the Second World War

Notwithstanding a decline, internationalist thought and action continued into the 1950s and beyond; and science and technology, broadly conceived, continued to remain important for much of this thinking. In that sense Lilienthal was wrong. Although the notion of over-arching global peace through technology had disappeared, internationalists continued to hope and believe that modern science and its applications could act as powerful motors for political and social transformation. Even Lilienthal himself could not entirely let go of these hopes. After his stint as Chairman of the Atomic Energy Commission he became a proponent of TVA-type projects abroad, and pushed for the creation of international dam projects modelled on the TVA in the Middle East, French Indo-China, and South Asia. These large, often transnational, water projects were seen not only as a means for industrial and social development but also as a way of reducing conflict and bringing neighbouring countries together.[13]

Such projects fitted neatly into a growing consensus that scientific planning by technocratic experts was the solution to developing poorer areas of the world and reducing conflict there. A new breed of technical experts, trained in the newly emergent Cold War disciplines of development economics and area studies, promised to alleviate poverty and modernize whilst also bringing international peace and security.[14] Nor were these technocratic hopes limited to the developing world. They were evident too in the growing movement for European integration in the 1940s and 1950s. Ideologues of European unification such as Robert Schuman, Jean Monnet, and Altiero Spinelli made the case for a technological and infrastructural-led political integration of Western Europe. These hopes were reflected in the neo-functionalist writings of political scientists such as Ernst Haas and Leon Lindberg in the 1950s and 1960s.[15]

---

[13] David Ekbladh, '"Mr. TVA": Grass-Roots Development, David Lilienthal, and the Rise and Fall of the Tennessee Valley Authority as a Symbol of U.S. Overseas Development, 1933–1973', *Diplomatic History* 26, no. 3 (Summer 2002): 335–374; David Ekbladh, *The Great American Mission: Modernization and the Construction of an American World Order* (Princeton, NJ: Princeton University Press, 2010), chapter 6.

[14] Nils Gilman, *Mandarins of the Future: Modernization Theory in Cold War America* (Baltimore, MD: Johns Hopkins University Press, 2003).

[15] John Pinder, 'Federalism and the Beginnings of European Union', in *A Companion to Europe Since 1945*, ed. Klaus Larres (Oxford: Wiley-Blackwell, 2009), 25–44; Philippe C. Schmitter, 'Ernst B. Haas and the legacy of neofunctionalism', *Journal of European Public Policy* 12, no. 2 (April 2005): 255–272; Johan Schot, 'Transnational

Science, technology, and internationalism remained prominent too in the continuing political activism of scientists. Whilst much diminished within the context of a deepening Cold War, increasing concerns about the destructiveness of hydrogen weapons and the effects of radioactive fallout led Western scientists to once again agitate for nuclear arms control. Although this rejuvenated scientific internationalism now incorporated non-Western scientists and included a wider range of political beliefs, it nevertheless retained a strong liberal internationalist aspect, the most prominent expression of which was the rise of the Pugwash scientists' movement in the late 1950s and early 1960s. These scientist-activists argued that scientists were uniquely placed to build bridges across national, political, and ideological divides. Science, they believed, could be used to reduce conflict between the Western and Eastern blocs, and point the world towards wider international cooperation, especially on nuclear weapons.[16] The internationalist rhetoric around nuclear matters was boosted by Eisenhower's Atoms for Peace Programme (announced 1953), and institutionalized in the International Atomic Energy Agency, founded 1957.[17]

The most radical expression of modern technology and international political integration continued in popular culture, beginning in juvenile fiction, but then making the jump to radio and television. British writer W. E. Johns, author since the 1930s of the popular Biggles adventure books, took his aviator hero in new directions after the war when his protagonist joined the Special Air Police division of Scotland Yard. Biggles ably wielded his authority as an aviator, and his machine, to chase international criminals across national boundaries.[18] Blackhawk spawned a radio show (1950–51) and a fifteen-chapter movie serial (beginning 1952); the comic continues to this day, though the aerial team increasingly fights extra-terrestrial threats.[19] Technological policing was taken to outer space in US science-fiction writer Robert A. Heinlein's best-selling 1948 novel *Space Cadet*, which saw its teenage hero join

Infrastructures and the Origins of European Integration', in Badenoch and Fickers, *Materializing Europe*, 82–109.

[16] Joseph Rotblat, 'Movements of Scientists against the Arms Race', in *Scientists, the Arms Race and Disarmament*, ed. Joseph Rotblat (London: Taylor & Francis, 1982), 115–157.

[17] Matthew Evangelista, *Unarmed Forces: The Transnational Movement to End the Cold War* (Ithaca, NY: Cornell University Press, 1999); Richard G. Hewlett and Jack M. Holl, *Atoms for Peace and War, 1953–1961: Eisenhower and the Atomic Energy Commission* (Berkeley: University of California Press, 1989).

[18] Christopher Routledge, 'Crime and Detective Literature for Young Readers', in *A Companion to Crime Fiction*, eds. Charles J. Rzepka and Lee Horsley (Oxford: Wiley-Blackwell, 2010), 321–331.

[19] Roger Hill, *Reed Crandall: Illustrator of the Comics* (Raleigh, NC: TwoMorrows Publishing, 2018), 130–131.

Earth's space police, the prestigious Space Patrol. Heinlein's future universe was modelled on the internationalist visions that did not come to pass: the future world government had given international control of all atomic weapons to its Space Patrol which used its monopoly to ensure peace and security on Earth and in space.[20] The popularity of *Space Cadet* led to the launch of an even more popular radio, television, book, comic book, and comic strip series in the 1950s. The 'Tom Corbett – Space Cadet' stories followed the exploits of cadets training to be members of the Solar Guard, the international force of the solar system's government, the Solar Alliance. The Solar Guard swore to 'uphold the Constitution of the Solar Alliance, to obey interplanetary law, to protect the liberties of the planets, to safeguard the freedom of space and to uphold the cause of peace throughout the universe' using their state-of-the-art spaceships. They used small arms which fired 'paralo rays' that did not kill but paralysed, though each ship was also equipped with four atomic bombs for emergency use. Atomic reactors powered most machines in the Alliance, and its capital was named Atom City.[21] By the 1960s the shine of nuclear energy had begun to fade, and it was replaced with even more fantastical machines in integrationist stories of the future. Most prominently Gene Roddenberry's *Star Trek* television series, which first aired in 1966, followed the exploits of an exploratory spaceship of a future intergalactic government, the United Federation of Planets, headquartered in San Francisco. The invention of the warp drive, which allowed faster-than-light travel, made the formation of the Federation possible, and the drive was used by the Federation's Starfleet to defend and police the Federation.[22]

These visions and movements declined in the 1970s. *Star Trek* ended in 1969. Pugwash lost much of its impetus as nuclear weapons continued to grow and arms control negotiations came under the close direction of states, with little room for intervention from external organizations of experts.[23] Neo-functionalist approaches to international integration melted away by the late 1960s as proponents of closer European union

[20] Robert A. Heinlein, *Space Cadet* (New York: Scribner's Sons, 1948). Also: Robert A. Heinlein, 'The Long Watch', in *Beyond Time and Space*, ed. August Derleth (New York: Pellegrini & Cudahy), 616–629.

[21] Carey Rockwell, *Stand by for Mars!* (New York: Grosset and Dunlap, 1952). See also the 50s TV series Space Patrol: Wheeler Winston Dixon, 'Tomorrowland TV: The Space Opera and Early Science Fiction Television', in *The Essential Science Fiction Television Reader*, ed. J. P. Telotte (Lexington, KY: University of Lexington Press, 2008), 93–110.

[22] Michael P. Scharf and Lawrence D. Robert, 'The Interstellar Relations of the Federation: International Law and "Star Trek: The Next Generation"', in *Star Trek Visions of Law and Justice*, eds. Robert H. Chaires and Bradley Stewart Chilton (Denton: University of North Texas Press, 2003), 73–113.

[23] Evangelista, *Unarmed Forces*, 144.

came to see the process as largely one of political negotiation.[24] Large dam projects continued into the 1980s and beyond, but shed the internationalist rhetoric that surrounded them by the 1980s. They were by then firmly seen as national rather than transnational projects, symbols of national might and nation-building. By the 2000s large dam projects, though still favoured by some developing countries, were increasingly questioned because of their massive ecological and human cost.[25]

Intellectuals and Modern Technology

It is perhaps in intellectual thought where the hopes and fears explored in this book remain the most enduring. Discourses on globalization, in particular, show striking continuities with the liberal internationalist thinking on science and technology explored in this book. Since its rise in the 1980s, the notion of globalization has been underpinned by the idea of the integrative nature of modern transportation and communication which, according to one recent article on globalization, have 'dissolved the barriers of time, distance, and lack of information that complicated all types of long-range relationships'.[26] In such techno-globalist narratives the current erasure of national boundaries and spread of global trade and communication is portrayed as the end-point of a centuries-long process of gradual global integration based on a succession of (modern-for-their-time) technologies.[27] In some histories the internationalization and technological development of these inventions march hand in hand from the late nineteenth century onwards.[28] Often just a small number of technologies, new in their day, are selected for their integrative effects. Typical are those which according to the *Companion to International History 1900–2001* have created 'A Shrinking World': the telegraph, the radio, aviation, cars, and the internet.[29]

---

[24] Paul Taylor, 'Introduction', in Mitrany, *The Functional Theory of Politics*, ix–xxv.

[25] Daniel Klingensmith, *One Valley and a Thousand: Dams, Nationalism, and Development* (New Delhi: Oxford University Press, 2007).

[26] Alfred E. Eckes, 'Globalization', in *Companion to International History 1900–2001*, ed. Gordon Martel (London: Blackwell Press, 2007), 408–421.

[27] For examples of globalization literature at its height, see: David Held, Anthony McGrew, David Goldblatt, and Jonathan Perraton, *Global Transformations: Politics, Economics and Culture* (Cambridge: Polity Press, 1999); Frank J. Lechner and John Boli, eds., *The Globalization Reader* (Oxford: Blackwell Publishing, 2000).

[28] For example: Anton A. Huurdeman, *A Worldwide History of Telecommunications* (New York: Wiley, 2003).

[29] Jeffrey Engel, 'A Shrinking World: Transport, Communication, and Towards a Global Culture', in Martel, *Companion to International History 1900–2001*, 52–64. For a longer perspective see: William H. McNeill, *The Rise of the West: A History of the Human Community* (Chicago: University of Chicago Press, 1963), 764–772, 794–795; and Ronald J. Deibert, *Parchment, Printing and Hypermedia: Communications in World Order Transformation* (New York: Columbia University Press, 1997).

Many modern techno-globalist accounts reach a crescendo with 'information technologies' and the 'Information Age'. Manuel Castells's three-volume *The Information Age* is perhaps the most extensive study of this kind. In his work, as in many others, we find echoes of earlier liberal internationalist dichotomies. Nation-states are relics of a now long-gone 'industrial era', rendered obsolete by the development of telecommunications, yet surviving 'beyond historical inertia' due to the 'defensive communalism' of their people.[30] For Jessica Mathews the growth of telecommunications was a transformative 'Power Shift' away from nations to non-state actors; states, argued Robert O. Keohane and Joseph S. Nye, Jr., needed new ways of exerting their power in the new interdependent information age. For Walter Lafeber this impact of the 'information revolution' on US foreign relations was as great as the impact of the previous two 'technological revolutions' (the first and second industrial revolutions).[31] Perhaps no one better captured this idea of the opposition between the integrative effects of modern technology (and in this case the international trade in modern high-tech goods) and the parochial traditional nation-state than journalist Thomas Friedman in his aptly titled best-selling account of globalization, *The Lexus and the Olive Tree*.[32]

The anxieties over convertibility and the militaristic perversion of modern technologies explored in this book also have echoes in modern-day discourses, particularly through the notion of dual-use technologies. Concerns over nuclear technologies in this regard remain, of course, but have been recently joined by a number of even more modern technologies. Drones first gained prominence in the early 2000s when they were seen as heralding a new era of intelligent warfare, but led soon to concerns over extra-judicial killings abroad, and illegal or illicit surveillance at home. Policymakers and activists then began to explore governance regimes that could counter their deleterious uses, whilst allowing them to fulfil their potential for legal surveillance, mapping, delivery, and transport.[33]

---

[30] Manuel Castells, *The Information Age: Economy, Society and Culture*, Vol. 2, *The Power of Identity* (Oxford: Blackwell, 1997), 308, 354.

[31] Jessica T. Mathews, 'Power Shift', *Foreign Affairs* 76, no. 1 (January/February 1997): 50–66; Robert O. Keohane and Joseph S. Nye, Jr., 'Power and Interdependence in the Information Age', *Foreign Affairs* 77, no. 5 (September/October 1998): 81–94; Walter Lafeber, 'Presidential Address: Technology and U.S. Foreign Relations', *Diplomatic History* 24, no. 1 (Winter 2000): 1–19.

[32] Thomas Friedman, *The Lexus and the Olive Tree* (London: Harper Collins, 2000), 18.

[33] See for example: Sarah Elizabeth Kreps, *Drones: What Everyone Needs to Know* (New York: Oxford University Press, 2016); Bart Custers, ed., *The Future of Drone Use: Opportunities and Threats from Ethical and Legal Perspectives* (Berlin: Springer, 2016); Kristin Bergtora Sandvik, Maria Gabrielsen Jumbert, eds., *The Good Drone* (Abingdon: Routledge, 2017).

Commentators now warn that, although Artificial Intelligence offers widespread economic growth, it also brings with it threats such as joblessness and arms races, and possibly the extinction of the human race.[34] Although the language used is different in many ways (talk of 'convertibility' and 'international control' has been replaced by 'dual-use', 'risk', 'proliferation', and 'governance'), the underlying notion, that there is a singular technology with radically positive and negative impacts, often reoccurs. As one author of the 2016 American Academy of Arts and Sciences report, *Governance of Dual-Use Technologies* notes: 'A gun is designed to have negative effects on objects and people. But in the hands of the good guys, (e.g., the police), its use is beneficial to society. Guns are misused for harmful purposes primarily when they are put into the hands of the bad guys (e.g., criminals). Similar comments apply to applications of IT with negative effects'.[35]

Technologies are, partially, discursive and ideological constructs. Our understandings of their essential nature are intertwined with our beliefs and assumptions about their effects and impact on society. But more than that, the urge to ascribe power to spectacular technologies appears to run deep in our modern increasingly mechanized and bureaucratized societies. A succession of technologies have been attached to our existential hopes and fears, and emerged as carriers of our dreams and nightmares. The technologies discussed in this book were thought of as machines of peace, capable of creating and sustaining liberal world orders. This belief was not limited to internationalists but was much more widespread, and nor was it divorced from real-world policymaking: it was embedded in political and diplomatic discussions on arms control, international organization, and regulation. It is only when we understand and explore the interconnectedness of these strands, the technological and the ideological, that we can hope to understand the material and ideational centrality of technology to the modern world.

---

[34] Elizer Yudkowsky, 'Artificial Intelligence as a Positive and Negative Factor in Global Risk', in *Global Catastrophic Risks*, eds. Nick Bostrom and Milan M. Ćirković (Cambridge: Cambridge University Press, 2008), 308–345; Nick Bostrom, *Superintelligence: Paths, Dangers, Strategies* (Oxford: Oxford University Press, 2014), chapter 8; Rosario Girasa, *Artificial Intelligence as a Disruptive Technology: Economic Transformation and Government Regulation* (Cham, Switzerland: Palgrave Macmillan, 2020).

[35] Elisa D. Harris, ed., *Governance of Dual-Use Technologies: Theory and Practice* (Cambridge, MA: American Academy of Arts & Sciences, 2016), 113.

Bibliography

Archives and Manuscripts

Archives and Special Collections, British Library of Political and Economic Science, London School of Economics and Political Science

Archives of the League of Nations, Geneva

Association of Oak Ridge Engineers and Scientists Records

Atomic Scientists of Chicago Records 1943–1955

British Library Manuscripts Collection, British Library

British Library of Political and Economic Science, London School of Economics and Political Science

Carnegie Endowment for International Peace Records, 1910–1954

Cecil of Chelwood Papers

Chatham House, London

Churchill Archives Centre, Churchill College, University of Cambridge

Clark M. Eichelberger Papers

Eugene I. Rabinowitch Papers

Federation of American Scientists Records

Leonard Woolf Papers

Lord Davies of Llandinam Papers

Manuscripts and Archives Division, The New York Public Library

Mass Observation Archives

Modern Records Centre, University of Warwick

National Archives and Records Administration, National Archives at College Park, College Park, MD

National Library of Wales

Papers of the United Nations Association of Great Britain and Northern Ireland

Paul Mantoux Papers

Quincy Wright Papers

Rare Books and Manuscript Library, Columbia University

Records of the Association of Scientific Workers

Records of the Council on Foreign Relations, 1921–1951

Records of the Royal Institute of International Affairs

Special Collections Research Center, University of Chicago Library

Special Collections, University of Sussex Library

The National Archives, UK
The Papers of Baron Noel-Baker
World Citizens Association Central Committee Records 1939–1953

Primary Sources: Newspapers and Periodicals

Advocate of Peace
Aeroplane
Air Force
Air Wonder Stories
American Magazine
Atlantic Monthly
Bookman
Chicago Tribune
Collier's
Daily Express
Daily Herald
Daily Mail
Economist
Evening Standard
Excelsior
Federal News
Flight
Flying
Flying and Popular Aviation
Harper's Magazine
La Paix
La République
La Revue des Vivants
L'Europe Nouvelle
Le Figaro
Liberal Magazine
Life
Listener
Manchester Guardian
News Chronicle
Newsweek
New York Herald Tribune
New York Times
New York Times Magazine
Notre Temps
Observer
Picture Post
Reader's Digest
Reynolds News
Rotarian
Saturday Evening Post

Saturday Review of Literature
Spectator
Survey Graphic
Times, The
Time Magazine
Tribune
Washington Post

Official Government Publications

Agreement between His Majesty's Government in the Commonwealth of Australia and His Majesty's Government in New Zealand, Cmnd. 6513. 1944.

Air Policy Commission. *Survival in the Air Age*. Washington, DC: Government Printing Office, 1948.

Barnard, Chester I., J. R. Oppenheimer, Charles A. Thomas et al. *A Report on the International Control of Atomic Energy*. Washington, DC: Department of State, 1946.

'Committee I: Multilateral Aviation Convention and International Aeronautical Body. Verbatim Minutes of Plenary Session, November 8', in Department of State, *Proceedings of the International Civil Aviation Conference*, vol. I, part 2, document 117, p. 540. Washington, DC: Government Printing Office, 1948, accessed 2 August 2020, www.icao.int/ChicagoConference/Pages/proceed.aspx.

Congressional Record.

Department of State. *United States Atomic Energy Proposals: Statement of the United States Policy on Control of Atomic Energy as Presented by Bernard M. Baruch, Esq., to the United Nations Atomic Energy Commission June 14, 1946*. Washington, DC: Government Printing Office, June 1946.

Foreign Relations of the United States.

Hansard.

International Air Transport Policy, Cmnd. 6561. October 1944.

Merriam, John C. 'The Relation of Science to Technological Trends', in *Technological Trends and National Policy*, ed. National Resources Committee, 91–92. Washington, DC: Government Printing Office, 1937.

'Memorandum [by Hugh R. Wilson] Arising from Conversations in Mr. Welles' Office, April 19 and 26', in *Postwar Foreign Policy Preparation 1939–1945*, ed. Harley Notter, 458–60. Washington, DC: Department of State, 1949.

Note on the Method of Employment of the Air Arm in Iraq, Cmnd. 2271. 1924.

'Verbatim Minutes of Opening Plenary Session, November 1', in Department of State, *Proceedings of the International Civil Aviation Conference*, vol. I, part 1, document 33, pp. 77–78, 80. Washington, DC: Government Printing Office, 1948, accessed 2 August 2020, www.icao.int/ChicagoConference/Pages/proceed.aspx.

US Congress, Senate Committee on Foreign Relations. *The Charter of the United Nations: Hearings before the Committee on Foreign Relations, United States Senate, Seventy-Ninth Congress, First Session, on the Charter of the United Nations for the Maintenance of International Peace and Security*, Submitted by the President of the United States on 2 July 1945: revised 9, 10, 11, 12, and 13 July 1945. Washington, DC: Government Printing Office, 1945.

League Of Nations Publications

Arbitration, *Security and Reduction of Armaments: Extracts from Debates of the Fifth Assembly C.708.1924.IX (C.C.O.1.)* Geneva: League of Nations, 1924.

Air Commission of the Conference for the Reduction and Limitation of Armaments. *Objective Study on the Internationalisation of Civil Aviation, Disarmament 1932.IX.43.* Geneva: League of Nations, 1932.

'Proposals of the French Delegation, Conf. D.56, Geneva, 5 February 1932', in League of Nations, *Conference for the Reduction and Limitation of Armaments. Conference Documents vol. 1, Disarmament 1932.IX.63*, pages 113–116. Geneva: League of Nations, 1932.

Records of the Conference for the Reduction and Limitation of Armaments. Series A Verbatim Records of Plenary Meetings vol. 1 February 2nd – July 23rd 1932, Disarmament 1932.IX.60. Geneva: League of Nations, 1932.

Other Published Primary Sources

Air League of the British Empire. *A Souvenir of the Great Empire Day of 1934.* London: Air League of the British Empire, 1934.

Allen, Clifford. *Britain's Political Future: A Plea For Liberty and Leadership.* London: Longman Green, 1934.

Allen, Clifford. 'Public Opinion and the Idea of International Government', *International Affairs* 13, no. 2 (March–April 1934): 186–207.

Allen, Stephen Haley. *International Relations.* Princeton, NJ: Princeton University Press, 1920.

Allenby, Edmund. *Allenby's Last Message: World Police for World Peace.* London: New Commonwealth Society, 1936.

Alexander, Edward. *Irving Howe: Socialist, Critic, Jew.* Bloomington: Indiana University Press, 1998.

Altman, Victor. *The International Police Force and World Security.* London: Alliance Press, 1945.

American Association for the United Nations. *Dumbarton Oaks Proposals.* New York: American Association for the United Nations, 1944.

American Association for the United Nations. *That These Dead Shall Not Have Died in Vain.* New York: American Association for the United Nations, 1945.

Anderson, Mosa. 'The Approach to World Government', *One World* 1 (June 1946): 3–6.

Anderson, R. Wherry. *The Romance of Air-Fighting.* New York: George H. Doran, 1917.

Angell, Norman. *Arms and Industry: A Study of the Foundations of International Polity.* London: GP Putnam & Sons, 1914.

Angell, Norman. *The Great Illusion: A Study of the Relation of Military Power to National Advantage*, 4th ed. New York: GP Putnam & Sons, 1913.

Angell, Norman. 'Introduction', in *Nationalism, War and Society*, by Edward Krehbiel, xiii–xxxv. New York: Macmillan: 1916.

Angell, Norman. *The Menace to our National Defence.* London: Hamish Hamilton, 1934.

Angell, Norman. *The World's Highway*. New York: George H. Doran, 1915.

Arlen, Michael. *Man's Mortality*. London: Heinemann, 1933.

Armstrong, Hamilton Fish. *Peace and Counterpeace: From Wilson to Hitler*. New York: Harper and Row, 1971.

Arnold-Forster, William. *The United Nations Charter Examined*. London: Labour Party, 1946.

Association of Los Alamos Scientists. 'The Policy of the A.L.A.S.', *Bulletin of the Atomic Scientists* 1, no. 1 (10 December 1945): 2.

'Atlantic Charter', The Avalon Project, accessed 2 August 2020, avalon .law.yale.edu/wwii/atlantic.asp.

'The Atomic Bomb and World Government Conference', *Rollins Alumni Record* 14, no. 1 (March 1946): 8–9.

'Atomic Power Control Problems. 2. General Analysis', *Bulletin of the Atomic Scientists* 1, no. 3 (10 January 1946): 2.

Atomic Scientists of Chicago. 'The Atomic Scientists of Chicago', *Bulletin of the Atomic Scientists* 1, no. 1 (10 December 1945): 1.

Attlee, Clement R. *An International Police Force*. London: New Commonwealth Society, 1934.

Attlee, Clement R. *The Labour Party in Perspective*. London: Victor Gollancz, 1937.

Attlee, Clement R. *Labour's Peace Aims*. London: Labour Party, 1939.

Ball, Joseph H. 'A Chance for Peace' in *A Chance for Peace and Responsibility for Peace*, by Joseph H. Ball and Harold C. Urey. Cincinnati, OH: Cincinnati Foreign Policy Association, November 1945.

Bennett, D. C. T. *Freedom from War*. London: Pilot Press, 1945.

Bennett, D. C. T. *Pathfinder: A War Autobiography*. London: Frederick Muller, 1958.

Berle, Jr., Adolph A. 'International Civil Aviation Conference', *Vital Speeches of the Day* 11, no. 4 (1 December 1944): 124–128.

Berle, Jr., Adolph A. *Navigating the Rapids, 1918–1971*, eds. Beatrice Bishop Berle and Travis Beal Jacobs. New York: Harcourt Brace Jovanovich, 1973.

Bernal, J. D. *The Social Function of Science*. London: George Routledge and Sons, 1939.

Blackburn, Raymond. *I am an Alcoholic*. London: Allan Wingate, 1959.

Blackett, P. M. S. *The Military and Political Consequences of Atomic Energy*. London: Turnstile Publications, 1948.

Blackett, P. M. S. 'The Military and Political Consequences of Atomic Energy', in *The Atomic Age*, by M. L. Oliphant et al., 32–51. London: George Allen and Unwin, 1949.

Blackett, P. M. S. *The Atom and the Charter*. London: Fabian Publications, 1946.

Bohr, Niels. 'Challenge to Civilization', *Science* 102, no. 2650 (12 October 1945): 363.

Bohr, Niels. 'Niels Bohr's Memorandum to President Roosevelt', Atomic Archive, July 1944, accessed 2 August 2020, www.atomicarchive.com/Docs/ ManhattanProject/Bohrmemo.shtml.

Bohr, Niels. 'Science and Civilization', in *One World or None*, eds. Dexter Masters and Katharine Way, ix–x. New York: McGraw Hill, 1946.

'The Bomb', *The New Commonwealth* (October 1945), 324–327.

Bonnet, Henri. *Outlines of the Future: World Organizations Emerging from the War.* Chicago: World Citizens Association, 1943.

Bonnet, Henri. *The United Nations: What They Are and What They May Become.* Chicago: World Citizens Association, 1942.

Boorse, Henry A. 'Two International Scientific Meetings in England', *Bulletin of the Atomic Scientists* 2, nos. 5–6 (1 September 1946): 5–6.

Boorse, Henry A. 'A World Federation of Scientific Workers', *Bulletin of the Atomic Scientists* 2, nos. 5–6 (1 September 1946): 7.

Bourquin, Maurice, ed. *Collective Security: A Record of the Seventh and Eighth International Studies Conferences, Paris 1934-London 1935.* Paris: International Institute of Intellectual Cooperation, 1936.

Bowett, D. W. *United Nations Forces: A Legal Study of United Nations Practice.* London: Stevens and Sons, 1964.

Bowles, F. G. and Major W. F. Vernon, *World Airways: What We Want.* London: World Airways Joint Committee, 1946.

Brailsford, H. N. *A League of Nations*, 2nd ed. New York: Macmillan, 1917.

Brailsford, H. N. 'A Socialist Foreign Policy' in *Problems of a Socialist Government,* by C. Addison et al., 252–286. London: Victor Gollancz, 1933.

Brittain, Vera. 'Peace and the Public Mind', in *Challenge to Death*, by P. J. Noel Baker et al., 40–66. London: Constable, 1934.

Brodie, Bernard. 'New Techniques of War and National Policies', in *Technology and International Relations*, ed. William F. Ogburn, 144–173. Chicago: University of Chicago Press, 1949.

Brown, Harrison. *Must Destruction be our Destiny? A Scientist Speaks as a Citizen.* New York: Simon and Schuster, 1946.

Bruner, Jerome S. 'Public Opinion and America's Foreign Policy', *American Sociological Review* 9, no. 1 (February 1944): 50–56.

Buell, Raymond Leslie. *International Relations.* London: Sir Isaac Pitway & Sons, 1926.

'Building the Peace', *Education for Victory* 3, no. 19 (3 April 1945): 4.

Bulwer-Lytton, Victor. *Can We Survive the Atomic Bomb?* London: Hutchinson, 1945.

Burney, Charles Dennistoun. *The World, The Air and the Future.* London: Alfred A. Knopf, 1929.

Burns, C. Delisle. *International Politics.* London: Methuen, 1920.

Burns, C. Delisle. *A Short History of International Intercourse.* London: George Allen & Unwin, 1924.

Burns, C. Delisle. *Modern Civilization on Trial.* New York: Macmillan, 1931.

Cadogan, Alexander. *Diaries of Sir Alexander Cadogan, 1938-1945*, ed. David Dilks. London: Cassell, 1972.

Carlton, D. 'The Problem of Civil Aviation in British Air Disarmament Policy, 1919-1934', *Royal United Services Institution Journal* 111, no. 664 (1966): 307–316.

Carnegie Endowment for International Peace. *The Challenge of Disarmament* 5 (10 June 1932).

Carr, E. H. *Conditions of Peace.* London: Macmillan, 1942.

Carr, E. H. *Nationalism and After*. London: Macmillan, 1945.

Carr, E. H. *The Twenty Years' Crisis 1919–1939: An Introduction to the Study of International Relations*. London: Macmillan, 1939.

Carter, W. Horsfall. *Peace Through Police*. London: New Commonwealth Society, 1934.

Caudwell, Christopher. *Studies in a Dying Culture*. London: Bodley Head, 1938.

Cecil, Robert. 'Disarmament and the League', in *The Way of Peace: Essays and Addresses*, 211–232. London: Philip Allan, 1928.

Cecil, Robert. *A Real Peace*. London: Hamish Hamilton, 1941.

Chafee, Zechariah, Jr. 'International Utopias', *Proceedings of the American Academy of Arts and Sciences* 75, no. 1 (October 1942): 39–53.

Churchill, Winston. 'The Soviet Danger: The Iron Curtain', in *The Speeches of Winston Churchill*, ed. David Cannadine, 295–308. London: Houghton Mifflin, 1989.

Churchill, Winston. *The World Crisis, vol. 3, 1916–1918*. London: Thornton Butterworth, 1927.

Churchill, Winston. *The World Crisis, vol. 4, 1918–1928: The Aftermath*. London: Thornton Butterworth, 1929.

Clark, G. N. *Unifying the World*. London: Swarthmore Press, 1920.

Clark-Hall, R. H. 'The Value of Civil Aviation as a Reserve to the Royal Air Force in the Time of War', *Journal of the Royal United Services Institution* 69, no. 475 (1924): 415–32.

Cole, G. D. H. and Margaret Cole. *The Intelligent Man's Review of Europe To-Day*. London: Victor Gollanz, 1933.

Commission to Study the Organization of Peace. 'Fifth Report: Security and Disarmament under the United Nations (June 1947)', in *Building Peace: Reports of the Commission to Study the Organization of Peace, 1939–1972 Vol. 1*, 185–215. Metuchen, NJ: Scarecrow Press, 1973.

Commission to Study the Organization of Peace. *For This We Fight: Program No. 2 Science and our Future*. New York: Commission to Study the Organization of Peace, 1943.

Commission to Study the Organization of Peace. *For This We Fight: Program No. 5 Making the World Secure*. New York: Commission to Study the Organization of Peace, 1943.

Commission to Study the Organization of Peace. 'Fourth Report of the Commission to Study the Organization of Peace', *International Conciliation* 22 (January 1944): 3–110.

Commission to Study the Organization of Peace. 'The General International Organization: Its Framework and Function', *International Conciliation* 22 (September 1944): 547–551.

Commission to Study the Organization of Peace. *Preliminary Report*. New York: Commission to Study the Organization of Peace, November 1940.

Commission to Study the Organization of Peace. 'Seventh Report: Collective Security under the United Nations (July 1951)', in *Building Peace: Reports of the Commission to Study the Organization of Peace, 1939–1972 Vol. 1*, 230–263. Metuchen, NJ: Scarecrow Press, 1973.

Commission to Study the Organization of Peace. *Study Outline on the Organization of Peace.* New York: Commission to Study the Organization of Peace, January 1940.

Commission to Study the Organization of Peace. *A Ten Year Record 1939–1949.* New York: Commission to Study the Organization of Peace, 1949.

Committee on Atomic Energy of the Carnegie Endowment for International Peace. *Utilization and Control of Atomic Energy: A Draft Convention Prepared by the Legal Subcommittee in consultation with other Legal, Political and Scientific Experts.* New York: Carnegie Endowment for International Peace, June 1946.

Cook, Don. *The Chicago Aviation Agreements: An Approach to World Policy.* New York: American Enterprise Association, 1945.

Cooper, John C. 'Air Transport and World Organization', *The Yale Law Journal* 55, no. 5 (August 1946): 1191–1213.

Cooper, John C. *The Right to Fly.* New York: Henry Holt, 1947.

Compton, Arthur H. 'Franklin Medal Lecture: Atomic Energy as a Human Asset', *Proceedings of the American Philosophical Society* 90, no. 1 (January 1946): 70–79.

Compton, Arthur H. 'The Social Implications of Atomic Energy', *American Journal of Physics* 14, no. 3 (May–June 1946): 173–178.

Corbett, Percy E. 'Effect on International Organization', in *The Absolute Weapon: Atomic Power and World Order*, ed. Bernard Brodie, 148–165. New York: Harcourt Brace, 1946.

Corbusier, Le. *Aircraft.* London: Studio Publications; Milan: Abitare Segesta, 1996.

Cot, Pierre. *Military Force or Air Police.* London: New Commonwealth Institute, 1935.

Council on Foreign Relations. *American Public Opinion and Postwar Security Commitments.* New York: Council on Foreign Relations, 1944.

Council on Foreign Relations. *The War and Peace Studies of the Council on Foreign Relations.* New York: Council on Foreign Relations, 1946.

Cousins, Norman. *Modern Man Is Obsolete.* New York: Viking, 1945.

'The Covenant of the League of Nations (Including Amendments adopted to December, 1924)', Avalon Project, accessed 2 August 2020, https://avalon.law.yale.edu/20th_century/leagcov.asp#art8.

Craig, F. W. S. *British General Election Manifestos 1918–1966.* Chichester: Political Reference Publications, 1970.

Cripps, Stafford. *The Struggle for Peace.* London: Victor Gollancz, 1936.

Crosby, Oscar T. 'International Force', *Advocate of Peace through Justice* 94, no.1 (March 1932): 44–48.

Crosby, Oscar T. 'An International Justice of the Peace and His Constable', *Advocate of Peace through Justice* 93, no. 2 (May 1931): 110–119.

Crosby, Oscar T. *International War: Its Causes and its Cure.* London: Macmillan, 1919.

Culbertson, Ely. *How to Control the Atomic Threat.* New York: Total Peace, 1945.

Culbertson, Ely. *Must We Fight Russia?* Philadelphia: John C. Winston, 1946.

Culbertson, Ely. 'Russia and World Security', *Russian Review* 4, no. 1 (Autumn 1944): 10–17

Culbertson, Ely. *The Strange Lives of One Man: An Autobiography*. Chicago: John C. Winston, 1940.

Culbertson, Ely. *Summary of the World Federation Plan: An Outline of a Practical and Detailed Plan for World Settlement*. New York: World Federation, 1943.

Culbertson, Ely. *Total Peace: What makes Wars and How to Organize Peace*. Garden City, NY: Doubleday, Doran, 1943.

Curtis, Lionel. *Civitas Dei: The Commonwealth of God*. London: Macmillan, 1938.

Davies, David. 'An International Police Force?', *International Affairs* 11, no. 1 (January 1932): 76–99.

Davies, David. 'Disarmament', *Transaction of the Grotius Society* 5 (1919): 109–118.

Davies, David. *Force and the Future* (London: New Commonwealth Society, 1934.

Davies, David. *An International Police Force: An Abridged Edition of 'The Problem of the Twentieth Century'*. London: Ernest Benn, 1932.

Davies, David. *An International Air Force: Its Functions and Organisation*. London: New Commonwealth Society, 1934.

Davies, David. *Nearing the Abyss: The Lesson of Ethiopia*. London: Constable, 1936.

Davies, David. *The Problem of the Twentieth Century: A Study in International Relationships*. London: Ernest Benn, 1930.

Davies, David. *Suicide or Sanity?* London: Williams and Norgate, 1932.

Davis, John W. et al. 'The President and Peace Forces: Letter to the New York Times', *International Conciliation* 22 (December 1944): 795–800.

Dean, Vera Micheles. *On the Threshold of World Order*. New York: Foreign Policy Association, January 1944.

Dewar, George A. B. *The Great Munition Feat 1914–1918*. London: Constable, 1921.

Dickinson, G. Lowes. *War: Its Nature, Cause and Cure*. London: George Allen and Unwin, 1923.

Dodd, Allen Robert. *Captain Gardiner of the International Police*. New York: Dodd, Mead, 1916.

Doman, Nicholas. *The Coming Age of World Control: The Transition to an Organized World Society*. New York: Harper & Bros., 1942.

'Donald C. Blaisdell Oral History Interview', Truman Library, accessed 2 August 2020, www.trumanlibrary.gov/library/oral-histories/blaisdel.

Dutt, Palme R. *World Politics 1918–1936*. London: Victor Gollancz, 1936.

Eagleton, Clyde. *Analysis of the Problem of War*. New York: Ronald Press, 1937.

Eagleton, Clyde. *The Forces that Shape Our Future*. New York: New York University Press, 1945.

Eagleton, Clyde. *International Government*. New York: Ronald Press, 1932.

Eagleton, Clyde. *International Government*, 2nd ed. New York: Ronald Press, 1948.

Eagleton, Clyde. *International Government*, 3rd ed. New York: Ronald Press, 1957.

Eagleton, Clyde. 'Peace Enforcement', *International Conciliation* 20 (April 1941): 499–504.

Eichelberger, Clark M. 'Next Steps in the Organization of the United Nations', *Annals of the American Academy of Political and Social Science* 228, no. 1 (July 1943): 34–39.

Eichelberger, Clark M. *Organizing for Peace: A Personal History of the Founding of the United Nations*. New York: Harper & Row, 1977.

Eichelberger, Clark M. *Proposals for the United Nations Charter: What was Done at Dumbarton Oaks*. New York: Commission to Study the Organization of Peace, 1944.

Eichelberger, Clark M. *Time Has Come for Action*, New York: Commission to Study the Organization of Peace, 1944.

Einstein, Albert. 'A Message to the World Congress of Intellectuals', *Bulletin of the Atomic Scientists* 4, no. 10 (October 1948): 295, 299.

Einstein, Albert. 'A Reply to the Soviet Scientists', *Bulletin of the Atomic Scientists* 4, no. 2 (February 1948): 35–37.

Einstein, Albert. 'The Way Out', in *One World or None*, eds. Dexter Masters and Katharine Way, 76–79. New York: McGraw Hill, 1946.

Eliot, George Fielding. *Bombs Bursting in Air: The Influence of Air Power on International Relations*. New York: Reynal & Hitchcock, 1939.

Enock, Arthur Guy. *The Problem of Armaments: A Book for Every Citizen of Every Country*. London: Macmillan, 1923.

'Extract from "The Brains Trust"', *Federal News* 130 (January 1946): 4–5.

Fabian Society. *Dumbarton Oaks: A Fabian Commentary*. London: Fabian Publications, December 1944.

Fahan R. [Irving Howe]. 'Struggle for Air Supremacy', *The New International* 9, no. 4 (April 1943): 103–106.

Federal Union. *Proposals for a Permanent United Nations Force*. London: Federal Union, 1957; 1960.

Federation of American Scientists. *The Federation of American Scientists*. Washington. DC: Federation of American Scientists, 1946.

Fleming, Denna. 'The Coming World Order, Closed or Free', *The Journal of Politics* 4, no. 2 (May 1942): 250–243.

Fosdick, Raymond Blaine. *The Old Savage in the New Civilization*. Garden City, NY: Doubleday, Doran, 1928.

Fox, William T. R. 'IV. Collective Enforcement of Peace and Security', *The American Political Science Review* 39, no. 5 (October 1945): 970–981.

Fox, William T. R. 'Atomic Energy and International Relations', in *Technology and International Relations*, ed. William F. Ogburn, 102–125. Chicago: University of Chicago Press, 1949.

Fox, William T. R. 'International Control of Atomic Weapons', in *The Absolute Weapon: Atomic Power and World Order*, ed. Bernard Brodie, 169–203. New York: Harcourt Brace, 1946.

Fradkin, Elvira K. *The Air Menace and the Answer*. New York: Macmillan, 1934.

Fradkin, Elvira K. *A World Airlift: The United Nations Air Police Patrol*. New York: Funk & Wagnalls, 1950.

Franck, James, et al. 'Appendix B: The Franck Report June 11, 1945', in *A Peril and a Hope: The Scientists' Movement in America: 1945–47*, by Alice Kimball Smith, 560–572. Chicago: University of Chicago Press, 1965.

Franck, James. 'The Social Task of the Scientist', *Bulletin of the Atomic Scientists* 3, no. 3 (March 1947): 70.

Franklin, William M. *Areas of Agreement in the Preparatory Disarmament Commission and the General Disarmament Conference.* New York: Council on Foreign Relations, 1940.

Fuller, J. F. C. *The Foundations of the Science of War.* London: Hutchinson, 1926.

Fuller, J. F. C. *The Reformation of War.* London: Hutchinson, 1923.

Fuller, J. F. C. *Tanks in the Great War, 1914–1918.* London: John Murray, 1920.

Fulljames, R. E. G. 'An International Air Police Force', *The Royal Air Force Quarterly* 6, no. 3 (July 1935): 245–250.

Fulljames, R. E. G. letter to the editor, 'International Force', *United Nations News* 1 (May 1946): 26.

Fulljames, R. E. G. *An International Police Force* Southampton: Shirley Press, 1946.

Fulljames, R. E. G. 'The Problem of Security from the Fear of War in the Atomic Age', *Fighting Forces* 24, no. 2 (June 1947): 84–86.

Fulljames, R. E. G. 'Security Under the United Nations', *World Affairs* 1 (October 1947): 402–408.

Garnett, Maxwell. *Freedom of the Air.* Letchworth: Garden City Press, 1933.

Garnett, Maxwell. 'The World Crisis and the Disarmament Conference', *Contemporary Review* 141 (Jan/June 1932): 144–154.

Goodrich, C. F. 'Wanted – an International Police', in *War Obviated by an International Police: A Series of Essays, Written in Various Countries,* by C. van Vollenhoven et al., 172. The Hague: Martinus Nijhoff, 1915.

Graham, Malbone W. 'Great Powers and Small States', in *Peace, Security and the United Nations,* ed. Hans J. Morgenthau, 57–82. Chicago: University of Chicago Press, 1946.

Greaves, H. R. G. *An International Police Board.* London: Allen & Unwin, 1935.

Greaves, H. R. G. *The League Committees and World Order: A Study of the Permanent Expert Committees of the League of Nations as an Instrument of International Government.* Oxford: Oxford University Press, 1931.

Greenwood, Arthur. 'International Economic Relations', in *Introduction to the Study of International Relations,* ed. A. J. Grant et al., 66–112. London: Macmillan, 1916.

Grey, C. G. *Bombers.* London: Faber and Faber, 1941.

Grey, C. G. *The Civil Air War.* London: Harborough, 1945.

Grey, C. G. *A History of the Air Ministry.* London: George Allen & Unwin, 1940.

Griessemer, Tom. 'An International Police Force', *World Affairs* 105, no. 4 (December 1942): 262–266.

Griffin, Jonathan. *Britain's Air Policy: Present and Future.* London: Gollancz, 1935.

Griffin, Jonathan. *World Airways – Why Not? A Practical Scheme for the Safeguarding of Peace.* London: Gollancz, 1934.

Groves, P. R. C. *Behind the Smoke Screen.* London: Faber and Faber, 1934.

Groves, P. R. C. *Our Future in the Air: A Survey of the Vital Question of British Air Power.* London: Hutchinson, 1922.

Groves, P. R. C. 'The Influence of Aviation in International Affairs', *Journal of the Royal Institute of International Affairs* 8, no. 4 (July 1929): 289–317.

Groves, P. R. C. 'The Influence of Aviation in International Relations', *Journal of the Royal Institute of International Affairs* 6, no. 3 (May 1927): 133–152.

Gundry, Ronald E. 'Letters: Two Roads – No Goal', *Federal News* 129 (December 1945): 12.

Haas, Ernst B. *Beyond the Nation-State: Functionalism and International Organization.* Stanford: Stanford University Press, 1964.

Haas, Ernst B. *The Uniting of Europe: Political, Social and Economic Forces 1950–1957.* Stanford: Stanford University Press, 1958.

Haldane, J. B. S. *Callinicus: A Defence of Chemical Warfare.* London: Kegan Paul, Trench, Trubner, 1925.

Hamilton, Mary Agnes. 'No Peace Apart from International Security: An Answer to Extreme Pacifists', in *Challenge to Death*, by P. J. Noel Baker et al., 261–274. London: Constable, 1934.

Handley Page, F. 'International Air Transport. By Sir Osborne Mance, assisted by J.E. Wheeler', *International Affairs* 20, no. 2 (April 1944): 281–282.

Hankey, Maurice. 'The Control of External Affairs', *International Affairs* 22, no. 2 (March 1946): 161–73.

Hankey, Maurice. 'The Human Element', in *Atomic Energy. Its International Implications. A Discussion by a Chatham House Study Group*, by Royal Institute of International Affairs, 112–124. London: Royal Institute of International Affairs, 1948.

Hankey, Maurice. 'International Forces', in *Diplomacy by Conference: Studies in Public Affairs 1920–1946*, 120–130. London: Ernest Benn, 1946.

Hankey, Maurice. 'The Study of Disarmament', in *Diplomacy by Conference: Studies in Public Affairs 1920–1946*, 105–119. London: Ernest Benn, 1946.

Hart, Hornell. 'Atomic Cultural Lag: I. The Value Frame', *Sociology and Social Research* 32 (March 1948): 768–75.

Hart, Hornell. 'Atomic Cultural Lag: II. Its Measurement', *Sociology and Social Research* 32 (May 1948): 845–55.

Hart, Hornell. 'Some Cultural-Lag Problems Which Social Science Has Solved', *American Sociological Review* 16, no. 2 (April 1951): 223–227.

Hart, Hornell. 'Technological Acceleration and the Atomic Bomb', *American Sociological Review* 11, no. 3 (June 1946): 277–293.

Hart, Hornell. 'Technology and the Growth of Political Areas', in *Technology and International Relations*, ed. William F. Ogburn, 29–57. Chicago: University of Chicago Press, 1949.

Hart, B. H. Liddell. 'An International Force', *Journal of the Royal Institute of International Affairs 1931–1939* 12, no. 2 (March 1932): 205–223.

Heinlein, Robert A. 'The Long Watch', in *Beyond Time and Space*, ed. August Derleth, 616–629. New York: Pellegrini & Cudahy.

Heinlein, Robert A. *Space Cadet.* New York: Scribner's Sons, 1948.

Henderson, Arthur. *Labour's Way to Peace.* London: Methuen, 1935.

Herz, John. 'Idealist Internationalism and the Security Dilemma', *World Politics* 2, no. 2 (January 1950): 157–80.

Herz, John. *International Politics in the Atomic Age.* New York: Columbia University Press, 1959.

Herz, John. *Political Realism and Political Idealism: A Study in Theories and Realities.* Chicago: University of Chicago Press, 1951.

Herz, John. 'The Rise and Demise of the Territorial State', *World Politics* 9, no. 4 (July 1957): 473–93.

Hill, Norman. *International Administration.* New York: McGraw-Hill, 1931.

Hirst, Francis W. *Armaments: The Race and the Crisis.* London: Cobden-Sanderson, 1937.

Hobson, J. A. *Democracy after the War.* London: George Allen & Unwin, 1917.

Hobson, J. A. 'Force Necessary to Government', *Hibbert Journal* 33, no. 3 (April 1935): 331–342.

Hobson, J. A. *Problems of a New World.* London: George Allen & Unwin, 1921.

Hobson, J. A. *Towards International Government.* London: George Allen & Unwin, 1915.

Holt, George C. 'The Conference on World Government', *Journal of Higher Education* 17, no. 5 (May 1946): 227–235.

Holt, Hamilton. 'How the Conference Came About', *World Affairs* 109, no. 2 (1946): 87–90.

Howard-Ellis, C. *The Origin, Structure, and Working of the League of Nations.* London: George Allen & Unwin, 1928.

Hull, Cordell. *The Memoirs of Cordell Hull,* 2 vols. London: Hodder & Stoughton, 1948.

Hutchison, Keith. *Freedom of the Air.* New York: Public Affairs Committee, 1944.

Huxley, Julian. *Scientific Research and Social Needs.* London: Watts, 1934.

Jeffries, Zay et al. 'Appendix A: Prospectus on Nucleonics (the Jeffries Report)', in *A Peril and a Hope: The Scientists' Movement in America: 1945–47*, by Alice Kimball Smith, 539–559. Chicago: University of Chicago Press, 1965.

Jennings, W. Ivor. *A Federation for Western Europe.* Cambridge: Cambridge University Press, 1940.

Johnsen, Julia E., ed. *International Police Force.* New York: H. W. Wilson, 1944.

Josephson, Matthew. *Empire of the Air: Juan Trippe and the Struggle for World Airways.* New York: Harcourt, Brace, 1944.

Josephy, F. L. 'The Federal Implications of the Atom', *Federal News* 126 (September 1945): 1–2, 11.

Josephy, F. L. 'Federate or Perish', *Federal News* 126 (September 1945): 3.

Jouvenel, Henry de. *Aviation for World Service.* London: New Commonwealth Society, 1933.

Joyce, James Avery and Michael Young. *Chicago Commentary: The Truth about the International Air Conference.* London: n.p., 1945.

Kaempffert, Waldemar. *The Airplane and Tomorrow's World.* New York: Public Affairs Committee, 1943.

Keeton, George. *National Sovereignty and International Order.* London: Peace Book, 1939.

Kerr, Philip. 'Address of the Marquess of Lothian', *International Conciliation* 20 (January 1941): 9–19.

Kerr, Philip. 'The Place of Britain in the Collective System', *International Affairs* 13, no. 5 (September–October 1934): 622–650.

Kerr, Philip and Lionel Curtis. *The Prevention of War*. New Haven, CT: Yale University Press, 1923.

Kenworthy, J. M. *Will Civilization Crash?* London: Ernest Benn, 1927.

King-Hall, Stephen. *Britain's Third Chance: A Book about Post-War Problems and the Individual*. London: Faber and Faber, 1943.

King-Hall, Stephen. *Total Victory*. London: Faber and Faber, 1941.

King-Hall, Stephen. *The World Since the War*. London: Thomas Nelson and Sons, 1937.

Kinkaid, T. W. [untitled], *in War Obviated by an International Police: A Series of Essays, Written in Various Countries*, by C. van Vollenhoven et al., 158–162. The Hague: Martinus Nijhoff, 1915.

Kirk, Grayson. *The Atlantic Charter and Postwar Security*. New York: Council on Foreign Relations, December 1941.

Kirk, Grayson. *International Policing (A Survey of Recent Proposals)*. New York: Council on Foreign Relations, October 1941.

Kirk, Grayson. *International Politics and International Policing*. New Haven, CT: Yale Institute of International Studies, 1944.

Kirk, Grayson, et al. *Some Problems of International Policing*. New York: Council on Foreign Relations, January 1944.

Kirk, Grayson. 'Wings over the Pacific', *Foreign Affairs* 20, no. 2 (January 1942): 293–302.

Kirk, Grayson and Walter R. Sharp, *Contemporary International Politics*. New York: Farrar & Rinehart, 1940.

Kirk, Grayson and Walter R. Sharp, *United Today for Tomorrow: The United Nations in War and Peace*. New York: Foreign Policy Association, 1942.

Labour Party. *For Socialism and Peace: The Labour Party's Programme of Action*. London: Labour Party, 1934.

Labour Party. *Labour and the Nation*. London: Labour Party, 1928.

Labour Party. *Labour Party Annual Conference Report*. London: Labour Party, 1933.

Labour Party. *The Old World and the New Society: A Report on the Problems of War and Peace Reconstruction*. London: Labour Party, 1942.

Labour Party. *Report of the Annual Conference of the Labour Party*. London: Labour Party, 1937.

Labour Party. *Wings for Peace: Labour's Post-War Policy for Civil Flying*. London: Labour Party, April 1944.

Langer, William L. 'Some Recent Books on International Relations', *Foreign Affairs* 11, no. 4 (July 1933): 720–732.

Laski, Harold J. *Reflections on the Revolution of Our Time*. London: Allen and Unwin, 1943.

Laski, Harold J. 'The Theory of an International Society', in *Problems of Peace 6th Series*, by H. J. Laski et al., 188–209. London: George Allen & Unwin, 1932.

Lauterbach, Albert. 'The Changing Nature of War', *International Conciliation* 20 (April 1941): 214–221.

Lawson, Neale. *A Plan for the Organisation of a European Air Service*. London: Constable, 1935.

League of Nations Union. *Annual Report for 1944*. London: League of Nations Union, 1945.

League of Nations Union. *The Problem of the Air*. London: League of Nations Union, 1935.

Lefebure, Victor. 'Chemical Warfare: The Possibility of Control', *Transactions of the Grotius Society* 7 (1921): 153–166.

Lefebure, Victor. *Common Sense about Disarmament*. London: Victor Gollancz, 1932.

Lefebure, Victor. *The Riddle of the Rhine: Chemical Strategy in Peace and War*. London: Collins, 1921.

Lefebure, Victor. *Scientific Disarmament*. London: Mundanus, 1931.

Liberal Party Organization. *20 Point Manifesto of the Liberal Party*. London: Liberal Party Organization, 1945.

Liberal Party Organization. *Final Agenda for the Meeting of the Assembly of the Liberal Party*. London: Liberal Party Organization, 1942.

Lilienthal, David E. *Change, Hope, and the Bomb*. Princeton, NJ: Princeton University Press, 1963.

Lilienthal, David E. *Journals of David E. Lilienthal Vol. II: The Atomic Energy Years, 1945–1950*. New York: Harper & Row, 1964.

Lippmann, Walter. 'International Control of Atomic Energy', in *One World or None*, eds. Dexter Masters and Katharine Way, 66–75. New York: McGraw Hill, 1946.

Lissitzyn, Oliver J. 'The Diplomacy of Air Transport', *Foreign Affairs* 19, no. 1 (October 1940): 156–170.

Lissitzyn, Oliver J. *International Air Transport and National Policy*. New York: Council on Foreign Relations, 1942.

Longmore, Arthur. *From Sea to Sky: 1910–1945*. London: Geoffrey Bles, 1946.

MacClure, Victor. *Ultimatum: A Romance of the Air*. London: George G Harrap, 1924.

MacReady, Gordon N. 'The Provision of Military Forces for Use by the World Organisation', *Proceedings of the American Academy of Political Sciences* 21, no. 3 (May 1945): 72–78.

Madariaga, Salvador de. *Disarmament*. New York: Coward-McCann, 1929.

Madariaga, Salvador de. *Morning without Noon: Memoirs*. Farnborough: Saxon House, 1973.

Madariaga, Salvador de. 'The Preparation of the First General Disarmament Conference', in *Problems of Peace 2nd Series: Lectures Delivered at the Geneva Institute of International Relations*, by William E. Rappard et al., 124–142. London: Oxford University Press, 1928.

Madariaga, Salvador de. *The World's Design*. London: Allen and Unwin, 1938.

Mance, H. O. 'Air Transport and the Future: Some International Problems', *Bulletin of International News* 19 (26 December 1942): 1173–1178.

Mance, H. O. *Frontiers, Peace Treaties, and International Organization*. London: Oxford University Press, 1946.

Mance, H. O. 'The Influence of Communications on the Regulation of International Affairs', *Journal of the British Institute of International Affairs* 1, no. 3 (May 1922): 78–89.

Mance, H. O. *International Air Transport*. London: Oxford University Press, 1943.

Mander, Geoffrey Le Mesurier. 'Military Disarmament', *Forward View* 3, no. 32 (September 1929).

Mander, Linden A. *Foundations of Modern World Society*. Stanford: Stanford University Press, 1941.

Maurice, Frederick. 'Disarmament', *Journal of the Royal Institute of International Affairs* 5, no. 3 (May 1926): 117–133.

Maurice, Frederick. 'The International Police Force', *The New Commonwealth Research Bureau Research Materials No. 1* (November 1934): 4–6.

Meyer, Jr., Cord. *The Search For Security*. New York: United World Federalists, 1947.

Millikan, Robert A. *The Autobiography of Robert A. Millikan*. London: MacDonald, 1951.

Millikan, Robert A. *Electrons (+ and -), Protons, Photons, Neutrons, Mesotrons, and Cosmic Rays*. Chicago: University of Chicago Press, 1947.

Millikan, Robert A. *Science, War and Human Progress: Address by Dr. Robert A. Millikan*. New York: The National Association of Manufacturers of the United States of America, 1943.

Mitrany, David. 'The Growth of World Organisation', *Common Wealth Review* 3, no. 8 (June 1946): 12–13.

Mitrany, David. *A Working Peace System: An Argument for the Functional Development of International Organization*. London: Royal Institute of International Affairs, 1943.

Mitrany, David. 'A Working Peace System (1943)' in *The Functional Theory of Politics*, 123–132. London: LSE & Political Science, 1975.

Molter, Bennett A. *Knights of the Air*. New York: D. Appleton, 1918.

Moore, John Bassett. 'An Appeal to Reason', *Foreign Affairs* 11, no. 4 (1933): 547–588.

Moore-Brabazon, J. 'Ad Astra', *Aeronautical Journal* 46, no. 382 (October 1942): 247–260.

Moore-Brabazon, J. 'Past, Present and Future of Aviation', *Nature* 150, no. 3796 (1 August 1942): 132–133.

Morgenthau, Hans. *Politics Among Nations*. New York: Alfred A. Knopf, 1948.

Morgenthau, Hans. *Scientific Man Versus Power Politics*. Chicago: University of Chicago Press, 1946.

Mosley, Philip E. *Alternatives to Absolute National Sovereignty of the Airspace*. New York: Council on Foreign Relations, June 1940.

Mott, N. F. 'The Atomic Bomb and World Affairs', *London Quarterly of World Affairs* 11 (October 1945): 187–190.

Moulton, John Fletcher. *Science and War: The Rede Lecture*. Cambridge: Cambridge University Press, 1919.

Mower, Edmund C. *International Government*. Boston: D. C. Heath, 1931.

Muir, Ramsay. *The Interdependent World and its Problems*. London: Constable, 1932.

Muir, Ramsay. *Political Consequences of the Great War*. London: Thornton Butterworth Limited, 1930.

National Council of Labour. *International Policy and Defence*. London: National Council of Labour, 1937.

National Democratic Committee. *The Campaign Text Book of the Democratic Party of the United States, For the Presidential Election of 1888*. New York: Brentanos, 1888.

National Executive Committee of the Labour Party. *Labour's Immediate Programme*. London: Labour Party, 1937.

National Executive Committee of the Labour Party. *The Old World and the New Society: A Report on the Problems of War and Peace Reconstruction*. London: Labour Party, 1942.

New Commonwealth Society. *The Functions of an International Air Police*. London: New Commonwealth Society, 1936.

New Commonwealth Society. *An International Air Force: its Functions and Organisation / being a Memorandum submitted by the Executive Committee of the New Commonwealth to the International Congress in Defence of Peace, Brussels, February, 1934*. London: New Commonwealth, 1934.

New Commonwealth Society. *The New Commonwealth Declaration*. London: New Commonwealth Society, February 1940.

New Commonwealth Society. *World Security Force*. London: New Commonwealth Society, 1949.

Next Five Years Group. *The Next Five Years: An Essay in Political Agreement*. London: Macmillan, 1935.

Next Five Years Group. *A Summary of the Book 'The Next Five Years': An Essay in Political Agreement*. London: Next Five Years Group, 1936.

Noel Baker, P. J. *Disarmament*. London: Hogarth Press, 1926.

Noel Baker, P. J. 'Disarmament', *Economica* 6, no. 16 (March 1926): 1–15.

Noel Baker, P. J. 'Disarmament', *International Affairs* 13, no.1 (January–February 1934): 3–25.

Noel Baker, P. J. *Disarmament and the Coolidge Conference*. London: Hogarth Press, 1927.

Noel Baker, P. J. 'The Future of the Collective System', in *Problems of Peace* 10th series Anarchy of World Order, by R. B. Mowat et al., 178–198. London: George Allen & Unwin, 1936.

Noel Baker, P. J. *The Geneva Protocol for the Pacific Settlement of International Disputes*. London: PS King & Son, 1925.

Noel Baker, P. J. 'The Growth of International Society', *Economica* 4, no. 12 (November 1924): 262–277.

Noel Baker, P. J. 'The International Air Police Force', in *Challenge to Death*, by P. J. Noel Baker et al., 206–239. London: Constable, 1934.

Noel Baker, P. J. 'Peace and the Official Mind', in *Challenge to Death*, by P. J. Noel Baker et al., 67–89. London: Constable, 1934.

Ogburn, William F. 'Introductory Ideas on Inventions and the State', in *Technology and International Relations*, ed. William F. Ogburn, 1–15. Chicago: University of Chicago Press, 1949.

Ogburn, William F. *On Culture and Social Change*. Chicago: University of Chicago Press, 1964.

Ogburn, William F. *Social Change with Respect to Culture and Original Nature*. New York: B.W. Huebsch, 1922.

Ogburn, William F. *The Social Effects of Aviation*. Boston: Houghton Mifflin, 1946.

Ogburn, William F. 'Sociology and the Atom', *The American Journal of Sociology* 51, no. 4 (January 1946): 267–275.

Oppenheimer, J. Robert. 'Atomic Weapons', *Proceedings of the American Philosophical Society* 90, no. 1 (January 1946):7–10.

Oppenheimer, J. Robert. 'A Future for Atomic Weapons', in *Science and Civilization*, by A. V. Hill et al., vol. 1, *The Future of Atomic Energy*, 77–90. New York: McGraw-Hill, 1946.

McKinzie, Richard D. and Harold Stassen. 'Oral History Interview with Harold Stassen', Truman Library, accessed 2 August 2020, www.trumanlibrary.gov/library/oral-histories/stassen.

Orwell, George. 'You and the Atom Bomb', in *The Collected Essays, Journalism and Letters of George Orwell*, eds. Sonia Orwell and Ian Angus, vol. 4, *In Front of Your Nose 1945–1950*, 6–10. London: Secker & Warburg, 1968.

Osterhout, Howard. 'A Review of the Recent Chicago International Air Conference', *Virginia Law Review* 31, no. 2 (March 1945): 376–386.

Palmer, Norman D. and Howard Perkins. *International Relations: The World Community in Transition*. Boston: Houghton Mifflin, 1953.

Pasvolsky, Leo. *Dumbarton Oaks Proposals*. Washington, DC: Department of State, 1944.

Perillier, Louis. *De La Limitation des Armements par la méthode budgétaire et du contrôle de cette limitation*. Paris: Rousseau, 1932.

Phillips, Hubert et al. *Whither Britain? A Radical Answer*. London: Faber & Faber, 1932.

Pigou, A. C. *The Political Economy of War*. London: Macmillan, 1921.

Pogue, L. Welch. 'The Next Ten Years in Air Transportation', *Proceedings of the Academy of Political Science* 21, no. 2 (January 1945): 17–27.

Political and Economic Planning. 'International Air Transport', in *Building Peace out of War: Studies in International Reconstruction*, 90–110. London: Political and Economic Planning, 1944.

Portal, C. F. A. 'Air Force Co-operation in Policing the Empire', *Royal United Services Institution Journal* 82, no. 526 (1937): 343–358.

Potter, Pitman B. *An Introduction to the Study of International Organization*. New York: Century, 1922.

Potter, Pitman B. *An Introduction to the Study of International Organization*, 4th ed. New York: D. Appleton-Century, 1935.

Potter, Pitman B. *The World of Nations: Foundations, Institutions, Practices*. New York: Macmillan, 1929.

Potter, Pitman B. and Roscoe L. West. *International Civics: The Community of Nations*. New York: Macmillan, 1927.

'Prominent Britons Urge Attlee to call Big Three Meeting', *Bulletin of the Atomic Scientists* 3, no. 8 (August 1947): 218, 220.

'The Quarter's Polls', *Public Opinion Quarterly* 10, no. 1 (Spring 1946): 104.

Rabinowitch, Eugene. Editorial, 'Scientists and World Government', *Bulletin of the Atomic Scientists* 3, no. 12 (December 1947): 345–346.

Rabinowitch, Eugene. 'Technical Feasibility of Atomic Energy Controls', *Bulletin of the Atomic Scientists* 1, no. 2 (24 December 1945): 1.

Raleigh, Walter. *The War in the Air: Being the Story of the Part Played in the Great War by the Royal Air Force*, vol. 1. Oxford: Clarendon Press, 1922.

Randall, John Herman. *A World Community: The Supreme Task of the Twentieth Century*. New York: Frederick A. Stokes, 1930.

Rappard, William E. 'The Evolution of the League of Nations', *The American Political Science Review* 21, no. 4 (November 1927): 792–826.

Rappard, William E. 'The League of Nations as a Historical Fact', *International Conciliation* 11 (June 1927): 270–322.

Rappard, William E. *International Relations as Viewed from Geneva*. New Haven, CT: Yale University Press, 1925.

Reinsch, Paul S. *Public International Unions*. Boston: Ginn, 1911.

Renner, George T. *Human Geography in the Air Age*. New York: Macmillan, 1942.

'Report by the Military Staff Committee to the Security Council on the General Principles Governing the Organization of the Armed Forces Made Available to the Security Council by Member Nations of the United Nations, April 30, 1947', *International Organization* 1, no. 3 (September 1947): 561–574.

Reves, Emery. *The Anatomy of Peace*. New York: Harper Brothers, 1945.

Reves, Emery. *The Anatomy of Peace*, 8th ed. New York: Harper & Brothers, 1946.

Reves, Emery. 'Emery Reves on World Government', *Bulletin of the Atomic Scientists* 1, no. 4 (1 February 1946): 6.

Reves, Emery. 'National Sovereignty – the Road to the Next War', *World Affairs* 109, no. 2 (June 1946): 109–116.

Rockwell, Carey. *Stand by for Mars!* New York: Grosset and Dunlap, 1952.

Roosevelt, Theodore. 'Acceptance Speech', The Nobel Prize, accessed 2 August 2020, www.nobelprize.org/prizes/peace/1906/roosevelt/acceptance-speech.

Rollins College Conference on World Government. 'An Appeal to the Peoples of the World: Issued by the Rollins College Conference on World Government (March 11–16, 1946, Winter Park, Fla.)', *World Affairs* 109, no. 2 (1946): 83–86.

Roucek, Joseph S. 'King-Hall, Stephen, Total Victory', *Annals of the American Academy of Political and Social Science* 222, no. 1 (July 1942): 199–200.

Rowan-Robinson, G. A. *An International General Staff*. London: New Commonwealth Society, 1948.

Royal Institute of International Affairs. *Atomic Energy. Its International Implications. A Discussion by a Chatham House Study Group*. London: Royal Institute of International Affairs, 1948.

Russell, Bertrand. 'The Atomic Bomb and the Prevention of War', *Bulletin of the Atomic Scientists* 2, nos. 7–8 (1 October 1946): 19–21.

Russell, Bertrand. *Autobiography*. London: Routledge, 2010.

Russell, Bertrand. 'The Bomb and Civilization', *Glasgow Forward* 39 (18 August 1945): 1, 3.

Russell, Bertrand. 'International Government', in *Towards World Government*, 3–12. London: New Commonwealth, 1948.

Russell, Bertrand. 'International Government', *The New Commonwealth* 9 (January 1948): 77–80.

Russell, Bertrand. *The Scientific Outlook*. London: George Allen & Unwin, 1931.

Russell, Bertrand. *The Scientific Outlook*, 2nd ed. London: George Allen & Unwin, 1949.

Russell, Bertrand. 'Values in the Atomic Age', in *The Atomic Age*, by M. L. Oliphant et al., 81–104. London: George Allen and Unwin, 1949.

Russell, Bertrand. *Which Way to Peace?* London: Michael Joseph, 1936.

Salter, Arthur. *The Framework for an Ordered Society*. Cambridge: Cambridge University Press, 1933.

Salter, Arthur. 'The Future of Economic Nationalism', *Foreign Affairs* 11, no. 1 (October 1932): 8–20.

Salter, Arthur. *Modern Mechanization and its Effects on the Structure of Society*. London: Oxford University Press, 1933.

Salter, Arthur. 'The United Nations and the Atomic Bomb', *International Conciliation* 24 (January 1946): 40–48.

Salter, Arthur. 'World government', in *The Modern State*, ed. Mary Adams, 253–316. London: George Allen & Unwin, 1933.

Schneider, Joseph. 'Cultural Lag: What Is It?', *American Sociological Review* 10, no. 6 (December 1945): 786–791.

Schuman, Frederick L. *International Politics: An Introduction to the Western State System*. New York: McGraw Hill, 1933.

Schuman, Frederick L. *International Politics: The Destiny of the Western State System*, 4th ed. New York: McGraw-Hill, 1948.

Schuman, Frederick L. *International Politics: The Western State System in Transition*, 3rd ed. New York: McGraw-Hill, 1941.

Schmitt, Carl. 'The Turn to the Discriminating Concept of War (1937)', in *Writings on War*, eds. Carl Schmitt and Timothy Nunan, 30–74. Cambridge: Polity Press, 2011.

Schwarzenberger, Georg. *The League of Nations and World Order*. London: Constable, 1936.

Editorial, 'A Scientific Approach to Peace', *Nature* 134, no. 3394 (17 November 1934): 749–751.

Scott, James Brown. 'International Organization – Executive and Administrative', *Advocate of Peace Through Justice* 83, no. 4 (April 1921): 133–136.

Seversky, Alexander P. de. *Victory Through Air Power*. New York: Simon and Schuster, 1942.

Sharp, Walter R. *National Sovereignty and the International Tasks of the Postwar World*. New York: Council on Foreign Relations, August 1942.

Shils, Edward. 'Some Political Implication of the State Dep't Report', *Bulletin of the Atomic Scientists* 1, no. 9 (15 April 1946): 7–9, 19.

Shotwell, James T. 'After the War', *International Conciliation* 21 (January 1942): 31–35.

Shotwell, James T. 'The Atomic Bomb and International Organization', *Bulletin of the Atomic Scientists* 1, no. 7 (15 March 1946): 8–9.

Shotwell, James T. 'The Control of Atomic Energy under the Charter', *Proceedings of the American Philosophical Society* 90, no. 1 (January 1946): 59–64.

Shotwell, James T. *The Great Decision*. New York: Macmillan, 1944.

Shotwell, James T. 'History', in *The Encyclopaedia Britannica: A Dictionary of Arts, Sciences, Literature and General Information 11th ed. vol. 13*, ed. Hugh Chisholm, 527–533. New York: Encyclopaedia Britannica, 1910.

Shotwell, James T. 'Implementing and Amending the Charter: An Address by James T. Shotwell, November 11, 1945', *International Conciliation* 23 (December 1945): 811–823.

Shotwell, James T. 'The International Implications of Nuclear Energy', *American Journal of Physics* 14, no. 3 (May–June 1946): 179–185.

Shotwell, James T. 'Introduction', in *The Study of International Relations in the United States: Survey for 1934*, ed. Edith W. Ware, 3–20. New York: Columbia University Press, 1934.

Shotwell, James T. 'Mechanism and Culture', in *Science and Man*, ed. Ruth Nanda Anshen, 151–162. New York: Harcourt, Brace, 1942.

Shotwell, James T. 'Mechanism and Culture', *The Historical Outlook* 16 (January 1925): 7–11.

Shotwell, James T. *On the Rim of the Abyss*. New York: Macmillan, 1936.

Shotwell, James T. *The Religious Revolution of Today*. Boston: Houghton Mifflin, 1913.

Shotwell, James T. *War as an Instrument of National Policy and its Renunciation in the Pact of Paris*. New York: Harcourt, Brace, 1929.

Shotwell, James T. 'War as an Instrument of Politics', *International Conciliation* 20 (April 1941): 205–213.

Simpson, Smith. 'The Commission to Study the Organization of Peace', *The American Political Science Review* 35, no.2 (April 1941): 317–324.

Sinclair, Archibald. *The People's Party*. London: Liberal Party, 1942.

Smith, Rennie. *General Disarmament or War?* London: George Allen & Unwin, 1927.

Spaight, J. M. *Air Power in the Next War*. London: Geoffrey Bles, 1938.

Spaight, J. M. *The Atomic Problem*. London: Arthur Barron, 1948.

Spaight, J. M. *An International Air Police*. London: Gale and Polden, 1932.

Spaight, J. M. *Pseudo-Security*. London: Longmans, 1928.

Spaight, J. M. 'Self Defence and International Air Power', *Journal of Comparative Legislation and International Law*, 3rd series, 14, no. 1 (1932): 20–29.

Spencer, Herbert. *Political Institutions, being Part V of the Principles of Sociology (The Concluding Portion of Vol. II)*. London: Williams and Norgate, 1882.

Stamp, Josiah. 'The Impact of Science Upon Society', *Science* 84, no. 2176 (September 1936), 235–239.

Stamp, Josiah. *The Science of Social Adjustment*. London: Macmillan, 1937.

Stassen, Harold E. 'Atomic Control', *Proceedings of the Academy of Political Science* 21, no. 4 (January 1946): 103–112.

Stassen, Harold E. 'From War to Peace: An Address by Harold E. Stassen, November 8, 1945', *International Conciliation* 23 (December 1945): 801–810.

Stassen, Harold E. *A Proposal of a Definite United Nations Government*. St. Paul, MN: Foreign Policy Association, 1943.

Stassen, Harold E. *United Nations: A Working Paper for Restructuring*. Minneapolis, MN: Lerner Publication, 1994.

Steed, H. Wickham. 'The Search for Security Against War', *The Journal of the Royal United Services Institution* 81, no. 524 (November 1936): 707–733.

Streit, Clarence K. *Union Now! A Proposal for a Federal Union of the Democracies of the North Atlantic*. New York: Harper & Brothers, 1939.

Swanwick, H. M. *Collective Insecurity*. London: Jonathan Cape, 1937.

Swanwick, H. M. *New Wars for Old*. London: Women's International League, 1934.

Swanwick, H. M. *Pooled Frankenstein and His Monster: Aviation for World Service*. London: Women's International League, 1934.

Swanwick, H. M. *Pooled Security: What Does it Mean?* London: Women's International League, 1934.

Sykes, F. H. *Aviation in Peace and War*. London: Edward Arnold, 1922.

Sykes, F. H. 'Memorandum by the Chief of the Air Staff on Air-Power Requirements of the Empire', *From Many Angles: An Autobiography*, 558–574. London: Harrap, 1942.

Tangye, Nigel, ed. *The Air is Our Concern*. London: Methuen, 1935.

Tangye, Nigel. *Britain in the Air*. London: William Collins, 1944.

Teller, Edward. 'Atomic Scientists have Two Responsibilities', *Bulletin of the Atomic Scientists* 3, no. 12 (December 1947): 355–356.

Things to Come, directed by William Cameron Menzies. London Film Production and United Artists, 1936.

A Third World War Can be Prevented Now! New York: True Comics Magazine, 1945.

Thomas, Elbert D. *The Four Fears*. Chicago: Ziff Davis, 1944.

Thomson, Christopher Birdwood. *Air Facts and Problems*. London: John Murray, 1927.

Tolédano, André Daniel. *Ce qu'il faut savoir sur le désarmement (avec en annexe le projet de convention du désarmement)*. Paris: A. Pedone, 1932.

Traxler, Arthur E. 'International Air Transport Policy of the United States', *International Conciliation* 21 (December 1943): 616–636.

Tree, Ronald et al., *Air Transport Policy: By Four Conservative Members of Parliament*. London: n.p., 1943.

Tunstall, Brian. *Eagles Restrained*. London: Allen & Unwin, 1936.

'UNA's First General Council', *International Outlook* 1 (March 1946): 7–12.

United Nations Information Office. *Towards Freedom in the Air*. New York: United Nations Information Office, 1944.

Universities Committee on Post-War International Problems. 'Summaries of Reports of Cooperating Groups: Problem XIII-International Air Traffic After the War', *International Conciliation* 23 (April 1945): 249–260.

Universities Committee on Post-war International Problems. 'Summaries of Reports of Cooperating Groups: XVIII-The Dumbarton Oaks Proposals: The Enforcement of Peace', *International Conciliation* 23 (October 1945): 671–682.

Urey, Harold C. 'An Alternative Course for the Control of Atomic Energy', *Bulletin of the Atomic Scientists* 3, no. 6 (June 1947): 139–142.

Urey, Harold C. 'The Atom and Humanity: An Address by Harold C. Urey, October 21, 1945', *International Conciliation* 23 (December 1945):790–800.

Urey, Harold C. 'Atomic Energy and World Peace', *Bulletin of the Atomic Scientists* 2, no. 10 (1 November 1946): 2–4.

Urey, Harold C. 'Atomic Energy, Aviation, and Society', *Air Affairs* 1, no.1 (September 1946): 21–29.

Urey, Harold C. 'Atomic Energy, Master or Servant?', *World Affairs* 109, no. 2 (June 1946): 99–108.

Usborne, Henry. *Towards World Government – The Role of Britain*. London: National Peace Council, 1946.

Usborne, Henry. 'What Holds us Back', *Federal News* 135 (June 1946): 2, 5.

Van Zandt, J. Parker. *Civil Aviation and Peace*. Washington, DC: Brookings Institution, 1944.

Vavilov, Sergei, A. N. Frumkin, A. F. Loffe, and N. N. Semyonov, 'Open Letter to Dr. Einstein – From Four Soviet Scientists', *Bulletin of the Atomic Scientists* 4, no. 2 (February 1948): 34, 37–38.

Vigilantes [Konni Zilliacus], *The Dying Peace*. London: New Statesmen and Nation, 1933.

Vollenhoven, C. van. 'International Police and Pacifism', in *War Obviated by an International Police: A Series of Essays, Written in Various Countries*, 73–82. The Hague: Martinus Nijhoff, 1915.

Vollenhoven, C. van. 'The Opportunity of this Solution', in *War Obviated by an International Police: A Series of Essays, Written in Various Countries*, 59–73. The Hague: Martinus Nijhoff, 1915.

Vollenhoven, C. van, et al. *War Obviated by an International Police: A Series of Essays, Written in Various Countries*. The Hague: Martinus Nijhoff, 1915.

Wallace, Henry. 'Business Measures', in *Democracy Reborn*, 208–214. New York: Reynal & Hitchcock, 1944.

Wallace, Henry. *The Century of the Common Man*. New York: Reynal & Hitchcock, 1943.

Wallace, Henry. 'Industrialization of the World', *Vital Speeches of the Day* 11, no. 19 (15 July 1945): 603–605.

Wallace, Henry. *The Price of Vision: The Diary of Henry A. Wallace, 1942–1946*, ed. John Morton Blum. Boston: Houghton Mifflin, 1973.

Walsh, Edmund A., ed. *History and Nature of International Relations*. New York: Macmillan, 1922.

Warner, Edward P. 'Airways for Peace', *Foreign Affairs* 22, no. 1 (October 1943): 11–27.

Warner, Edward P. 'Chicago Air Conference: Accomplishments and Unfinished Business', *Foreign Affairs* 23, no. 3 (April 1945): 406–421.

Warner, Edward P. 'Douhet, Mitchell, Seversky: Theories of Air Warfare', in *Makers of Modern Strategy: Military Thought from Machiavelli to Hitler*, ed. Edward Mead Earle, 485–503. Princeton, NJ: Princeton University Press, 1941.

Warner, Edward P. 'Future Controls Over German Aviation', *Foreign Affairs* 21, no. 3 (April 1943): 427–439.

Warner, Edward P. *Possibilities of Controlling or Limiting Aircraft Suitable for Offense against Ground Objectives*. New York: Council on Foreign Relations, 1940.

Warner, Edward P. 'Postwar Aviation', in *Proceedings of the Section of International and Comparative Law*, 51–57. Chicago: American Bar Association, September 1944.

Webster, Charles. 'The Making of the Charter of the United Nations', in *The Art and Practice of Diplomacy*, 70–91. London: Chatto & Windus, 1961.

Webster, Charles and Sydney Herbert. *The League of Nations in Theory and Practice*. Boston: Houghton Mifflin, 1933.

Webster, Charles, Arthur Salter and Oliver Franks. "The Control of Nuclear Energy and the Development of International Institutions", in *Atomic Energy. Its International Implications. A Discussion by a Chatham House Study Group* by Royal Institute of International Affairs, 91–101. London: Royal Institute of International Affairs, 1948.

Wehberg, Hans. 'La Police Internationale', *Recueil des Cours* 48 (1934 part II): 1–132.

Welles, Sumner. 'Under Secretary Of State Memorial Day Address at the Arlington National Amphitheater', Ibiblio, accessed 2 August 2020, www.ibiblio.org/pha/policy/1942/420530a.html.

Wells, H. G. 'Foreword', in J. M. Kenworthy, *Peace or War?*, vii–xvii. New York: Boni & Liveright, 1927.

Wells, H. G. *In the Fourth Year: Anticipations of a World Peace*. New York: Macmillan, 1918.

Wells, H. G. *The Idea of a League of Nations*. Boston: Atlantic Monthly Press, 1919.

Wells, H. G. *The New World Order: Whether it is Attainable, How it Can be Attained, and What Sort of World a World at Peace will have to be*. London: Secker and Warburg, 1940.

Wells, H. G. *The Open Conspiracy*. London: Victor Gollancz, 1928.

Wells, H. G. *The Outline of History: Being a Plain History of Life and Mankind* 3rd ed. New York: Macmillan, 1921.

Wells, H. G. *The Shape of Things to Come*. London: Hutchinson, 1933; London: Penguin, 1993.

Wells, H. G. *The War in the Air*. London: George Bell and Sons, 1908.

Wells, H. G. *Washington and the Riddle of Peace*. New York: Macmillan, 1922.

Wells, H. G. *What Are We To Do with Our Lives?* London: William Heinemann, 1931.

Wells, H. G. *The World of William Clissold*. London: Ernest Benn, 1926.

Welles, Sumner. *Where Are We Heading?* London: Hamish Hamilton, 1947.

Wheeler-Bennett, John W. *Information on the Problem of Security*. London: George Allen & Unwin, 1927.

Wheeler-Bennett, John W. *Pipe Dream of Peace: The Story of the Collapse of Disarmament*. New York: William Morrow, 1935.

Wild, Payson S. *American Participation in International Policing*. New York: Council on Foreign Relations, July 1943.

Wild, Payson S. *The Distinction between War and International Policing*. New York: Council on Foreign Relations, June 1943.

Willkie, Wendell. 'Economic Freedom for the World', in *Representative American Speeches: 1942–1943*, ed. A. Craig Baird, 105–115. New York: HW Wilson.

Willkie, Wendell. *One World*. New York: Limited Editions Club, 1944.

Wilson, Eugene P. *Air Power for Peace*. New York: McGraw-Hill, 1945.

Wilson, Eugene P. 'Fundamentals of Freedom: Initiative and Enterprise Protect Real Security', Ibiblio, 28 May 1942, accessed 2 August 2020, www.ibiblio.org /pha/policy/1942/1942-05-28b.html.

Wilson, Robert R. and Philip Morrison, 'Half a World ... and None: Partial World Government Criticized', *Bulletin of the Atomic Scientists* 3, no. 7 (July 1947): 181–182.

Wimperis, H. E. 'International Co-operation in the Development of Atomic Energy', in *Atomic Energy. Its International Implications. A Discussion by a Chatham House Study Group* by Royal Institute of International Affairs, 102–111. London: Royal Institute of International Affairs, 1948.

Wimperis, H. E. *World Power and Atomic Energy: The Impact on International Relations*. London: Royal Institute of International Affairs, 1946.

Woolf, Leonard. *International Government: Two Reports by L.S. Woolf Prepared for the Fabian Research Department, Together with a Project by a Committee for Supernational Authority that will Prevent War*. London: Fabian Society, 1916.

Woolf, Leonard, *The International Post-War Settlement*. London: Fabian Publications, 1944.

Woolf, Leonard. 'Review of *The League Committees and World Order*, by H. R. G. Greaves', *Economica* 11, no. 34 (November 1931): 485–487.

Woolf, Leonard. *The War for Peace*. London: George Routledge & Sons, 1940.

Wolfers, Arnold. 'The Atomic Bomb in Soviet-American Relations', in *The Absolute Weapon: Atomic Power and World Order*, ed. Bernard Brodie, 111–147. New York: Harcourt Brace, 1946.

World Airways Joint Committee. *Civil Aviation and World Unity: A Manifesto*. London: World Airways Joint Committee, 1944.

World Citizens Association. *The World's Destiny and the United States: A Conference of Experts in International Relations*. Chicago: World Citizens Association, 1941.

World Government Committee of the Association of Oak Ridge Engineers and Scientists. *Primer for Peace: World Government for the Atomic Age – An Analysis with Recommendations for Action*. Oak Ridge, TN: Association of Scientists for Atomic Education, November 1947.

Wright, Quincy. *The Causes of War and the Conditions of Peace*. London: Longmans, Green, 1935.

Wright, Quincy. 'Draft for a Convention on Atomic Energy', *Bulletin of the Atomic Scientists* 1, no. 8 (1 April 1946): 11–13.

Wright, Quincy. 'Fundamental Problems of International Organization', *International Conciliation* 20 (April 1941): 468–492.

Wright, Quincy. 'How an International Police Force Would Work', in *Winning the Peace*, eds. Benjamin Akzin et al., 2nd ed., 14–17. St Louis, MS: St Louis Star Times, 1944.

Wright, Quincy. *Human Rights and the World Order*. New York: Commission to Study the Organization of Peace, 1943.

Wright, Quincy. 'National Security and International Police', *The American Journal of International Law* 37, no. 3 (July 1943): 499–505.

Wright, Quincy. 'Peace Problems of Today and Yesterday', *American Political Science Review* 38, no. 3 (June 1944): 512–521.

Wright, Quincy. 'Political Conditions of the Period of Transition', *International Conciliation* 21 (April 1942): 264–278.

Wright, Quincy. 'The International Regulation of the Air', *The American Economic Review* 35, no. 2 (May 1945): 243–248.

Wright, Quincy. 'The Law of the Nuremberg Trial', *The American Journal of International Law* 41, no. 1 (January 1947): 38–72.

Wright, Quincy. *The Study of International Relations*. New York: Appleton-Century-Crofts, 1955.

Wright, Quincy. *A Study of War 2 vols*. Chicago: University of Chicago Press, 1942.

Wright, Quincy. 'United Nations-Phrase or Reality?', *Annals of the American Academy of Political and Social Science* 228, no.1 (July 1943): 1–10.

Wright, Quincy and Edward Warner. 'Enforcement of International Law', *Proceedings of the American Society of International Law at Its Annual Meeting* 38 (April 1944): 77–97.

Wright, T. P. *Aircraft Production and National Defense*. New York: Farrar & Rinehart, 1941.

Wright, T. P. 'Aviation's Place in Civilization', *The Journal of the Royal Aeronautical Society* 49, no. 414 (June 1945): 299–342.

Young, Michael Dunlop. *Civil Aviation*. London: Pilot Press, 1944.

Young, Michael Dunlop. *Labour's Plan for Plenty*. London: Victor Gollancz, 1947.

Ziff, Sr., William B. *The Gentlemen Talk of Peace*. London: John Lane the Bodley Head, 1945.

Zimmern, Alfred. 'German Culture and the British Commonwealth', in *The War and Democracy*, by R. W. Seton-Watson et al., 348–382. London: Macmillan, 1914.

Zimmern, Alfred. *The League of Nations and the Rule of Law 1918–1935*. London: Macmillan, 1936.

Zimmern, Alfred. 'The Prospects of Democracy', *Journal of the Royal Institute of International Affairs* 7, no. 3 (May 1928): 153–191.

Zimmern, Alfred. *Quo Vadimus?* London: Oxford University Press, 1934.

Zimmern, Alfred. 'Review of The Interdependent World and its Problems, by Ramsay Muir', *International Affairs* 12, no. 2 (March 1933): 247–248.

Zimmern, Alfred. *Spiritual Values and World Affairs*. Oxford: Clarendon Press, 1939.

Secondary Sources

Aaserud, Finn. 'The Scientist and the Statesmen: Niels Bohr's Political Crusade during World War II', *Historical Studies in the Physical and Biological Sciences* 30, no. 1 (1999): 1–47.

'Air Cargo Matters', IATA, accessed 2 August 2020, www.iata.org/whatwedo/cargo/sustainability/Pages/benefits.aspx.

Anderson, David G. 'British Rearmament and the "Merchants of Death": The 1935–36 Royal Commission on the Manufacture of and Trade in Armaments', *Journal of Contemporary History* 29, no.1 (January 1994): 5–37.

Anker, Peder. *Imperial Ecology: Environmental Order in the British Empire, 1895–1945.* Cambridge, MA: Harvard University Press, 2001.

Ashkenazi, Ofer. 'Transnational Anti-war Activity in the Third Reich: The Nazi Branch of the New Commonwealth Society', *German History* 36, no. 2 (June 2018): 207–228.

Ashworth, Lucian M. *Creating International Studies: Angell, Mitrany and the Liberal Tradition.* London: Taylor and Francis, 2017.

Ashworth, Lucian M. 'Democratic Socialism and International Thought in Interwar Britain', in *Radicals and Reactionaries in Twentieth-Century International Thought*, ed. Ian Hall, 75–100. New York: Palgrave Macmillan, 2015.

Ashworth, Lucian M. 'Did the Realist-Idealist Great Debate Really Happen? A Revisionist History of International Relations', *International Relations* 16, no. 1 (2002), 33–51.

Ashworth, Lucian M. 'Feminism, War and the Prospects for Peace: Helena Swanwick (1864–1939) and the Lost Feminists of Inter-War International Relations', *International Feminist Journal of Politics* 13, no. 1 (2011): 25–43.

Badenoch, Alexander and Andreas Fickers, eds., *Materializing Europe: Transnational Infrastructures and the Project of Europe.* London: Palgrave Macmillan, 2010.

Balogh, Brian. *Chain Reaction: Expert Debate and Public Participation in American Commercial Nuclear Power, 1945–1975.* Cambridge: Cambridge University Press, 1991.

Bankwitz, Philip Charles Farwell. *Maxime Weygand and Civil-Military Relations in Modern France.* Cambridge, MA: Harvard University Press, 1967.

Baratta, Joseph Preston. 'The International Federalist Movement: Toward Global Governance', *Peace & Change* 24, no. 3 (July 1999): 340–372.

Baratta, Joseph Preston. *The Politics of World Federation: From World Federalism to Global Governance*, vol. 1. Westport, CT: Praeger, 2004.

Bartel, Fritz. 'Surviving the Years of Grace: The Atomic Bomb and the Specter of World Government, 1945–1950', *Diplomatic History* 38, no. 20 (2015): 275–302.

Bartlett, Ruhl Jacob. *The League to Enforce Peace.* Chapel Hill: University of North Carolina Press, 1944.

Bauernfeind, Neil D. *'Lord Davies and the New Commonwealth Society 1932–1944'.* MPhil diss., University of Wales at Aberystwyth, 1990.

Baylis, John. *Ambiguity and Deterrence: British Nuclear Strategy, 1945–1964.* Oxford: Clarendon Press, 1995.

Beaumont, Roger. *Right Backed by Might: The International Air Force Concept.* Westport, CT: Praeger, 2001.

Bell, Duncan. 'Dissolving Distance: Technology, Space, and Empire in British Political Thought, 1770–1900', *The Journal of Modern History* 77, no.3 (September 2005): 523–562.

Bell, Duncan. *Reordering the World: Essays on Liberalism and Empire.* Princeton, NJ: Princeton University Press, 2016.

Bell, Duncan. 'Victorian Visions of Global Order: An Introduction', in *Victorian Visions of Global Order: Empire and International Relations in Nineteenth-Century Political Thought*, 1–25. Cambridge: Cambridge University Press, 2007.

Bialer, Uri. *The Shadow of the Bomber: The Fear of Air Attack and British Politics 1932–1939.* London: Royal Historical Society, 1980.

Biddle, Tami Davis. *Rhetoric and Reality in Air Warfare: The Evolution of British and American Ideas About Strategic Bombing, 1914–1945.* Princeton, NJ: Princeton University Press, 2002.

Bilstein, Roger E. 'Edward Pearson Warner and the New Air Age', in *Aviation's Golden Age: Portraits from the 1920s and 1930s*, ed. William M. Leary, 113–126. Iowa City, IA: University of Iowa Press, 1989.

Bingham, A. '"The Monster"? The British Popular Press and Nuclear Culture, 1945–early 1960s', *British Journal of the History of Science* 45, no. 2 (2012): 609–625.

Blower, Brooke L. 'From Isolationism to Neutrality: A New Framework for Understanding American Political Culture, 1919–1941', *Diplomatic History* 38, no. 2 (2014): 345–376.

Borg, Stefan. *European Integration and the Problem of the State: A Critique of the Bordering of Europe.* Houndmills: Palgrave Macmillan, 2015.

Borgwardt, Elizabeth. *A New Deal for the World: America's Vision for Human Rights.* Cambridge, MA: Belknap Press, 2005.

Bosco, Andrea. *June 1940, Great Britain and the First Attempt to Build a European Union.* Cambridge: Cambridge Scholars Publishing, 2016.

Bosco, Andrea. 'Lothian, Curtis, Kimber and the Federal Union Movement (1938–40)', *Journal of Contemporary History* 23, no. 2 (July 1988): 465–502.

Bostrom, Nick. *Superintelligence: Paths, Dangers, Strategies.* Oxford: Oxford University Press, 2014.

Bottelier, Thomas. 'Associated Powers: Britain, France, the United States and the Defence of World Order, 1931–1943'. PhD diss., King's College London, 2018.

Boyer, Paul. *By the Bomb's Early Light: American Thought and Culture at the Dawn of the Atomic Age.* New York: Pantheon Books, 1985.

Brewin, Christopher. 'British Plans for International Operating Agencies for Civil Aviation, 1941–1945', *International History Review* 4, no. 1 (1982): 91–110.

Briggs, Philip J. 'Congress and Collective Security: The Resolutions of 1943', *World Affairs* 132, no. 4 (March 1970): 332–344.

Butts, Dennis. 'Imperialists of the Air-Flying Stories', in *Imperialism and Juvenile Literature*, ed. Jeffrey Richards, 126–143. Manchester: Manchester University Press, 1989.

Call, Steve. *Selling Air Power: Military Aviation and American Popular Culture.* College Station: Texas A&M University Press, 2009.

Callaghan, John and Mark Phythian, 'Intellectuals of the Left and the Atomic Dilemma in the Age of the US Atomic Monopoly, 1945–1949', *Contemporary British History* 29, no. 4 (2015): 441–463.

Callahan, Michael D. *The League of Nations, International Terrorism, and British Foreign Policy, 1934–1938*. Cham, Switzerland: Palgrave Macmillan, 2018.

Cantril, Hadley. *Public Opinion 1935–1946*. Princeton, NJ: Princeton University Press, 1951.

Castells, Manuel. *The Information Age: Economy, Society and Culture, Vol. 2, The Power of Identity*. Oxford: Blackwell, 1997.

Ceadel, Martin. *Semi-detached Idealists: The British Peace Movement and International Relations, 1854–1945*. Oxford: Oxford University Press, 2000.

Chabot, Jean-Luc. *Aux Origines Intellectuelles de L'Union Européenne: L'idée d'Europe unie de 1919 à 1939*. Grenoble: Presses Universitaires de Grenoble, 2005.

Clarke, Sabine. *Science at the End of Empire: Experts and the Development of the British Caribbean, 1940–62*. Manchester: Manchester University Press, 2018.

Clavin, Patricia. 'Men and Markets: Global Capital and the International Economy', in *Internationalisms: A Twentieth Century History*, eds. Glenda Sluga and Patricia Clavin, 85–112. Cambridge: Cambridge University Press, 2017.

Clavin, Patricia. *Securing the League of Nations: The Reinvention of the League of nations, 1920–1946*. Oxford: Oxford University Press, 2013.

Coker, Christopher. *The Future of War: The Re-Enchantment of War in the Twenty-First Century*. Oxford: Blackwell, 2014.

Collins, Harry and Robert Evans. *Rethinking Expertise*. Chicago: University of Chicago, 2007.

Converse III, Elliott V. *Circling the Earth: United States Plans for a Postwar Overseas Military Base System, 1942–1948*. Maxwell Air Force Base, AL: Air University Press, 2005.

Corn, Joseph J. *The Winged Gospel: America's Romance with Aviation, 1900–1950*. New York: Oxford University Press, 1983.

Corthorn, Paul. 'The Labour Party and the League of Nations: The Socialist League's Role in the Sanctions Crisis of 1935', *Twentieth Century British History* 13, no. 1 (2002): 62–85.

Corum, James S. 'The Myth of Air Control: Reassessing the History', *Aerospace Power Journal* 14, no. 4 (2000): 61–77.

Cotton, James. '"The Standard Work in English on the League" and Its Authorship: Charles Howard Ellis, an Unlikely Australian Internationalist', *History of European Ideas* 42, no. 8 (2016): 1089–1104.

Craft, Stephen G. *V.K. Wellington Koo and the Emergence of Modern China*. Lexington: University Press of Kentucky, 2004.

Crockatt, Richard. *Einstein and Twentieth-Century Politics: 'A Salutary Moral Influence'*. Oxford: Oxford University Press, 2016.

Crompton, Teresa. 'British Imperial Policy and the Indian Air Route'. PhD diss., Sheffield Hallam University, 2014.

Custers, Bart, ed., *The Future of Drone Use: Opportunities and Threats from Ethical and Legal Perspectives*. Berlin: Springer, 2016.

Dallek, Robert. *Franklin D. Roosevelt and American Foreign Policy 1932–1945*, 2nd ed. New York: Oxford University Press, 1995.

Davies, Thomas Richard. *The Possibilities of Transnational Activism: The Campaign for Disarmament between the Two World Wars*. Leiden: Martinus Nijhoff, 2007.

Deibert, Ronald J. *Parchment, Printing and Hypermedia: Communications in World Order Transformation*. New York: Columbia University Press, 1997.

Denton, Peter. *Bertrand Russell on Science, Religion, and the Next War, 1919–1938*. New York: State University of New York Press, 2001.

Dierikx, Marc. *Clipping the Clouds: How Air Travel Changed the World*. Westport, CT: Praeger, 2008.

Dierikx, Marc. 'Shaping World Aviation: Anglo-American Civil Aviation Relations, 1944–1946', *Journal of Air Law and Commerce* 57, no. 4 (1992): 795–840.

Divine, Robert A. *Second Chance: A Triumph of Internationalism in America During World War II*. New York: Atheneum, 1967.

Dixon, Wheeler Winston. 'Tomorrowland TV: The Space Opera and Early Science Fiction Television', in *The Essential Science Fiction Television Reader*, ed. J. P. Telotte, 93–110. Lexington, KY: University of Lexington Press, 2008.

Dobson, Alan P. 'The Other Air Battle: The American Pursuit of Post-War Civil Aviation Rights', *The Historical Journal* 28, no.2 (June 1985): 429–439.

Doenecke, Justus D. 'Edwin M. Borchard, John Bassett Moore, and Opposition to American Intervention in World War II', *The Journal of Libertarian Studies* 6, no. 1 (1982): 1–34.

Douglas, R. M. *The Labour Party, Nationalism and Internationalism, 1939–1951*. London: Routledge, 2004.

Dubin, Martin David. 'Elihu Root and the Advocacy of a League of Nations, 1914–1917', *The Western Political Quarterly* 19, no.3 (September 1966): 439–455.

Duggan, John and Henry Cord Meyer. *Airships in International Affairs, 1890–1945*. Houndmills: Palgrave, 2001.

Dunne, Gerald T. *Grenville Clark: Public Citizen*. New York: Farrar, Strauss, Giroux, 1986.

Eckes, Alfred E. 'Globalization', in *Companion to International History 1900–2001*, ed. Gordon Martel, 408–421. London: Blackwell Press, 2007.

Edgerton, David. 'British Scientific Intellectuals and the Relations of Science, Technology and War', in *National Military Establishments and the Advancement of Science and Technology Studies in the 20th Century*, eds. P. Forman and J. M. Sanchez-Ron, 1–35. Dordrecht: Kluwer Academic, 1996.

Edgerton, David. *England and the Aeroplane: Militarism, Modernity and Machines*, 2nd ed. London: Penguin, 2013.

Edgerton, David. 'Liberal Militarism and the British State', *New Left Review* 185 (January–February 1991): 138–169.

Edgerton, David. *Warfare State: Britain 1920–1970*. Cambridge: Cambridge University Press, 2006.

Egerton, George. 'Conservative Internationalism: British Approaches to International Organization and the Creation of the League of Nations', *Diplomacy & Statecraft* 5, no. 1 (1994): 1–20.

Ekbladh, David. '"Mr. TVA": Grass-Roots Development, David Lilienthal, and the Rise and Fall of the Tennessee Valley Authority as a Symbol of U.S. Overseas Development, 1933–1973', *Diplomatic History* 26, no. 3 (Summer 2002): 335–374.

Ekbladh, David. *The Great American Mission: Modernization and the Construction of an American World Order*. Princeton, NJ: Princeton University Press, 2010.

Engel Jeffrey. 'A Shrinking World: Transport, Communication, and Towards a Global Culture', in *Companion to International History 1900–2001*, ed. Gordon Martel, 52–64. London: Blackwell Press, 2007.

Evangelista, Matthew. *Unarmed Forces: The Transnational Movement to End the Cold War*. Ithaca, NY: Cornell University Press, 1999.

Evanier, Mark. 'Point of View', Internet Archive Wayback Machine, accessed 2 August 2020, https://web.archive.org/web/20101217065650/http://povon line.com/cols/COL306.htm.

Freedman, Lawrence. 'The First Two Generations of Nuclear Strategists', in *Makers of Modern Strategy from Machiavelli to the Nuclear Age*, ed. Peter Paret, 735–778. Princeton, NJ: Princeton University Press, 1986.

Friedman, Thomas. *The Lexus and the Olive Tree*. London: Harper Collins, 2000.

Gieryn, Thomas F. *Cultural Boundaries of Science: Credibility on the Line*. Chicago: University of Chicago, 1999.

Gilbert, Martin. *Plough My Own Furrow: The Story of Lord Allen of Hurtwood as Told Through his Writings and Correspondence*. London: Longmans, 1965.

Gilpin, Robert. *American Scientists and Nuclear Weapons Policy*. Princeton, NJ: Princeton University Press, 1962.

Gilman, Nils. *Mandarins of the Future: Modernization Theory in Cold War America*. Baltimore, MD: Johns Hopkins University Press, 2003.

Girard, Marion. *A Strange and Formidable Weapon: British Responses to World War I Poison Gas*. Lincoln: University of Nebraska Press, 2008.

Girasa, Rosario, *Artificial Intelligence as a Disruptive Technology: Economic Transformation and Government Regulation*. Cham, Switzerland: Palgrave Macmillan, 2020.

Goebel, Stefan. *The Great War and Medieval Memory: War, Remembrance and Medievalism in Britain and Germany, 1914–1940*. Cambridge: Cambridge University Press, 2007.

Goodrich, Leland M., Edvard Hambro and Anne Patricia Simons, *Charter of the United Nations: Commentary and Documents*. New York: Columbia University Press, 1969.

Gorman, Daniel. *The Emergence of International Society in the 1920s*. Cambridge: Cambridge University Press, 2012.

Gormly, James L. 'The Washington Declaration and the 'Poor Relation': Anglo-American Atomic Diplomacy, 1945–46', *Diplomatic History* 8, no. 2 (1984): 125–143.

Gossling, Stefan and Paul Upham. *Climate Change and Aviation: Issues, Challenges and Solutions*. Abingdon: Earthscan, 2009.

Gowing, Margaret. *Independence and Deterrence: Britain and Atomic Energy, 1945–1952, Vol. 1, Policy Making*. London: Macmillan, 1974.

Gowing, Margaret. 'Britain, America and the Bomb', in *British Foreign Policy, 1945–56*, eds. Michael Dockrill and John W. Young, 31–46. New York: Palgrave Macmillan, 1989.

Grant, Matthew. 'The Imaginative Landscape of Nuclear War in Britain, 1945–65', in *Understanding the Imaginary War: Culture, Thought and Nuclear Conflict, 1945–90*, eds. Matthew Grant and Benjamin Ziemann, 92–115. Manchester: Manchester University Press, 2016.

Grayson, Richard S. *Liberals, International Relations and Appeasement: The Liberal Party 1919–1939*. London: Routledge, 1990.

Gross, Charles J. *Military Aviation: The Indispensable Arm*. College Station: Texas A&M University Press, 2002.

Gwinn, Kristen E. *Emily Greene Balch: The Long Road to Internationalism*. Urbana: University of Illinois Press, 2010.

Hall, Ian. *Dilemmas of Decline: British Intellectuals and World Politics, 1945–1975*. Berkeley: University of California Press, 2012.

Halliday, Fred. 'Three Concepts of Internationalism', *International Affairs* 64, no. 2 (Spring, 1988): 187–198.

Hamerlinck, Paul. *Fawcett Companion: The Best of FCA*. Raleigh, NC: TwoMorrows Publishing, 2001.

Harris, Elisa D., ed. *Governance of Dual-Use Technologies: Theory and Practice*. Cambridge, MA: American Academy of Arts & Sciences, 2016.

Hatch, F. J. *The Aerodrome of Democracy: Canada and the British Commonwealth Air Training Plan, 1939–1945*. Ottawa, ON: Department of National Defence, 1983.

Headrick, Daniel R. *Power over Peoples: Technology, Environments, and Western Imperialism, 1400 to the Present*. Princeton, NJ: Princeton University Press, 2010.

Hearden, Patrick J. *Architects of Globalism: Building a New World Order During World War II*. Fayetteville: University of Arkansas Press, 2002.

Hecht, Gabrielle. *Being Nuclear: Africans and the Global Uranium Trade*. Cambridge, MA: MIT Press, 2012.

Held, David, Anthony McGrew, David Goldblatt and Jonathan Perraton. *Global Transformations: Politics, Economics and Culture*. Cambridge: Polity Press, 1999.

Herken, Gregg. *The Winning Weapon: The Atomic Bomb in the Cold War 1945–1950*, 2nd ed. Princeton, NJ: Princeton University Press, 1988.

Hershberg, James G. *James B. Conant: Harvard to Hiroshima and the Making of the Nuclear Age*. New York: Alfred A. Knopf, 1993.

Hewlett, Richard G. and Jack M. Holl. *Atoms for Peace and War, 1953–1961: Eisenhower and the Atomic Energy Commission*. Berkeley: University of California Press, 1989.

Higham, Robin. *Britain's Imperial Air Routes 1918–1939*. London: G.T. Foulis, 1960.

Higham, Robin. *Speedbird: The Complete History of BOAC*. London: IB Tauris, 2013.

Hilderbrand, Robert C. *Dumbarton Oaks: The Origins of the United Nations and the Search for Postwar Security*. Chapel Hill: University of North Carolina Press, 2001.

Hill, Leonard et al. *Atomic Energy: Science v. Sovereignty.* London: World Unity Movement, 1946.

Hill, Roger. *Reed Crandall: Illustrator of the Comics.* Raleigh, NC: TwoMorrows Publishing, 2018.

Hippler, Thomas. *Governing from the Skies: A Global History of Aerial Bombing.* London: Verso, 2017.

Hodge, Joseph Morgan. *Triumph of the Expert: Agrarian Doctrines of Development and the Legacies of British Colonialism.* Athens: Ohio University Press, 2007.

Hogg, Jonathan. *British Nuclear Culture: Official and Unofficial Narratives in the Long Twentieth Century.* London: Bloomsbury, 2016.

Holman, Brett. 'The Air Panic of 1935: British Press Opinion between Disarmament and Rearmament', *Journal of Contemporary History* 46, no. 2 (April 2011): 288–307.

Holman, Brett. 'The Militarisation of Aerial Theatre: Air Displays and Airmindedness in Britain and Australia Between the World Wars', *Contemporary British History* 33, no. 4 (2019): 483–506.

Holman, Brett. *The Next War in the Air: Britain's Fear of the Bomber, 1908–1941.* Farnham: Ashgate, 2014.

Holman, Brett. 'The Shadow of the Airliner: Commercial Bombers and the Rhetorical Destruction of Britain, 1917–35', *Twentieth Century British History* 24, no. 4 (2013): 495–517.

Holmes, James R. *Theodore Roosevelt and World Order: Police Power in International Relations.* Washington, DC: Potomac Books, 2006.

Holthaus, Leonie and Jens Steffek, 'Experiments in international administration: The forgotten functionalism of James Arthur Salter', *Review of International Studies* 42, no. 1 (2016): 114–135.

Hoopes, Townsend and Douglas Brinkley. *FDR and the Creation of the U.N.* New Haven, CT: Yale University Press, 2000.

Howe, A. 'Free Trade and Global Order: The Rise and Fall of a Victorian Vision', in *Victorian Visions of Global Order: Empire and International Relations in Nineteenth-Century Political Thought*, ed. Duncan Bell, 26–46. Cambridge: Cambridge University Press, 2007.

Huurdeman, Anton A. *A Worldwide History of Telecommunications.* New York: Wiley, 2003.

Ikenberry, G. John, 'Liberal Internationalism 3.0: America and the Dilemmas of Liberal World Order', *Perspectives on Politics* 7, no. 1 (2009): 71–87.

Ikenberry, G. John. *Liberal Leviathan: The Origins, Crisis, and Transformation of the Liberal World Order.* Princeton, NJ: Princeton University Press, 2011.

Jackson, Julian. *The Popular Front in France: Defending Democracy, 1934–1938.* Cambridge: Cambridge University Press, 1990.

Jackson, Peter. *Beyond the Balance of Power: France and the Politics of National Security in the Era of the First World War.* Cambridge: Cambridge University Press, 2013.

Jackson, Simon and Alanna O'Malley, eds., *The Institution of International Order: From the League of Nations to the United Nations.* London: Routledge, 2018.

Jahn, Beate. *Liberal Internationalism: Theory, History, Practice.* Houndmills: Palgrave Macmillan, 2013.

Jasanoff, Sheila. 'Future Imperfect: Science, Technology, and the Imaginations of Modernity', in *Dreamscapes of Modernity: Sociotechnical Imaginaries and the Fabrication of Power*, eds. Sheila Jasanoff and Sang-Hyun Kim, 1–33. Chicago: University of Chicago Press, 2015.

Jenkins, Gwyn. 'Lord Davies, Howard Hughes and the "The Quinine Proposition": The Plan to Set up an International Air Force to Defend Chinese Cities from Japanese Air Raids, 1938', *The National Library of Wales Journal* 21, no. 4 (Winter 1980): 414–422.

Jerónimo, Miguel Bandeira. 'A League of Empires: Imperial Political Imagination and Interwar Internationalisms', in *Internationalism, Imperialism and the Formation of the Contemporary World*, eds. Miguel Bandeira Jerónimo and José Pedro Monteiro, 87–126. Cham, Switzerland: Palgrave Macmillan, 2018.

Johnson, Edward. 'British Proposals for a United Nations Force 1946–48', in *Britain and the First Cold War*, ed. Ann Deighton, 109–129. New York: Palgrave Macmillan, 1990.

Johnson, Gaynor. *Lord Robert Cecil: Politician and Internationalist*. Abingdon: Routledge, 2016.

Johnstone, Andrew. *Against Immediate Evil: American Internationalists and the Four Freedoms on the Eve of World War II*. Ithaca, NY: Cornell University Press, 2014.

Johnstone, Andrew. *Dilemmas of Internationalism: The American Association for the United Nations and US Foreign Policy, 1941–1948*. Farnham: Ashgate, 2009.

Johnstone, Andrew. 'Isolationism and Internationalism in American Foreign Relations', *Journal of Transatlantic Studies* 9, no. 1 (March 2011): 7–20.

Johnstone, Andrew. 'Shaping our Post-war Foreign Policy: The Carnegie Endowment for International Peace and the Promotion of the United Nations Organisation during World War II', *Global Society* 28, no. 1 (2014): 24–39.

Jones, Greta. *Science, Politics, and the Cold War*. London: Routledge, 1988.

Jones, Greta. 'The Mushroom-Shaped Cloud: British Scientists' Opposition to Nuclear Weapons Policy, 1945–57', *Annals of Science* 43, no. 1 (1986): 1–26.

Jones, Matthew. *The Official History of the UK Strategic Nuclear Deterrent, vol. I, From the V-Bomber Era to the Arrival of Polaris, 1945–1964*. Abingdon: Routledge, 2017, 1–12.

Jönsson, Christer. 'Classical Liberal Internationalism', in *International Organization and Global Governance*, eds. Thomas G. Weiss and Rorden Wilkinson, 105–117. Abingdon: Routledge, 2014.

Jordan, John M. *Machine-Age Ideology: Social Engineering & American Liberalism, 1911–1939*. Chapel Hill: University of North Carolina Press, 1994.

Josephson, Harold. *James T. Shotwell and the Rise of Internationalism in America*. Cranbury, NJ: Associated University Presses, 1975.

Joyce, Daniel. 'Liberal Internationalism', in *The Oxford Handbook of the Theory of International Law*, eds. Anne Orford, Florian Hoffmann, and Martin Clark, 471–487. Oxford: Oxford University Press, 2016.

Kaplan, Fred. *The Wizards of Armageddon*. Stanford: Stanford University Press, 1991.

Kaplan, Lawrence S. *Harold Stassen: Eisenhower, the Cold War, and the Pursuit of Nuclear Disarmament*. Lexington: University Press of Kentucky, 2017.

Keohane, Robert O. and Joseph S. Nye, Jr., 'Power and Interdependence in the Information Age', *Foreign Affairs* 77, no. 5 (September/October 1998): 81–94.

Keohane, Robert O. and Joseph S. Nye, Jr., *Power and Interdependence*, 2nd ed. New York: Harper Collins, 1989.

Kendall, Eric M. 'Liberal Internationalism, the Peace Movement, and the Ambiguous Legacy of Woodrow Wilson'. PhD diss., Case Western Reserve University, 2012.

Kitching, C. J. *Britain and the Geneva Disarmament Conference: A Study in International History*. Houndmills: Palgrave Macmillan, 2003.

Kitching, C. J. *Britain and the Problem of International Disarmament 1919–1934*. London: Routledge, 1999.

Kimball, Warren F. 'The Sheriffs: FDR's Postwar World', in *FDR's World: War, Peace, and Legacies*, eds. David B. Woolner, Warren F. Kimball, and David Reynolds, 91–122. New York: Palgrave Macmillan, 2008.

Klabbers, Jan. 'The Emergence of Functionalism in International Institutional Law: Colonial Inspirations', *The European Journal of International Law* 25, no. 3 (2014): 645–675.

Klingensmith, Daniel. *One Valley and a Thousand: Dams, Nationalism, and Development*. New Delhi: Oxford University Press, 2007.

Koskenniemi, Martti. *The Gentle Civilizer of Nations: The Rise and Fall of International Law 1870–1960*. Cambridge: Cambridge University Press, 2001.

Kott, Sandrine. 'Cold War Internationalism', in *Internationalisms: A Twentieth Century History*, eds. Glenda Sluga and Patricia Clavin, 340–362. Cambridge: Cambridge University Press, 2017.

Knepper, Paul. *International Crime in the 20th Century: The League of Nations Era, 1919–1939*. Houndmills: Palgrave Macmillan, 2011.

Kreps, Sarah Elizabeth. *Drones: What Everyone Needs to Know*. New York: Oxford University Press, 2016.

Kuehl, Warren F. 'Concepts of Internationalism in History', *Peace and Change* 11, no. 2 (1986): 1–10.

Kuehl, Warren F. 'Preface', in *Biographical Dictionary of Internationalists*, ed. Warren F. Kuehl, ix–xii. Westport, CT: Greenwood Press, 1983.

Kuehl, Warren F. 'Webs of Common Interests Revisited: Nationalism, Internationalism, and Historians of American Foreign Relations', *Diplomatic History* 10, no. 2 (April 1986): 107–120.

Kuehl, Warren F. and Lynne K. Dunn, *Keeping the Covenant: American Internationalists and the League of Nations, 1920–1939*. Kent, OH: The Kent State University Press, 1997.

Lafeber, Walter. 'Presidential Address: Technology and U.S. Foreign Relations', *Diplomatic History* 24, no. 1 (Winter 2000): 1–19.

Lagendijk, Vincent. *Electrifying Europe: The Power of Europe in the Construction of Electricity Networks*. Eindhoven: Aksant, 2009.

Laqua, Daniel. *The Age of Internationalism and Belgium, 1880–1930: Peace, Progress and Prestige*. Oxford: Oxford University Press, 2015.

Laucht, Christoph. "'Dawn – Or Dusk?" Britain's Picture Post Confronts Nuclear Energy', in *The Nuclear Age in Popular Media: A Transnational History, 1945–1965*, ed. Dick van Lente, 117–148. New York: Palgrave Macmillan, 2012.

Laucht, Christoph. *Elemental Germans: Klaus Fuchs, Rudolf Peierls and the Making of British Nuclear Culture 1939–59.* Houndmills: Palgrave Macmillan, 2012.

Laucht, Christoph. 'Transnational Professional Activism and the Prevention of Nuclear War in Britain', *Journal of Social History* 52, no. 2 (2018): 439–467.

Lavin, Deborah. *From Empire to International Commonwealth: A Biography of Lionel Curtis.* Oxford: Clarendon Press, 1995.

Lechner, Frank J. and John Boli, eds., *The Globalization Reader.* Oxford: Blackwell Publishing, 2000.

Leffler, Melvyn P. 'Strategy, Diplomacy, and the Cold War: The United States, Turkey, and NATO, 1945–1952', *Journal of American History* 71, no. 4 (March 1985): 807–825.

Levi, Michael A. and Michael E. O'Hanlon. *The Future of Arms Control.* Washington, DC: Brookings Institution Press, 2005.

Lewis, Julian. *Changing Direction: British Military Planning for Post-war Strategic Defence, 1942–1947.* London: Sherwood, 1988.

Lomas, Daniel W. B. *Intelligence, Security and the Attlee Governments, 1945–51: An Uneasy Relationship?* Manchester: Manchester University Press, 2017.

Long, David, *Towards a New Internationalism: The International Theory of J.A. Hobson.* Cambridge: Cambridge University Press, 1996.

Long, David and Brian C. Schmidt, eds., *Imperialism and Internationalism in the Discipline of International Relations.* Albany: State University of New York Press, 2005.

Long, David and Peter Wilson, eds., *Thinkers of the Twenty Years' Crisis: Inter-war Idealism Reassessed.* Oxford: Clarendon Press, 1995.

Lopez-Claros, Augusto, Arthur L. Dahl, and Maja Groff. 'Completing the Collective Security Mechanism of the Charter: Establishing an International Peace Force', in *Global Governance and the Emergence of Global Institutions for the 21st Century*, 145–180. Cambridge: Cambridge University Press, 2020.

Lyons, Justin D. 'Strength Without Mercy: Winston Churchill on Technology and the Fate of Civilization', *Perspectives on Political Science* 43, no. 2 (2014): 102–108

Mabee, Bryan. 'From "Liberal War" to "Liberal Militarism": United States Security Policy as the Promotion of Military Modernity', *Critical Military Studies* 2, no. 3 (2016): 242–261.

MacDonald, Sean. 'Radar the International Policeman', Writeups.org, accessed 2 August 2020, www.writeups.org/radar-the-international-policeman-fawcett-comics.

Macfadyen, David, et al. *Eric Drummond and His Legacies: The League of Nations and the Beginnings of Global Governance.* Cham, Switzerland: Palgrave Macmillan, 2019.

MacKenzie, David. 'An "Ambitious Dream": The Chicago Conference and the Quest for Multilateralism in International Air Transport', *Diplomacy and Statecraft* 2, no. 2 (1991): 270–293.

MacKenzie, David. 'Australia and Canada in the World of International Commercial Aviation', in *Parties Long Estranged: Canada and Australia in the Twentieth Century*, eds. Margaret MacMillan and Francine McKenzie, 99–123. Vancouver: University of British Columbia Press, 2003.

MacKenzie, David. *Canada and International Civil Aviation 1932–1948*. Toronto: University of Toronto Press, 1989.

MacKenzie, David. *ICAO: A History of the International Civil Aviation Organization*. Toronto: University of Toronto Press, 2010.

MacKenzie, David. *A World Beyond Borders: An Introduction to the History of International Organizations*. Toronto: University of Toronto Press, 2010.

Maddock, Shane J. *Nuclear Apartheid: The Quest for American Atomic Supremacy from World War II to the Present*. Chapel Hill: University of North Carolina Press, 2010.

Maiolo, Joseph. 'Introduction', in *Arms Races in International Politics: From the Nineteenth to the Twenty-First Century*, eds. Thomas Mahnken, Joseph Maiolo, and David Stevenson, 1–9. Oxford: Oxford University Press, 2016.

Manela, Erez. *The Wilsonian Moment: Self-Determination and the International Origins of Anticolonial Nationalism*. Oxford: Oxford University Press, 2007.

Manzione, Joseph. '"Amusing and Amazing and Practical and Military": The Legacy of Scientific Internationalism in American Foreign Policy, 1945–1963', *Diplomatic History* 24, no. 1 (Winter 2000): 21–55.

Marks, Sally. *The Illusion of Peace: International Relations in Europe, 1918–1933*, 2nd ed. Houndmills: Palgrave Macmillan, 2003.

Marwick, Arthur. 'Middle opinion in the thirties: planning, progress and political "agreement"', *The English Historical Review* 79, no. 311 (1964): 285–298.

Mathews, Jessica T. 'Power Shift', *Foreign Affairs* 76, no. 1 (January/February 1997): 50–66.

Mayne, Richard, John Pinder and John C. de V. Roberts. *Federal Union: The Pioneers. A History of Federal Union*. Houndmills: Palgrave Macmillan, 1990.

Mazower, Mark. *Governing the World: The History of an Idea*. New York: Penguin, 2012.

Mazower, Mark. *No Enchanted Palace: The End of Empire and the Ideological Origins of the United Nations*. Princeton, NJ: Princeton University Press, 2009.

McCarthy, Helen. *The British People and the League of Nations: Democracy, Citizenship and Internationalism, c.1918–45*. Manchester: Manchester University Press, 2012.

McNeill, William H. *The Rise of the West: A History of the Human Community*. Chicago: University of Chicago Press, 1963.

Mehta, Uday Singh. *Liberalism and Empire: A Study in Nineteenth-Century British Liberal Thought*. Chicago: University of Chicago Press, 1999.

Meskell, Lynn, *A Future in Ruins: UNESCO, World Heritage, and the Dream of Peace*. New York: Oxford University Press, 2018.

Miller, Karen S. *The Voice of Business: Hill & Knowlton and Postwar Public Relations*. Chapel Hill: University of North Carolina Press, 1999.

Miscamble, Wilson D. *From Roosevelt to Truman: Potsdam, Hiroshima, and the Cold War*. Cambridge: Cambridge University Press, 2007.

Mitchell, Kurt and Roy Thomas. *American Comic Book Chronicles: 1940–1944*. Raleigh, NC: TwoMorrows Publishing, 2019.

Mitoma, Glenn Tatsuya. 'Civil Society and International Human Rights: The Commission to Study the Organization of Peace and the Origins of the UN Human Rights Regime', *Human Rights Quarterly* 30, no. 3 (August 2008): 607–630.

Monk, Ray. *Bertrand Russell 1921–70. The Ghost of Madness*. London: Jonathan Cape, 2000.

Morefield, Jeanne. *Covenants without Swords: Idealist Liberalism and the Spirit of Empire*. Princeton, NJ: Princeton University Press, 2005.

Morefield, Jeanne. *Empires Without Imperialism: Anglo-American Decline and the Politics of Deflection*. Oxford: Oxford University Press, 2014.

Morrell, Peter S. *Moving Boxes by Air: The Economics of International Air Cargo*. London: Routledge, 2011.

Morrow Jr., John H. 'Knights of the Sky: Rise of Military Aviation' in *Authority, Identity and the Social History of the Great War*, eds. Frans Coetzee and Marilyn Shevin-Coetzee, 305–324. Oxford: Berghahn Books, 1995.

Müller, Simone M. *Wiring the World: The Social and Cultural Creation of Global Telegraph Networks*. New York: Columbia University Press, 2016.

Mundy, Martha. *The Strategies of the Coalition in the Yemen War: Aerial Bombardment and Food War*. London: World Peace Foundation, 2018.

Nehring, Holger. *Politics of Security: British and West German Protest Movements and the Early Cold War, 1945–1970*. Oxford: Oxford University Press, 2013.

Nehring, Holger. 'National Internationalists: British and West German Protests Against Nuclear Weapons, the Politics of Transnational Communication and the Social History of the Cold War, 1957–1964', *Contemporary European History* 14, no. 4 (2005): 559–582.

Navari, Cornelia. 'David Mitrany and International Functionalism', in *Thinkers of the Twenty Years' Crisis: Inter-war Idealism Reassessed*, eds. David Long and Peter Wilson, 214–246. Oxford: Clarendon Press, 1995.

Navari, Cornelia. 'James T. Shotwell and the Organization of Peace', in *Progressivism and US Foreign Policy between the World Wars*, eds. Molly Cochran and Cornelia Navari, 167–192. London: Palgrave Macmillan, 2017.

Ndiaye, Pap A. *Nylon and Bombs: DuPont and the March of Modern America*. Baltimore, MD: Johns Hopkins University Press, 2007.

'Nuclear Power in the World Today', World Nuclear Association, accessed 2 August 2020, www.world-nuclear.org/information-library/current-and-future-generation/nuclear-power-in-the-world-today.aspx.

Nye, Mary Jo. 'A Physicist in the Corridors of Power: P. M. S. Blackett's Opposition to Atomic Weapons Following the War', *Physics in Perspective* 1 (1999): 136–156.

Omissi, David E. *Air Power and Colonial Control: The Royal Air Force, 1919–1939*. Manchester: Manchester University Press, 1990.

Ostrower, Gary B. *The United Nations and the United States*. New York: Twayne, 1998.

O'Sullivan, Christopher D. *Sumner Welles, Postwar Planning, and the Quest for a New World Order, 1937–1943*. New York: Columbia University Press, 2008.

Overy, Richard. *The Bombing War: Europe 1939–1945*. London: Allen Lane, 2013.

Paris, Michael. 'Air Power and Imperial Defence 1880–1919', *Journal of Contemporary History* 24, no. 2 (April 1989): 209–225.

Paris, Michael. 'The Rise of the Airmen: The Origins of Air Force Elitism, c. 1890–1918', *Journal of Contemporary History* 28, no. 1 (January 1993): 123–141.

Paris, Michael. *Warrior Nation: Images of War in British Popular Culture, 1850–2000*. London: Reaktion Books, 2000.

Parmar, Inderjeet. *Foundations of the American Century: The Ford, Carnegie and Rockefeller Foundations in the Rise of American Power*. New York: Columbia University Press, 2012.

Parmar, Inderjeet. *Think Tanks and Foreign Policy: A Comparative Study of the Role and Influence of the Council on Foreign Relations and the Royal Institute of International Affairs*. Houndmills: Palgrave Macmillan, 2004.

Patterson, Ian. *Guernica and Total War*. London: Profile Books, 2007.

Paul, Septimus H. *Nuclear Rivals: Anglo-American Atomic Relations, 1941–1952*. Columbus: Ohio State University Press, 2000.

Peden, G. C. *Arms, Economics and British Strategy: From Dreadnoughts to Hydrogen Bombs*. Cambridge: Cambridge University Press, 2007.

Pedersen, Susan. *The Guardians: The League of Nations and the Crisis of Empire*. Oxford: Oxford University Press, 2015.

Pemberton, Jo-Anne. *The Story of International Relations, parts one and two, Cold-Blooded Idealists*. Cham, Switzerland: Palgrave Macmillan, 2019, 2020.

Perkins, Ray. 'Bertrand Russell and Preventative War: A Reply to David Blitz', *Russell: The Journal of Bertrand Russell Studies* 22, no. 2 (Winter 2002–03): 161–172.

Pietrobon, Allen. 'Peacemaker in the Cold War: Norman Cousins and the Making of a Citizen Diplomat in the Atomic Age'. PhD diss., Washington, DC: American University, 2016.

Pinder, John. 'Federalism and the Beginnings of European Union', in *A Companion to Europe Since 1945*, ed. Klaus Larres, 25–44. Oxford: Wiley-Blackwell, 2009.

Pirie, Gordon. *Air Empire: British Imperial Civil Aviation, 1919–39*. Manchester: Manchester University Press, 2009.

Pitts, Jennifer. *A Turn to Empire: The Rise of Imperial Liberalism in Britain and France*. Princeton, NJ: Princeton University Press, 2005.

Porter, Brian. 'David Davies and the Enforcement of Peace', in *Thinkers of the Twenty Years' Crisis: Inter-war Idealism Reassessed*, eds. David Long and Peter Wilson, 58–78. Oxford: Clarendon Press, 1995.

Price, Matt. 'Roots of Dissent: The Chicago Met Lab and the Origins of the Franck Report', *Isis* 86, no.2 (June 1995): 222–244.

Provence, Michael. *The Great Syrian Revolt and the Rise of Arab Nationalism*. Austin: University of Texas Press, 2005.

Pugh, Michael. *Liberal Internationalism: The Interwar Movement for Peace in Britain*. Houndmills: Palgrave Macmillan, 2012.

Pugh, Michael. 'Policing the World: Lord Davies and the Quest for Order in the 1930s', *International Relations* 16, no. 1 (2002): 97–115.

Ramos, Paulo. 'The Role of the Yale Institute for International Studies in the Construction of United States National Security Ideology, 1935–1951'. PhD diss., University of Manchester, 2003.

Randall, Mercedes M. *Improper Bostonian: Emily Greene Balch*. New York: Twayne, 1964.

Rasmussen, Anne. 'Science and Technology', in *A Companion to World War I*, ed. John Horne, 307–322. Chichester: Wiley-Blackwell, 2010.

Reinisch, Jessica. 'Introduction: Relief in the Aftermath of War', *Journal of Contemporary History* 43, no. 3 (2008): 371–404.

Reinisch, Jessica, 'What Makes an Expert? The View from UNRRA, 1943–47', in *Work in Progress. Economy and Environment in the Hands of Experts*, eds. Frank Trentmann, A. B. Sum, and M. Riviera, 103–130. Munich: Oekom, 2018.

Rich, Paul. 'Reinventing Peace: David Davies, Alfred Zimmern and Liberal Internationalism in Interwar Britain', *International Relations* 16, no. 1 (2002): 117–133.

Rieger, Bernhard. *Technology and the Culture of Modernity in Britain and Germany, 1890–1945*. Cambridge: Cambridge University Press, 2005.

Rietzler, Katharina. 'American Foundations and the 'Scientific Study' of International Relations in Europe, 1910–1940'. PhD diss., University College London, 2009.

Rietzler, Katharina. 'Fortunes of a Profession: American Foundations and International Law, 1910–1939', *Global Society* 28, no. 1 (2014): 8–23.

Rosenberg, Emily S. 'Transnational Currents in a Shrinking World', in *A World Connecting 1870–1945*, ed. Emily S. Rosenberg, 813–996. Cambridge, MA: Belknap Press, 2012.

Rosenboim, Or. *The Emergence of Globalism: Visions of World Order in Britain and the United States, 1939–1950*. Princeton, NJ: Princeton University Press, 2017.

Rotblat, Joseph. 'Movements of Scientists against the Arms Race', in *Scientists, the Arms Race and Disarmament*, ed. Joseph Rotblat, 115–157. London: Taylor & Francis, 1982.

Routledge, Christopher, 'Crime and Detective Literature for Young Readers', in *A Companion to Crime Fiction*, eds. Charles J. Rzepka and Lee Horsley, 321–331. Oxford: Wiley-Blackwell, 2010.

Rowe, David E. and Robert Schulmann, eds., *Einstein on Politics: His Private Thoughts and Public Stands on Nationalism, Zionism, War, Peace, and the Bomb*. Princeton, NJ: Princeton University Press, 2007.

Sandvik, Kristin Bergtora and Maria Gabrielsen Jumbert, eds., *The Good Drone*. Abingdon: Routledge, 2017.

Satia, Priya. *Spies in Arabia: The Great War and the Cultural Foundations of Britain's Covert Empire in the Middle East*. Oxford: Oxford University Press, 2008.

Scharf, Michael P. and Lawrence D. Robert, 'The Interstellar Relations of the Federation: International Law and "Star Trek: The Next Generation"', in *Star*

Trek Visions of Law and Justice, eds. Robert H. Chaires and Bradley Stewart Chilton, 73–113. Denton: University of North Texas Press, 2003.

Schatzberg, Eric. *Technology: Critical History of a Concept*. Chicago: Chicago University Press, 2018.

Schmitter, Philippe C. 'Ernst B. Haas and the legacy of neofunctionalism', *Journal of European Public Policy* 12, no. 2 (April 2005): 255–272.

Schot, Johan and Vincent Lagendijk, 'Technocratic Internationalism in the Interwar Years: Building Europe on Motorways and Electricity Networks', *Journal of Modern European History* 6, no. 2 (2008): 196–217.

Schot, Johan. 'Transnational Infrastructures and the Origins of European Integration', in *Materializing Europe: Transnational Infrastructures and the Project of Europe*, eds. Alexander Badenoch and Andreas Fickers, 82–109. London: Palgrave Macmillan, 2010.

Schrafstetter, Susanna. '"Loquacious . . . and Pointless as Ever"? Britain, the United States and the United Nations Negotiations on International Control of Nuclear Energy, 1945–48', *Contemporary British History* 16, no. 4 (Winter 2002): 87–108.

Schulten, Susan. *The Geographical Imagination in America, 1880–1950*. Chicago: University of Chicago Press, 2001.

Schulzinger, Robert D. *The Wise Men of Foreign Affairs: The History of the Council on Foreign Relations*. New York: Columbia University Press, 1984.

Sethi, Megan Barnhart. 'Information, Education, and Indoctrination: The Federation of American Scientists and Public Communication Strategies in the Atomic Age', *Historical Studies in the Natural Sciences* 42, no. 1 February (2012): 1–29.

Shepard, Elizabeth and Mike Farrelly. *Legendary Locals of Montclair*. Mount Pleasant, SC: Arcadia, 2013.

Sherry, Michael S. *The Rise of American Air Power: The Creation of Armageddon*. New Haven, CT: Yale University Press, 1987.

Sherry, Michael S. *In the Shadow of War: The United States Since the 1930s*. New Haven, CT: Yale University Press, 1995.

Sherwin, Martin J. 'Niels Bohr and the First Principles of Arms Control', in *Niels Bohr: Physics and the World*, eds. Herman Feshbach, Tetsuo Matsui, and Alexandra Oleson, 319–330. Abingdon: Routledge, 1998.

Shinohara, Hatsue. *US International Lawyers in the Interwar Years: A Forgotten Crusade*. Cambridge: Cambridge University Press, 2012.

Shoup Laurence H. and William Minter. *Imperial Brain Trust: The Council on Foreign Relations and United States Foreign Policy*. New York: Monthly Review Press, 1977.

Sims, Chris. 'Bizarro Back Issues: Boxing Day With Captain Marvel! (1944)', *Comics Alliance*, accessed 2 August 2020, https://comicsalliance.com/bizarro-back-issues-captain-marvel-boxing-radar-the-international-policeman.

Sinclair, Guy Fiti. *To Reform the World: International Organizations and the Making of Modern States*. Oxford: Oxford University Press, 2017.

Smith, Alice Kimball. *A Peril and a Hope: The Scientists' Movement in America: 1945–47*. Chicago: University of Chicago Press, 1965.

Smith, Malcolm. *British Air Strategy Between the Wars*. Oxford: Clarendon Press, 1984.

Smith, Neil. *American Empire: Roosevelt's Geographer and the Prelude to Globalization*. Berkeley: University of California Press, 2003.

Smith, Tony. *Why Wilson Matters: The Origin of American Liberal Internationalism and its Crisis Today*. Princeton, NJ: Princeton University Press, 2017.

Sluga, Glenda. *Internationalism in the Age of Nationalism*. Philadelphia: University of Pennsylvania Press, 2013.

Sluga, Glenda and Patricia Clavin, eds. *Internationalisms: A Twentieth Century History*. Cambridge: Cambridge University Press, 2017.

Sluga, Glenda and Patricia Clavin. 'Rethinking the History of Internationalism', in *Internationalisms: A Twentieth Century History*, 3–14. Cambridge: Cambridge University Press, 2017.

Soffer, Jonathan. 'All for One or All for All: The UN Military Staff Committee and the Contradictions within American Internationalism', *Diplomatic History* 21, no. 1 (Winter 1997): 45–69.

Stirk, Peter M. R., ed. *European Unity in Context: The Interwar Period*. London: Pinter, 1989.

Stoler, Mark A. *Allies and Adversaries: The Joint Chiefs of Staff, the Grand Alliance, and U.S. Strategies in World War II*. Chapel Hill: University of North Carolina Press, 2000.

Strachan, Hew. 'Essay and Reflection: On Total War and Modern War', *The International History Review* 22, no. 2 (2000): 341–370.

Suganami, Hidemi. *The Domestic Analogy and World Order Proposals*. Cambridge: Cambridge University Press, 1989.

Sylvest, Casper. *British Liberal Internationalism, 1880–1930: Making Progress?* Manchester: Manchester University Press, 2009.

Sylvest, Casper. 'Technology and Global Politics: The Modern Experiences of Bertrand Russell and John H. Herz', *The International History Review* 35, no. 1 (2013): 121–142.

Tannenwald, Nina. *The Nuclear Taboo: The United States and the Non-Use of Nuclear Weapons Since 1945*. Cambridge: Cambridge University Press, 2007.

Taylor, Paul. 'Introduction', in *The Functional Theory of Politics*, by David Mitrany, ix–xxv. London: LSE & Political Science, 1975.

Toye, Richard. *The Labour Party and the Planned Economy, 1931–1951*. London: The Royal Historical Society, 2003.

Trentmann, Frank, A. B. Sum, and M. Riviera, eds., *Work in Progress. Economy and Environment in the Hands of Experts*. Munich: Oekom, 2018.

Turchetti, Simone. 'Atomic Secrets and Governmental Lies: Nuclear Science, Politics and Security in the Pontecorvo Case', *British Journal for the History of Science* 36, no. 4 (December 2003): 389–415.

Turner, Frank M. 'Public Science in Britain, 1880–1919', *Isis* 71, no.4 (December 1980): 589–608.

'United States aircraft production during World War II', Wikipedia, last modified 6 February 2020, https://en.wikipedia.org/wiki/United_States_aircraft_production_during_World_War_II.

Vaïsse, Maurice. *Sécurité D'abord: la Politique Française en Matière de Désarmement, 9 Décembre 1930 – 17 Avril 1934*. Paris: Pedone, 1981.

Vleck, Jenifer Van. *Empire of the Air: Aviation and the American Ascendancy.* Cambridge, MA: Harvard University Press, 2013.

Vleck, Jenifer Van. 'The "Logic of the Air": Aviation and the Globalism of the "American Century"', *New Global Studies* 1, no. 1 (2007): 1–37.

Waltz, Kenneth. *Theory of International Politics.* New York: Addison-Wesley, 1979.

Wang, Jessica. *American Science in the Age of Anxiety: Scientists, Anticommunism, and the Cold War.* Chapel Hill: University of North Carolina, 1999.

Weart, Spencer. *Nuclear Fear: A History of Images.* Cambridge, MA: Harvard University Press, 1988.

Webster, Andrew. '"Absolutely Irresponsible Amateurs": The Temporary Mixed Commission on Armaments, 1921–1924', *Australian Journal of Politics and History* 54, no. 3 (September 2008): 373–388.

Webster, Andrew. 'From Versailles to Geneva: The Many Forms of Interwar Disarmament', *The Journal of Strategic Studies* 29, no. 2 (April 2006): 225–246.

Webster, Andrew. 'The League of Nations, Disarmament and Internationalism', in *Internationalisms: A Twentieth Century History*, eds. Glenda Sluga and Patricia Clavin, 139–169. Cambridge: Cambridge University Press, 2017.

Weindling, Paul, ed. *International Health Organisations and Movements, 1918–1939.* Cambridge: Cambridge University Press, 1995.

Wendt, Alexander. *Social Theory of International Politics.* Cambridge: Cambridge University Press, 1999.

Wertheim, Stephen Alexander. 'Tomorrow, the World: The Birth of U.S. Global Supremacy in World War II'. PhD diss., Columbia University, 2015.

White, Donald W. 'The "American Century" in World History', *Journal of World History* 3, no. 1 (Spring 1992): 105–127.

Wilson, Peter. *The International Theory of Leonard Woolf: A Study in Twentieth-Century Idealism.* London: Palgrave Macmillan, 2003.

Wilson, Peter. 'Introduction: The Twenty Years Crisis and the Category of "Idealism" in International Relations', in *Thinkers of the Twenty Years' Crisis: Inter-war Idealism Reassessed*, eds. David Long and Peter Wilson, 1–24. Oxford: Clarendon Press, 1995.

Wilson, Peter. 'The Myth of the "First Great Debate"', in *The Eighty Years' Crisis: International Relations 1919–1999*, eds. T. Dunne, M. Cox, and K. Booth, 1–15. Cambridge: Cambridge University Press, 1998.

Wilson, Peter. 'The New Europe Debate in Wartime Britain', in *Visions of European Unity*, eds. Philomena Murray and Paul Rich, 39–62. Boulder, CO: Westview, 1996.

Winkler, Allan M. *Life Under a Cloud: American Anxiety about the Atom*, 2nd ed. Urbana: University of Illinois Press, 1999.

Winkler, Henry Ralph. *British Labour seeks a Foreign Policy, 1900–1940.* New Brunswick: Transaction Publishers, 2005.

Wittner, Lawrence S. *The Struggle Against the Bomb, Vol. 1, One World or None: A History of the World Nuclear Disarmament Movement Through 1953.* Stanford: Stanford University Press, 1993.

Wohl, Robert. *A Dream of Wings: Aviation and the Western Imagination, 1908–1918.* New Haven, CT: Yale University Press, 1994.

Wohl, Robert. *The Spectacle of Flight: Aviation and the Western Imagination, 1920–1950*. New Haven, CT: Yale University Press, 2005.

Wood, James. 'Anglo-American Liberal Militarism and the Idea of the Citizen Soldier', *International Journal* 62, no. 2 (2007): 403–422.

Wright, Bradford W. *Comic Book Nation: The Transformation of Youth Culture in America*. Baltimore, MD: Johns Hopkins Press, 2001.

Yearwood, Peter J. *Guarantee of Peace: The League of Nations in British Policy 1914–1925*. Oxford: Oxford University Press, 2009.

Yudkowsky, Elizer. 'Artificial Intelligence as a Positive and Negative Factor in Global Risk', in *Global Catastrophic Risks*, eds. Nick Bostrom and Milan M. Ćirković, 308–345. Cambridge: Cambridge University Press, 2008.

Zaidi, Waqar H. '"Aviation Will Either Destroy or Save Our Civilization": Proposals for the International Control of Aviation, 1920–45', *Journal of Contemporary History* 46, no. 150 (2011): 150–178.

Zaidi, Waqar H. 'Stages of War, Stages of Man: Quincy Wright and the Liberal Internationalist Study of War', *The International History Review* 40, no. 2 (2018): 416–435.

Zaidi, Waqar H. '"Wings for Peace" versus "Airopia": Contested Visions of Postwar European Aviation in World War Two Britain', in *Linking Networks: The Formation of Common Standards and Visions for Infrastructure Development*, eds. Martin Schiefelbusch and Hans-Liudger Dienel, 151–168. London: Ashgate, 2014.

Zeman, Scott C. '"To See . . . Things Dangerous to Come to": Life Magazine and the Atomic Age in the United States, 1945–1965', in *The Nuclear Age in Popular Media: A Transnational History, 1945–1965*, ed. Dick Van Lente, 53–77. New York: Palgrave Macmillan, 2012.

Zipp, Samuel. 'Dilemmas of World-Wide Thinking: Popular Geographies and the Problem of Empire in Wendell Willkie's Search for One World', *Modern American History* 1, no. 3 (2018): 295–319.

Index

Printed in the United States
by Baker & Taylor Publisher Services